Springer Theses

Recognizing Outstanding Ph.D. Research

For further volumes:
http://www.springer.com/series/8790

Aims and Scope

The series "Springer Theses" brings together a selection of the very best Ph.D. theses from around the world and across the physical sciences. Nominated and endorsed by two recognized specialists, each published volume has been selected for its scientific excellence and the high impact of its contents for the pertinent field of research. For greater accessibility to non-specialists, the published versions include an extended introduction, as well as a foreword by the student's supervisor explaining the special relevance of the work for the field. As a whole, the series will provide a valuable resource both for newcomers to the research fields described, and for other scientists seeking detailed background information on special questions. Finally, it provides an accredited documentation of the valuable contributions made by today's younger generation of scientists.

Theses are accepted into the series by invited nomination only and must fulfill all of the following criteria

- They must be written in good English.
- The topic of should fall within the confines of Chemistry, Physics and related interdisciplinary fields such as Materials, Nanoscience, Chemical Engineering, Complex Systems and Biophysics.
- The work reported in the thesis must represent a significant scientific advance.
- If the thesis includes previously published material, permission to reproduce this must be gained from the respective copyright holder.
- They must have been examined and passed during the 12 months prior to nomination.
- Each thesis should include a foreword by the supervisor outlining the significance of its content.
- The theses should have a clearly defined structure including an introduction accessible to scientists not expert in that particular field.

Juan Pedro Ochoa-Ricoux

A Search for Muon Neutrino to Electron Neutrino Oscillations in the MINOS Experiment

Doctoral Thesis accepted by California Institute of Technology, Pasadena, California, USA

Springer

Author
Dr. Juan Pedro Ochoa-Ricoux
Lawrence Berkeley National Laboratory
1 Cyclotron Rd.
Ms 50A2161
Berkeley, CA 94720
USA
e-mail: jpochoa@lbl.gov

Supervisor
Harvey B. Newman
California Institute of Technology
1200 East California Boulevard
Pasadena, CA 91125
USA

ISSN 2190-5053

e-ISSN 2190-5061

ISBN 978-1-4419-7948-3

e-ISBN 978-1-4419-7949-0

DOI 10.1007/978-1-4419-7949-0

Springer New York Dordrecht Heidelberg London

Cover design: eStudio Calamar, Berlin/Figueres

Printed on acid-free paper

Springer is part of Springer Science+Business Media (www.springer.com)

Supervisor's Foreword

"Search for Muon Neutrino to Electron Neutrino Oscillations in the MINOS Experiment" by Juan "Pedro" Ochoa, is a tour de force that summarizes, clearly and in some depth, much of his vast body of work which has had a major impact on MINOS. The centerpiece of the thesis is the search for ν_μ to ν_e oscillations which would indicate a non-zero mixing angle between the first and third neutrino generations θ_{13}, currently the "holy grail" of neutrino physics. The optimal extraction of the ν_e oscillation signal is based on the novel "library event matching" (LEM) method which Ochoa developed and implemented together with colleagues at Caltech and at Cambridge, which improves MINOS reach for establishing an oscillation signal over any other method.

LEM will now be the basis for MINOS' final results, and will likely keep MINOS at the forefront of this field until it completes its data taking in 2011. The LEM method has been further refined in the past year, to maximize MINOS' reach in this domain.

Ochoa and his colleagues also developed the plan to run MINOS with a beam tuned for antineutrinos, which has recently proven to be successful, leading to a sensitive test of CPT symmetry by comparing the inter-generational mass splitting for neutrinos and antineutrinos. The results of the antineutrino oscillation analysis have provided tantalizing evidence of a possible difference, and so this most fundamental symmetry of particle physics will be further tested as MINOS doubles its antineutrino dataset within the next 12 months, and possibly sooner.

Ochoa's in-depth, creative approach to the solution of a variety of complex experimental problems is an outstanding example for graduate students and longtime practitioners of experimental physics alike. Some of the most exciting results in this field to emerge in the near future may find their foundations in this thesis, for which Ochoa was awarded the National Prize for Youth of Mexico in 2008.

It was a great pleasure for me to take part in supervising the work in this thesis, and to contribute to Pedro's growth as a physicist, together with my colleagues at Caltech, most notably Professor Ryan Patterson and Dr. Caius Howcroft in recent years. I am therefore very glad that the publication of this thesis will give a wider audience the chance to appreciate and learn from Ochoa's work.

It is important to mention that this work would not have been possible without the founding contributions to the study of neutrinos by Professors Barry Barish and Charles Peck at Caltech's Lauritsen Lab over the last 40 years, the invention of this class of massive neutrino detector and the many contributions to the MINOS experiment by the late Dr. Doug Michael, and the strong support of the Department of Energy Office of High Energy Physics.

Pasadena, August 2010 Harvey B. Newman

Acknowledgments

There are many people to whom I am indebted and without whom this work would not have reached its fruitful completion.

First I would like to thank my advisor, Harvey B. Newman. Despite his many occupations, he always found the time to meet every week and constantly took an active interest in my progress. In addition, he always provided me with every available opportunity to benefit my work. I owe the successful completion of this part of my life and the opportunities that have opened up as a result to his constant support and guidance.

When I first started my graduate career, my knowledge of experimental neutrino physics was very limited. I was fortunate enough to have three extremely talented postdocs working with me at Caltech who were always willing to go well beyond the call of duty. Those postdocs are Chris Smith, Caius Howcroft, and Ryan Patterson. Almost everything I have learned about computer programming, particle physics, and statistics has been directly from them.

I was privileged enough to spend two years working with Doug Michael. His love for physics and his boldness when attacking problems has continued to inspire me since his tragic death in 2005. I have also been fortunate to work with very talented people at Caltech such as Alex Himmel, Mhair Orchanian and Hai Zheng. I would like to thank Leon Mualem who, along with Harvey Newman and Ryan Patterson, dedicated a lot of his time to ensure that this thesis was the best it could be.

Needless to say, the work presented here could not have been completed without the hard work of everybody in the MINOS collaboration and in the ν_e group in particular. The strong leadership of Mayly Sanchez and Trisha Vahle was key to the success of the analysis. It was truly a pleasure to work with such dedicated people such as Josh Boehm, Anna Holin, Greg Pawloski and Tingjun Yang. I want to thank David Jaffe for his encouragement during the work that we did together. I also want to thank Mark Thomson for his support throughout this time and for including me in the development of the ν_e selection described in this thesis that he originally proposed.

I could not have completed this thesis without the support of my dear wife. Carol, I am lucky to have you, and I am looking forward to many more adventures together. I am also grateful to all my friends from church for their constant encouragement and fellowship. I thank God for you, as well as for allowing me to complete this stage of my life, through which I hope to honor Him.

Last but not least, I want to thank my family in Mexico. My sister Sofia has always been a source of joy and a motivation for me to go visit as often as possible. My parents Lucia and Juan Manuel have provided me with every opportunity and have never failed to teach me about the importance of hard work and dedication through their own example. I dedicate this thesis to them.

Pasadena, October 2009 Juan Pedro Ochoa-Ricoux

Contents

Chapter 1
Introduction

Neutrinos are unique subatomic particles. The fact that they display extraordinary properties that are unmatched by any other particle makes the field of neutrino physics full of both challenges and opportunities. For instance, due to their extremely low interaction rate, neutrinos have the ability to travel several light-years through a solid material without interacting. On the one hand, this property makes them elusive and difficult to study. On the other hand, it is precisely the fact that they interact so differently from other particles that has led the way to the formulation of groundbreaking physics theories and a deeper understanding of our universe.

This thesis begins with a historical look at the neutrino in Chap. 2. Since they were first proposed in 1930, until their incorporation in the Standard Model, neutrinos have been at the heart of many major discoveries and breakthroughs in the field of particle physics. So far, every piece of evidence suggests that this trend will continue. In 1998 the discovery that neutrinos change flavor as they travel revolutionized their place in physics and in our universe. The fact that neutrinos have mass currently stands as the only confirmed piece of experimental evidence that lies outside the realm of the Standard Model. Just as the proposal of the neutrino led the way to the formulation and discovery of weak interactions, the study of neutrino oscillations will not only augment our understanding of the universe, but has great potential as a beacon that leads the way to new physics.

There has been substantial progress in our understanding of neutrinos in the last decade, particularly through the study of neutrino oscillations. Nevertheless, as explained at the end of Chap. 2, neutrinos remain the least understood fundamental particles to have been experimentally observed. Among the many initiatives in neutrino physics currently being undertaken, measuring the amount of mixing between the first and third generations, as described by the θ_{13} mixing angle, is of particular importance. Not only is the determination of θ_{13} needed to complete our picture of neutrino oscillations, but it also constitutes a gateway to other discoveries. For instance, the phenomenon of CP violation in the leptonic sector, which

J. P. Ochoa Ricoux, *A Search for Muon Neutrino to Electron Neutrino Oscillations in the MINOS Experiment*, Springer Theses, DOI: 10.1007/978-1-4419-7949-0_1,

has important consequences in cosmology, is inextricably linked to the value of θ_{13}. Also, it may be possible to determine the mass hierarchy of the neutrino sector through the study of θ_{13} driven oscillations.

An aggressive experimental program, consisting of two new reactor experiments and two new accelerator experiments, has been set in motion in order to measure θ_{13}. Before these next generation experiments come online, the MINOS experiment has a chance of making the first measurement of a nonzero θ_{13} if this angle lies close to the current experimental limit. MINOS is a long-baseline neutrino oscillation experiment, as described in Chap. 3. As such, it consists of a beam of muon neutrinos that is sampled at two locations. The MINOS Near Detector allows us to make a direct measurement of the beam's composition and energy spectrum before any oscillations take place. The MINOS Far Detector, located at a distance of approximately 735 km from the Near Detector, is used to search for evidence of oscillations.

The MINOS experiment is thus capable of searching for $\nu_\mu \rightarrow \nu_e$ oscillations, which is the main topic of this thesis. As explained in Chap. 4, the observation of ν_e charged-current (CC) event appearance at the Far Detector would constitute an unmistakeable signature of a nonzero θ_{13}. The search for ν_e appearance is challenging however, as the MINOS detectors were not designed for ν_e CC identification. The consequence is that when ν_e CC events are selected in MINOS, an even larger number of background events are also selected. The reach in θ_{13} thus depends primarily on the ability to separate the signal from the backgrounds.

In order to maximize our reach, we have developed a novel ν_e CC identification technique called the Library Event Matching (LEM) selection. As opposed to other techniques, where reconstructed variables are combined through a multivariate approach, LEM performs event identification based on each event's pattern of energy deposition. This is achieved by comparing each input event to a large library of simulated signal and background events. A discriminant is then constructed from the information of the N best matches. By making better use of all the available information in each event, LEM gives the best sensitivity to θ_{13} among all the other selection methods ever attempted in MINOS. The mechanism and the performance of LEM are described in Chap. 5. The technical details are presented in Appendix B.

While LEM achieves a higher signal-to-background ratio than other methods, the number of background events is still larger than the number of signal events expected for a value of θ_{13} at the current experimental limit. The backgrounds must thus be accurately predicted in the Far Detector in order to maximize the reach in θ_{13}. The total number of background events can be directly measured in the Near Detector data. This is not sufficient however, given that the prediction of the Far Detector rates from the measured Near Detector rates (also referred to as the "extrapolation") depends on the event type considered. It is thus necessary to decompose the Near Detector spectrum using the two methods which are described in Chap. 8.

The prediction of the Far Detector signal and background rates is described in Chap. 7. The signal rates are predicted from the tuned simulation, while background rates are predicted from the Near Detector data. By using two detectors the systematic uncertainties that affect the selected events equally in the two detectors cancel to first order. The largest uncertainties in the simulation, such as the ones in the hadronic shower simulation, fall into this category. Nevertheless, it is still necessary to account for the small differences between the detectors as well as any other sources of systematic error. This is done as shown in Chap. 8.

A blind analysis procedure was followed, meaning that the estimation of the Far Detector predictions and associated systematic errors was finalized before looking at the Far Detector data. Also, a number of sidebands were analyzed in order to ensure the correctness of the different aspects of the analysis. The results from these sidebands, as well as from the ν_e appearance search itself, can be found in Chap. 9. The confidence intervals are derived using a Feldman-Cousins method described in Appendix D.

Chapter 10 ends with the prospects for future analyses in MINOS. The outlook for the next ν_e appearance analysis is very promising, given that the amount of data will have doubled and that much was learned with this first analysis. We find also that it is beneficial for MINOS to focus on the area of antineutrino physics after a second ν_e appearance analysis. MINOS has a unique advantage with respect to other neutrino experiments in that it can separate neutrinos and antineutrinos on an individual basis. In the second half of Chap. 10 we analyze the potential of two antineutrino physics analyses in MINOS. We also present the proposal, that we developed at Caltech, to run in antineutrino mode, where the beam consists mostly of antineutrinos instead of neutrinos. Our proposal has been accepted by the collaboration.

Chapter 2
Neutrino Physics

Neutrinos are truly fascinating particles that, due to their extraordinary properties, have played and continue to play a crucial role in our understanding of the universe. In the first part of this chapter we review the basic properties of neutrinos from a historical perspective until the discovery of neutrino mass, approximately a decade ago. We then devote one section to the topic of neutrino oscillations, the study of which constitutes the main purpose of the MINOS experiment. We end with an assessment of the open questions in neutrino physics, with a particular emphasis on those addressed by this thesis.

2.1 A Historical Look at the Neutrino

2.1.1 Proposal and Discovery

The twentieth century saw a number of developments in physics unparalleled in human history. In the area of particle physics, neutrinos were frequently at the heart of these developments, which is why studying their history is an enriching task.

After the discovery of radioactivity [1] by the French physicist Henri Becquerel in 1896 it was thought that the process of β-decay was fully explained by the emission of an electron from the radioactive nucleus. In 1914 however the English physicist James Chadwick conclusively demonstrated [2] that the electron energy spectrum from β-decay is continuous. The theoretical interpretation of this fact was troublesome. One possibility was that it was due directly to how electrons are emitted from the nucleus, which seemed to violate the principle of energy conservation. Another possibility was that secondary processes were at work, such as electrons losing a fraction of their energy to the surrounding atoms depending on the source's thickness traversed.

J. P. Ochoa Ricoux, *A Search for Muon Neutrino to Electron Neutrino Oscillations in the MINOS Experiment*, Springer Theses, DOI: 10.1007/978-1-4419-7949-0_2,

In 1927 a crucial experiment was performed by Ellis and Wooster [3], where they measured the average absolute heat in the absorption of β-electrons and found it to be in excellent agreement with the mean energy of the β-spectrum. This meant that no energy was being lost in secondary processes and that the electrons were indeed being emitted with varying energies from the nucleus.

Although these results were conclusive, they were very hard to interpret at the time. Some distinguished physicists, such as Niels Bohr, even suggested that energy conservation could be violated in individual decays, although perhaps not statistically. Momentum and angular momentum also appeared to not be conserved. It was not until 1930 that the solution was stumbled upon, after being suggested by Wolfgang Pauli as a "desperate remedy" in an open letter [4] to Geiger and Meitner who were in a physics meeting in Tübingen. Pauli postulated that another particle was being emitted alongside the electron during β-decay which carried a portion of the total energy and momentum. The hypothetical particle had to be neutral in order to ensure conservation of electric charge (as well as to explain why it had not been detected), and spin 1/2 in order to ensure conservation of angular momentum. It also had to have a very small mass in order to be consistent with the endpoint of the β-spectrum and its apparent lack of interactions with other particles. Pauli called this particle the "neutron."

In 1932 Chadwick discovered [5] what we now refer to as the neutron, which was too heavy to be the particle predicted by Pauli. In 1933 Enrico Fermi renamed Pauli's particle as "neutrino," which in Italian means "little neutral one." Fermi then gave the neutrinos a solid theoretical framework in his theory of β-decay [6]. Fermi's theory was so successful in its explanation of most of the important features of β-decay that the majority of physicists accepted the reality of the neutrino. Based on Fermi's theory, Hans Bethe and Rudolf Peierls showed in 1934 [7] that the cross section between neutrinos and matter should be extremely small, billions of times smaller than the corresponding cross section for an electron. This meant that neutrinos could go through the earth without deviation.

By 1950, there was thus a solid theoretical basis for the neutrino as well as compelling indirect evidence coming from the study of pion decays [8]. Using the photographic emulsion method, pions could be seen decaying into muons but at distinct angles from the original pion direction, suggesting that other particles that were not being detected were being emitted as well. The study of muon decays reached similar conclusions. No neutrinos had been directly detected however, due to their extremely small interaction rate.

The first direct detection of a neutrino came with the development of nuclear technology, which made very large fluxes of man-made neutrinos available. Physicists Cowan and Reines set up two tanks of approximately 200 l each of a mixture of water and cadmium chloride next to the Savannah River Plant's nuclear reactor [9]. The (anti-)neutrinos originating from the nuclear reactor interacted via the inverse β-decay reaction:

$$\overline{\nu} + p^{+} \rightarrow n + e^{+}. \tag{2.1}$$

The positrons then annihilated with electrons of the surrounding material, emitting two photons. The neutrons were captured by cadmium nuclei, giving a second emission of photons some 5 μs after the positron annihilation. While the flux of neutrinos in the detector was 5×10^{13} neutrinos per second per square centimeter, Cowan and Reines observed only approximately three neutrino interactions per hour, with a signal-to-background ratio of approximately 4 to 1. To have absolute certainty that reactor neutrinos were being detected they turned off the reactor and observed a lower rate of events. Reines was honored with the Nobel Prize in 1995 for this discovery.[1]

At around the same time, physicists such as R. Davis and L. Alvarez also tried to detect neutrinos next to nuclear power plants. They used carbonate chloride solutions, hoping for chlorine to transform into radioactive argon under the action of neutrinos. The argon could then be chemically separated and counted. Unfortunately for them however, a nuclear power plant delivers only antineutrinos. Nevertheless, this led to the confirmation that neutrinos are different from antineutrinos [10] and to the famous Homestake experiment, described later on.

The neutrinos detected by Reines and Cowan were electron antineutrinos $\overline{\nu}_e$, although the neutrino type was not known then. By the late fifties, the law of lepton conservation number introduced by Konopinski and Mahmoud [11] had proven successful in every instance, except in the non-observation of the decay of a muon into an electron plus a photon,

$$\mu^- \nrightarrow e^- + \gamma, \tag{2.2}$$

which seemed consistent with conservation of charge and lepton number. The absence of this decay suggested a conservation of muon-type and electron-type leptons. If that was the case however, the observation of the decay of a muon into an electron and two neutrinos,

$$\mu^- \rightarrow e^- + \overline{\nu} + \nu, \tag{2.3}$$

could only be explained if the neutrino and the antineutrino were really of two different kinds, one associated with the muon and one with the electron. The first experimental test of this hypothesis was conducted in 1962 by Leon M. Lederman, Melvin Schwartz, and Jack Steinberger at Brookhaven National Laboratory [12]. The experiment used a beam of protons impinging on a beryllium target to produce a shower of π mesons, which travelled 70 ft towards a 5,000-ton steel wall made of old battleship plates. This ensured that only the neutrinos from pion decay[2] reached a 10-ton aluminum spark chamber where 56 neutrino-like interactions were recorded during the lifetime of the experiment. In all these interactions a

[1] Unfortunately Cowan died in 1974.

[2] It should be mentioned that the principle for the production of a muon neutrino beam first introduced in this experiment is still the one used today in experiments like MINOS.

muon was produced instead of an electron, confirming that the muon neutrino ν_μ and the electron neutrino ν_e were indeed different particles. Lederman, Schwartz, and Steinberger received the 1988 Nobel Prize in physics for this discovery.

In 1975 the tau lepton was discovered [13] at the Stanford Linear Accelerator Center (SLAC) by Martin Perl using the SPEAR e^+e^- collider. This discovery expanded the number of lepton families to three, and hinted at the existence of a respective tau neutrino ν_τ. The tau-neutrino was not observed until 2001, due to the experimental challenges associated with its detection. The difficulties arise because tau-neutrinos are identified by the creation of the tau-lepton, which decays very quickly (after traveling only for $\sim$1 mm). By colliding 800 GeV protons with a large block of tungsten (referred to as a "beam dump"), the DONUT collaboration was able to create D_S mesons that subsequently decayed into tau-leptons which then produced tau-neutrinos. These decays produced other particles that were deflected and stopped in magnetized iron shielding, leaving only neutrinos downstream. The resulting tau-neutrino beam then interacted in a detector consisting of emulsion targets followed by a spectrometer. With an initial observation of 4 tau-neutrino events, the DONUT experiment made the first detection of the tau-neutrino [14]. This completed the experimental observation of all the neutrinos that are currently thought to exist.

2.1.2 Neutrinos in the Standard Model

2.1.2.1 Basic Ingredients of the Standard Model

In order to continue with our description of the neutrino, it is necessary to understand its place in the most complete and successful theory to date of all the known particle interactions with the exception of gravity: the Standard Model.

While Fermi's theory provides an excellent description of observed low energy weak interactions, it is flawed when considered as a field theory. One of the main reasons for this is that the calculation of higher-order weak interaction corrections to any lowest order weak process yields divergent results. For instance, the neutrino-electron scattering cross section grows without limit as a function of increasing energy. Fermi's description of the weak interaction simply breaks down above a certain energy. Historically, this prompted the search for a renormalizable Lagrangian and culminated in the discovery of gauge theories, which are a fundamental ingredient of the Standard Model.

The three forces considered by the Standard Model are the electromagnetic, weak, and strong forces. The electromagnetic force can be attractive or repulsive, is long range, and acts only between particles carrying electric charge. The weak force is responsible for radioactive decay and is very short range. Finally, the strong force is responsible for holding the nuclei of atoms together, is very short range (acts at distances in the order of 10^{-13} cm), and acts only on particles that carry color.

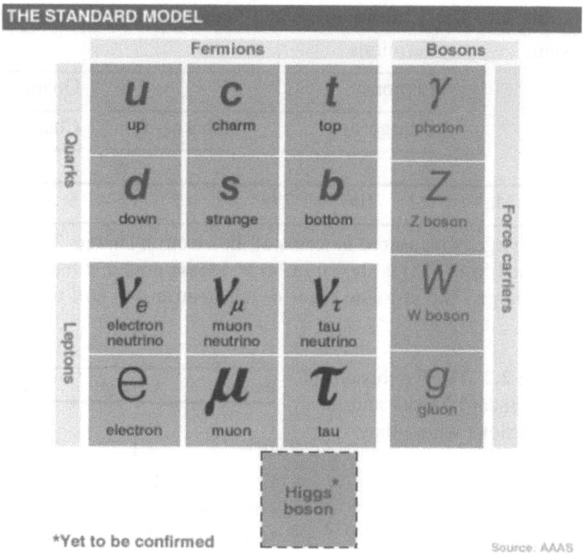

Fig. 2.1 A summary of the particle content of the Standard Model, which is composed of 12 fermions and 4 force carrier gauge bosons. There is also a Higgs boson, whose mechanism is thought to be responsible for the origin of mass. Source: AAAS

All the particles that make up the Standard Model of Particle Physics are shown in Fig. 2.1. Neutrinos are part of the Standard Model, and they belong to the group of 12 fundamental particles of spin-$1/2$ known as fermions. These 12 particles make up all the visible matter in the universe, and they are divided between leptons and quarks. Leptons are stable by themselves and interact only via the electromagnetic and weak forces, whereas quarks always exist in combination with others (they are "confined") and interact also through the strong force. The quarks and the leptons with the same charge come in groups of three, and each fermion has its own antiparticle. Table 2.1 shows the type, charge and mass of each fermion. The absolute mass of the neutrinos is still unknown, as will be emphasized later.

The Standard Model also contains bosons, which can have spin 0 or 1.[3] Four bosons are the mediators of the three fundamental forces described in the Standard Model: the electromagnetic force, the weak force and the strong force. The bosons and their corresponding interactions are shown in Table 2.2. Also in the Standard Model is the Higgs boson, through which the origin of mass in the universe is explained. The Higgs boson has not been observed to this date, although its detection is one of the goals [16] of the experiments taking place in the Large Hadron Collider (LHC) at CERN in Switzerland.

---

[3] The graviton, which is the force carrier for gravity, is thought to have spin-2. Gravity however is not included in the Standard Model.

Table 2.1 The 12 fermions of the Standard Model, separated by type and charge

Name	Fermions											
	Leptons						Quarks					
	Charge: −1			Charge: 0			Charge: −1/3			Charge: +2/3		
	e	μ	τ	ν_e	ν_μ	ν_τ	d	s	b	u	c	t
Mass (MeV)	0.5	106	1,777	?	?	?	5	105	4,200	3	1,270	171,300

The charge is measured in terms of the electron charge. The approximate mass of each fermion as obtained from [15] is shown in MeV. Please note that the uncertainty in the mass of each quark is quite large. The absolute mass of the neutrinos is still unknown.

Table 2.2 The five bosons of the Standard Model and the interactions which they mediate

Boson	Charge	Spin	Mass (GeV)	Force
γ	0	1	0	Electromagnetic
$W^{\pm}$	±1	1	80.4	Weak
Z	0	1	91.2	Weak
Gluon(8)	0	1	0	Strong
Higgs	0	0	?	-

2.1.2.2 Weak Interactions in the Standard Model

Theoretical formalism

In the Standard Model, all of the fundamental interactions between the fundamental particles are derived from the requirement of local gauge invariance. The Standard Model is a gauge theory of the electroweak and strong interactions with the gauge group $SU(3) \otimes SU(2) \otimes U(1)$. The theory of strong interactions, Quantum Chromodynamics (QCD), is a gauge theory of the $SU(3)$ gauge group by taking the color charge to define a local symmetry. The electromagnetic and weak interactions are unified under an $SU(2)_L \otimes U(1)_Y$ gauge group. Neutrinos interact only through the weak interaction, mediated by the $W^{\pm}$ and Z bosons. Consequently, we concentrate our attention on the electroweak portion of the Standard Model.

The unification of the electromagnetic and weak forces was achieved by Glashow, Weinberg and Salam (GWS) in the 1960s [17–19]. Experiments by Wu [20], Goldhaber [21] and others had demonstrated the left-handedness of the neutrino, showing that the charged current weak interactions couple to particular chirality states. The SU(2) group allowed for a natural implementation of this property by allowing that only the left-handed fermion fields (and the right-handed anti-fermion fields) have a non-trivial representation in the group. In other words, the left-handed states must transform non-trivially under the $SU(2)$ algebra, and they are consequently arranged in doublets:

$$E_e = \begin{pmatrix} \nu_e \\ e^- \end{pmatrix}_L, \quad E_\mu = \begin{pmatrix} \nu_\mu \\ \mu^- \end{pmatrix}_L, \quad E_\tau = \begin{pmatrix} \nu_\tau \\ \tau^- \end{pmatrix}_L. \tag{2.4}$$

Given that the right-handed states do not couple to the $W^{\pm}$, they are expressed as $SU(2)$ singlets:

$$e_e = e_R, \quad e_\mu = \mu_R, \quad e_\tau = \tau_R. \tag{2.5}$$

No right-handed neutrinos are considered given that none have been observed to this date.[4] We make similar assignments in the quark sector:

$$Q_u = \begin{pmatrix} u \\ d \end{pmatrix}_L, \quad Q_c = \begin{pmatrix} c \\ s \end{pmatrix}_L, \quad Q_t = \begin{pmatrix} t \\ b \end{pmatrix}_L, \tag{2.6}$$

$$u_u = u_R, \quad u_c = c_R, \quad u_t = t_R, \quad d_d = d_R, \quad d_s = s_R, \quad d_b = b_R. \tag{2.7}$$

In order to derive the weak interactions in the Standard Model we follow the procedure and the notation of [22]. A generalized position dependent rotation in the $SU(2)_L \otimes U(1)_Y$ of a field ϕ requires the existence of four vector gauge fields. We denote the $SU(2)$ and $U(1)$ gauge bosons by A^a_μ and B_μ respectively. The complete gauge transformation of ϕ is given by

$$\phi \rightarrow e^{ig\tau^a A^a_\mu} e^{\frac{i}{2}g' B_\mu} \phi, \tag{2.8}$$

where $\tau^a = \sigma^a/2$, and σ^a denotes the Pauli matrices. g and g' are the different coupling constants of the $SU(2)$ and $U(1)$ factors of the gauge group. The covariant derivative of ϕ is thus given by

$$D_\mu \phi = \left(\partial_\mu - igA^a_\mu \tau^a - \frac{i}{2} g' B_\mu \right) \phi. \tag{2.9}$$

In order to have massive gauge bosons, the symmetry must be spontaneously broken. In the GWS theory this is accomplished through the Higgs Mechanism [23–25], where a scalar field H is introduced and assumed to acquire a vacuum expectation value of the form

$$\langle H \rangle = \frac{1}{\sqrt{2}} \begin{pmatrix} 0 \\ v \end{pmatrix}. \tag{2.10}$$

The gauge boson mass terms come from the square of Eq. 2.8 evaluated at $<H>$:

$$\Delta\mathcal{L} = \frac{1}{2} (0 \quad v) \left(gA^a_\mu \tau^a + \frac{1}{2} g' B_\mu \right) \left(gA^{b\mu} \tau^b + \frac{1}{2} g' B^\mu \right) \begin{pmatrix} 0 \\ v \end{pmatrix}. \tag{2.11}$$

Evaluating the previous expression we find

---

[4] As we will show later, this implies that neutrinos are non-massive in the minimal Standard Model.

$$\Delta\mathcal{L} = \frac{v^2}{8}\left[g^2(A_\mu^1)^2 + g^2(A_\mu^2)^2 + (-gA_\mu^3 + g'B_\mu)^2\right]. \tag{2.12}$$

If we define a new set of gauge fields by

$$W_\mu^\pm = \frac{1}{\sqrt{2}}(A_\mu^1 \mp iA_\mu^2), \tag{2.13}$$

$$Z_\mu^0 = \frac{1}{\sqrt{g^2 + g'^2}}(gA_\mu^3 - g'B_\mu), \tag{2.14}$$

$$A_\mu = \frac{1}{\sqrt{g^2 + g'^2}}(g'A_\mu^3 + gB_\mu), \tag{2.15}$$

then we can re-express the relevant terms of Eq. 2.12 as

$$\Delta\mathcal{L} = \frac{1}{2}\frac{v^2}{4}\left(g^2 W_\mu^+ W^{-\mu} + (g^2 + g'^2)Z_\mu^0 Z^{0\mu}\right), \tag{2.16}$$

where the $W^\pm$ and Z^0 fields are the three massive vector bosons. From the previous expression we can see[5] that their masses are given by

$$m_{W^\pm} = g\frac{v}{2}, \tag{2.17}$$

$$m_{Z^0} = \frac{v}{2}\sqrt{g^2 + g'^2}. \tag{2.18}$$

The A_μ field, which is orthogonal to Z_μ^0, is massless. This is the electromagnetic vector potential.

To complete the description of the electroweak interactions, we consider a fermion field belonging to a general $SU(2)$ representation, with $U(1)$ charge Y. The covariant derivative, in terms of the mass eigenstate fields, becomes

$$\begin{aligned} D_\mu = \partial_\mu - i\frac{g}{\sqrt{2}}(W_\mu^+ T^+ + W_\mu^- T^-) - i\frac{1}{\sqrt{g^2 + g'^2}}Z_\mu(g^2T^3 - g'^2Y) \\ - i\frac{gg'}{\sqrt{g^2 + g'^2}}A_\mu(T^3 + Y), \end{aligned} \tag{2.19}$$

where $T^\pm = (T^1 \pm iT^2)$. In the spinor representation of $SU(2)$, $T^\pm = \sigma^\pm$.

From Eq. 2.19, it can be seen that the electromagnetic vector potential A_μ couples to the gauge generator $T^3 + Y$. We can thus identify the electric charge quantum number as $Q = T^3 + Y$. Moreover, we can identify the coefficient of electromagnetic interaction as the electron charge e,

---

[5] Section 2.1.2 includes a discussion of what the Dirac mass terms in the Lagrangian look like.

$$e = \frac{gg'}{\sqrt{g^2 + g'^2}}. \tag{2.20}$$

To further simplify Eq. 2.19, we introduce the weak mixing angle defined by

$$\cos\theta_w = \frac{g}{\sqrt{g^2 + g'^2}}, \quad \sin\theta_w = \frac{g'}{\sqrt{g^2 + g'^2}}. \tag{2.21}$$

Then, using the following manipulation in the Z^0 coupling,

$$g^2 T^3 - g'^2 Y = (g^2 + g'^2) T^3 - g'^2 Q, \tag{2.22}$$

we rewrite the covariant derivative (2.19) as

$$D_\mu = \partial_\mu - i\frac{g}{\sqrt{2}}(W_\mu^+ T^+ + W_\mu^- T^-) - i\frac{g}{\cos\theta_{\theta_w}} Z_\mu (T^3 - \sin^2\theta_w Q) - ieA_\mu Q, \tag{2.23}$$

where $g = e/\sin\theta_w$. Ignoring fermion masses and using the terminology introduced in Eqs. 2.4–2.7, the Lagrangian for the weak interactions of quarks and leptons is given by

$$\mathcal{L}_{ew} = \overline{E}_i(i\not{D})E_i + \overline{e}_i(i\not{D})e_i + \overline{Q}_j(i\not{D})Q_j + \overline{u}_j(i\not{D})u_j + \overline{d}_k(i\not{D})d_k, \tag{2.24}$$

where $i = e, \mu, \tau$ and $j = u, c, t$ and $k = d, s, b$. Rewriting this expression using the form of the covariant derivative of Eq. 2.23 we get

$$\begin{aligned} \mathcal{L}_{ew} = {} & \overline{E}_i(i\not{\partial})E_i + \overline{e}_i(i\not{\partial})e_i + \overline{Q}_j(i\not{\partial})Q_j + \overline{u}_j(i\not{\partial})u_j + \overline{d}_k(i\not{\partial})d_k \\ & + g(W_\mu^+ J_W^{\mu+} + W_\mu^- J_W^{\mu-} + Z_\mu^0 J_Z^\mu) + eA_\mu J_{EM}^\mu, \end{aligned} \tag{2.25}$$

where

$$J_W^{\mu+} = \frac{1}{\sqrt{2}}(\overline{\nu}_L \gamma^\mu e_L + \overline{u}_L \gamma^\mu d_L); \tag{2.26}$$

$$J_W^{\mu-} = \frac{1}{\sqrt{2}}(\overline{e}_L \gamma^\mu \nu_L + \overline{d}_L \gamma^\mu u_L); \tag{2.27}$$

$$\begin{aligned} J_Z^\mu = \frac{1}{\cos\theta_w} \Bigg[& \overline{\nu}_L \gamma^\mu \left(\frac{1}{2}\right) \nu_L + \overline{e}_L \gamma^\mu \left(-\frac{1}{2} + \sin^2\theta_w\right) e_L + \overline{e}_R \gamma^\mu (\sin^2\theta_w) e_R \\ & + \overline{u}_L \gamma^\mu \left(\frac{1}{2} - \frac{2}{3}\sin^2\theta_w\right) u_L + \overline{u}_R \gamma^\mu \left(-\frac{2}{3}\sin^2\theta_w\right) u_R. \\ & + \overline{d}_L \gamma^\mu \left(-\frac{1}{2} + \frac{1}{3}\sin^2\theta_w\right) d_L + \overline{d}_R \gamma^\mu \left(\frac{1}{3}\sin^2\theta_w\right) d_R \Bigg]; \end{aligned} \tag{2.28}$$

$$J_{EM}^\mu = -\overline{e}\gamma^\mu e + \frac{2}{3}\overline{u}\gamma^\mu u - \frac{1}{3}\overline{d}\gamma^\mu d, \tag{2.29}$$

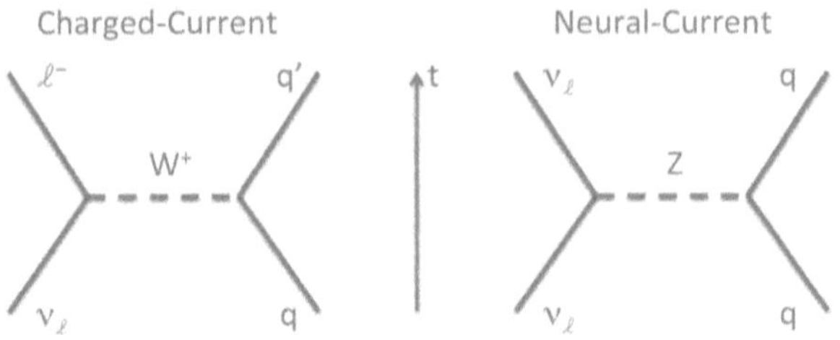

Fig. 2.2 Feynman diagrams of the charged-current and neutral-current neutrino interactions in MINOS. In the case of antineutrinos the sign of the W boson and of the outgoing lepton are reversed. Time flows in the upward direction, as indicated

and where we have dropped the i, j and k indices in the e, ν, u and d symbols for simplicity, although they are still there.[6] Also, we are ignoring quark mixing.

Interactions in MINOS. From the preceding equations we confirm that neutrinos interact only via the weak interaction. We see also that there are two possible types of weak interactions with neutrinos: the one where the neutrino couples to the $W^{\pm}$ bosons, and the one where it couples to the Z boson. We refer to these two as the Charged-Current (CC) and the Neutral-Current (NC) interactions respectively. In the MINOS experiment the primary neutrino interactions consist of the exchange of a $W^{\pm}$ or a Z boson with an iron nucleus in the detector, in which case the CC reactions for neutrinos and antineutrinos are described by

$$\nu_l + q \rightarrow l^- + q', \tag{2.30}$$

$$\overline{\nu}_l + q \rightarrow l^+ + q', \tag{2.31}$$

and the NC reactions by

$$\nu_l + q \rightarrow \nu_l + q, \tag{2.32}$$

$$\overline{\nu}_l + q \rightarrow \overline{\nu}_l + q. \tag{2.33}$$

Figure 2.2 shows the corresponding Feynman diagrams.

In the previous equations, q and q' are the two members of a quark family, as given by Eq. 2.6. For instance, if $q = u$, then $q' = d$. Nevertheless, due to quark mixing as described by the Cabbibo–Kobayashi–Maskawa (CKM) matrix [26, 27], there is a small probability that if $q = u$ then $q' = s$ or $q' = b$, and similarly for the other families. In any case, the quark cannot be identified experimentally, as it hadronizes almost immediatley through the production of other quarks and gluons. The same occurs for a NC interaction, where the energy

[6] In other words, without those indices Eqs. 2.26–2.29 are correct for the first family of leptons and quarks. Nevertheless, all of the terms still apply to the other two families as well.

transferred to the quark initiates a hadronic shower.[7] This is explained by color confinement [28], which states that quarks cannot exist individually. When two quarks become separated, at some point it is more energetically favorable for a new quark/anti-quark pair to spontaneously appear out of the vacuum [29]. CC and NC interactions thus appear in the detector as hadronic showers. In the case of CC interactions the outgoing lepton is also visible, which allows for the identification of the original neutrino flavor. A complete description of event topologies in MINOS is given in Chap. 4.

2.1.2.3 Neutrino Mass in the Standard Model

The Standard Model also makes a prediction concerning the mass of neutrinos. A Lagrangian can contain either Dirac mass terms or Majorana mass terms. A Dirac mass term for a neutrino v is of the form

$$-\mathcal{L}_{Dirac} = M\overline{v}v = M(\overline{v}_L v_R + \overline{v}_R v_L), \tag{2.34}$$

where M is the neutrino mass. In the Standard Model only the left-handed helicity state of the neutrino v_L is present at each generation. Consequently the Standard Model does not contain any Dirac mass terms for neutrinos.

Alternatively, a Majorana mass term is of the form

$$-\mathcal{L}_{Majorana} = \frac{1}{2}M(\overline{v}_L v_R^C + \overline{v}_R^C v_L) = \frac{1}{2}m(\overline{v}_L C\overline{v}_L^T + v_L^T C^\dagger, v_L) \tag{2.35}$$

where $v = v + v_R^C$ is a self-conjugate two component state [30] satisfying $v = v^C = C\overline{v}^T$. C is the charge conjugation matrix and $\dagger$ denotes the hermitian conjugate. A Majorana mass term thus requires only one helicity state of the particle and uses the opposite helicity state of the anti-particle. However, since v_L is part of the $SU(2)_L$ doublet (shown in Eq. 2.4) and has lepton number +1, the above neutrino mass term transforms as a $SU(2)_L$ triplet, which is not gauge invariant. Consequently, it is not allowed by the symmetries of the Standard Model. Moreover, the terms in Eq. 2.35 create (destroy) two neutrinos (antineutrinos) simultaneously and therefore do not conserve any quantum numbers. This is, for instance, why a charged particle cannot have a Majorana mass term. Neutrinos are neutral, but the Majorana mass term still breaks lepton number conservation by two units. The Standard Model however has perfect lepton number symmetry, even after symmetry breaking and to all orders in perturbation theory. Because of these reasons the Standard Model contains no Majorana mass terms and incorporates neutrinos as massless particles.

---

[7] We refer to the shower initiated by the energy transfer to the struck quark as a hadronic shower, even though it can have an electromagnetic component to it. This is discussed in Chap. 4.

2.1.3 Beyond the Standard Model

2.1.3.1 Vindication of the Standard Model

Soon after the proposal of the GWS theory, and after a tight race between groups at Fermilab and CERN, the first great vindication of the Standard Model was obtained with the Gargamelle experiment. Gargamelle was a $\sim 12\ m^3$ bubble chamber exposed to a neutrino beam from the CERN Proton Synchrotron. In 1973, the Gargamelle group reported the first observation of neutrino events inducing hadronic showers but with no outgoing lepton (i.e., neutral-current events) [31], thus providing the first experimental evidence of the existence of the Z boson.[8] The direct observation of the W and Z bosons had to wait another decade, for the construction of an accelerator powerful enough to produce them. Using the stochastic cooling technique developed by Simon van der Meer [32], the Super Proton Synchrotron (SPS) at CERN was used as a proton/anti-proton collider, achieving a center of mass energy of 540 GeV per collision. This was enough to produce the W and Z bosons, whose masses were estimated to be around 80 and 90 GeV at the time. In 1983, the UA1 and UA2 experiments at CERN announced their discovery [33, 34] to the world. In both cases the observation was made through the bosons' leptonic decays (a lepton and a neutrino in the case of the W and a lepton–antilepton pair in the case of the Z). C. Rubbia and S. van der Meer received the 1984 Nobel Prize in Physics for this discovery.

Perhaps the most comprehensive validation of the Standard Model was carried out at the Large Electron–Positron (LEP) collider at CERN. LEP had a circumference of 27 km and had four experiments built around four collision points in underground caverns. LEP allowed for the precise determination of many precise quantities of the Standard Model, most importantly the masses of the W and Z bosons which were determined to within 500 and 20 ppm respectively. Of particular interest to this thesis is the precision measurement of the shape of the Z boson mass peak done at LEP. The invisible width of the Z boson was determined by subtracting the measured visible width from the total width, then compared to the expectation due to N_ν light neutrino species as given by the Standard Model. The combined results from the four LEP experiments, shown in Fig. 2.3, set N_ν at 2.984 ± 0.008 [15, 35], thus confirming the Standard Model prediction of three neutrino families with impressive precision. This means that, if more neutrinos exist, they would have to be sterile or heavier than the Z boson.

The Standard Model continues to be tested at facilities such as SLAC, Fermilab and CERN, among others. The Standard Model currently stands as one of the most successful theories in the history of physics, as nearly every high-energy physics experiment to date has found results that are consistent with its predictions [36].

---

[8] As an interesting detail, it should be mentioned that the Gargamelle group scanned approximately 83,000 events, out of which only 102 neutral-current interactions were identified.

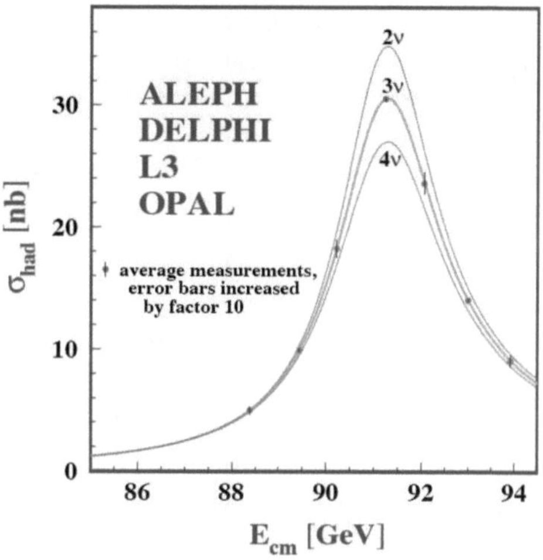

Fig. 2.3 Hadron production cross-section measurements around the Z resonance, done at LEP. The curves indicate the predictions with two, three and four neutrino species with Standard Model couplings and negligible mass. Image obtained from [35]

There is only one area where the Standard Model is known to be inadequate: neutrino masses.

2.1.3.2 The Discovery of Neutrino Oscillations

The first indication of nonzero neutrino masses came with the so-called solar neutrino problem. Starting in 1969, and using the same technique with which he had attempted the detection of neutrinos from a nuclear reactor, R. Davis set up a 100,000 gallon tank of perchloroethylene some 4,850 ft beneath the surface, in the Homestake gold mine, in South Dakota. Until its end in 1994, the experiment observed only $\sim 1/3$ of the expected flux of solar neutrinos [37] as predicted by John Bahcall's calculations [38]. This meant that either there was a problem with the neutrino flux predictions from the sun, or that electron neutrinos were somehow being lost or transformed on their way to the earth. These results were later confirmed by other experiments, such as SAGE [39], GALLEX [40] and Kamiokande [41].

The solar neutrino problem was not solved until approximately 30 years after the start of the Homestake experiment. Super-Kamiokande, a detector 15 times heavier than its predecessor Kamiokande, started operations in 1996. Super-Kamiokande is a cylindrical stainless steel tank containing 50,000 tons of ultrapure water and surrounded by 11,146 photomultipliers (PMTs) that remains in operation to this day. Neutrinos are identified by the Cherenkov light cones that charged particles originating in neutrino interactions produce while traveling in the water. In 1998, the Super-Kamiokande experiment produced solid evidence for the disappearance of muon-neutrinos produced in the atmosphere [42]. An atmospheric neutrino deficit had been seen previously by other experiments such as IMB [43],

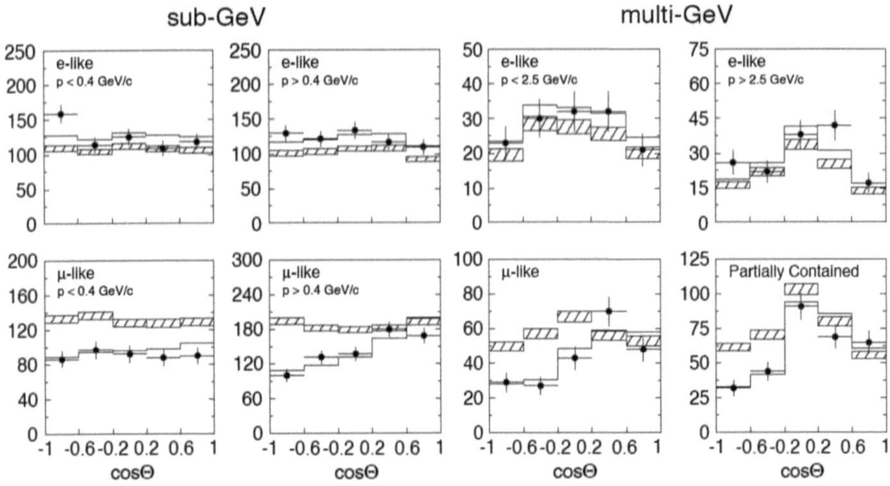

Fig. 2.4 Zenith angle distributions of μ-like and e-like events in sub-GeV and multi-GeV data sets in Super-Kamiokande. The hatched region shows the MC expectation in the absence of oscillations. The *bold line* is the best fit under the $\nu_\mu \rightarrow \nu_\tau$ oscillations hypothesis. Image obtained from [42]

MACRO [44] and Soudan-2 [45]. It was not until Super-Kamiokande however that the zenith angle dependence of the deficit was clearly established, as shown in Fig. 2.4. This meant that atmospheric ν_μ's were disappearing as a function of the distance traveled from their creation point in the atmosphere to the detector. Given that the e-like events showed no deficit, the preferred explanation was that ν_μ's were transitioning to ν_τ's on their way to the detector, as explained by the phenomenon of neutrino oscillations which is described in the next section. The observed disappearance was found to be in excellent agreement with the $\nu_\mu \rightarrow \nu_\tau$ oscillations hypothesis.

The final resolution of the solar neutrino problem came with the Sudbury Neutrino Observatory (SNO) experiment in the year 2000 [46], which showed that solar neutrinos were undergoing the same process of neutrino oscillations that had just been observed by Super-Kamiokande. The original SNO experiment consisted of approximately 1,000 tons of heavy water contained in a 20 ft acrylic vessel. The heavy water was seen by approximately 9,600 PMTs mounted on a geodesic sphere at a radius of about 850 cm. The presence of the deuteron allowed for different detection techniques. First, an incoming electron neutrino could undergo a charged-current (CC) interaction:

$$\nu_e + d \rightarrow p + p + e^- . \tag{2.36}$$

In this case the electron was detected through its Cherenkov radiation, which gives information about its energy and direction. Second, an incoming neutrino (of any flavor) could also undergo a neutral-current (NC) reaction:

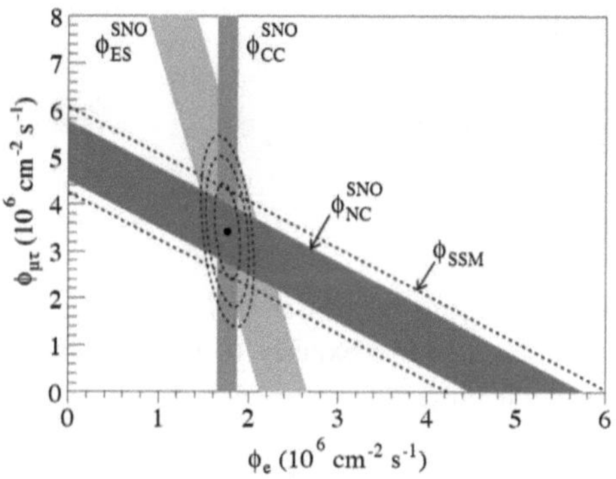

Fig. 2.5 Flux of boron-8 solar neutrinos which are μ or τ flavor vs. flux of electron neutrinos deduced from the three neutrino reactions in SNO. The diagonal bands show the total boron-8 flux as predicted by the Solar Standard Model (SSM), while the diagonal solid bands represents the SNO NC measurement. The intercept of these bands with the axes represent the $\pm 1\sigma$ errors. The bands intersect at the fit values for ϕ_e and $\phi_{\mu\tau}$, indicating that the combined flux results are consistent with neutrino flavor transformation with no distortion in the boron-8 neutrino energy spectrum. Image obtained from [47]

$$\nu_l + d \rightarrow \nu_l + n + p. \tag{2.37}$$

In this case the neutron was detected through its capture by a deuteron to form a tritium nucleus, a process that results in the emission of a gamma ray of roughly 6 MeV. Third, a neutrino of any flavor could also interact through electron elastic scattering (ES) reaction,

$$\nu_l + e^- \rightarrow \nu_l + e^-, \tag{2.38}$$

where the neutrino imparts some of its energy to an atomic electron. The NC and ES reactions are sensitive to all neutrino flavors, while the CC reaction is only sensitive to electron neutrinos. Through the CC reaction, a deficit of ν_e solar neutrinos consistent with the one observed by Homestake was seen in SNO. Nevertheless, the total neutrino flux (measured from the NC and ES reactions) was found to be in perfect agreement with the solar model predictions. These results provided unambiguous evidence that solar neutrinos transition to other flavors on their way to the earth, and also vindicated the solar model. Figure 2.5 illustrates these two points by comparing the observed ν_e and $\nu_\mu + \nu_\tau$ fluxes derived from the rate of CC, NC and ES events observed in the SNO data with the predictions of the Standard Solar Model (SSM) [47].

The preferred mechanism to account for the SNO and Super-Kamiokande observations is neutrino oscillations. The implications are enormous however, given that neutrino oscillations can only be explained with massive neutrinos, as shown next. After reviewing neutrino oscillations from the theoretical and experimental points of view in the next section, we address the questions that this discovery has raised at the end of the chapter.

2.2 Neutrino Oscillations

2.2.1 Theoretical Formalism

2.2.1.1 General Case

The underlying principle responsible for neutrino oscillations is that the neutrino weak eigenstates do not correspond exactly to the neutrino mass eigenstates, but rather are mixtures of each other:

$$|\nu_\alpha\rangle = \sum_{i=1}^{N} U^*_{\alpha i}|\nu_i\rangle, \tag{2.39}$$

where U is a $N \times N$ unitary matrix known as the Pontecorvo–Maki–Nakagawa–Sakata (PMNS) mixing matrix, in honor of the pioneering contributions of these physicists to neutrino mixing and oscillations [48, 49]. In the Standard Model, $N = 3$. The weak (flavor) eigenstates, denoted by $|\nu_\alpha\rangle$ in the previous equation, are the familiar ν_e, ν_μ and ν_τ. They are simply the states in which neutrinos interact through the exchange of the W and Z bosons. The mass eigenstates on the other hand, denoted $|\nu_1\rangle, |\nu_2\rangle$ and $|\nu_3\rangle$, are the neutrino states that have definite mass.

A mass eigenstate $|\nu_i\rangle$ with mass m_i and momentum p evolves in time according to

$$|\nu_i(t)\rangle = e^{-iE_it}|\nu_i\rangle, \tag{2.40}$$

where $E = \sqrt{p^2 + m_i^2}$ and where we use $c = 1$ units. The time evolution of a weak eigenstate $|\nu_\alpha\rangle$ is given by

$$|\nu_\alpha(t)\rangle = \sum_{i=1}^{N} U^*_{\alpha i}|\nu_i(t)\rangle. \tag{2.41}$$

Combining Eq. 2.40 with Eq. 2.39, we find

$$|\nu_\alpha(t)\rangle = \sum_{i=1}^{N} U^*_{\alpha i}e^{-iE_it}|\nu_i\rangle \tag{2.42}$$

$$= \sum_{i=1}^{N}\sum_{\beta=1}^{N} U_{\beta i}U^*_{\alpha i}e^{-iE_it}|\nu_\beta\rangle, \tag{2.43}$$

where we have used the unitarity of U to invert Eq. 2.39,

$$|\nu_i\rangle = \sum_{\beta=1}^{N} U_{\beta i}|\nu_\beta\rangle. \tag{2.44}$$

We can thus determine the probability that a neutrino created as a ν_α weak eigenstate will be detected as a ν_β weak eigenstate after time t:

$$\begin{aligned} P(\nu_\alpha \to \nu_\beta)(t) &= |\langle \nu_\beta | \nu_\alpha(t) \rangle|^2 \\ &= \left| \sum_{i=1}^{N} U_{\beta i} U^*_{\alpha i} e^{-iE_i t} \right|^2 \\ &= \sum_{i=1}^{N} \sum_{j=1}^{N} U^*_{\alpha i} U_{\alpha j} U_{\beta i} U^*_{\beta j} e^{-it(E_i - E_j)}. \end{aligned} \tag{2.45}$$

A highly relativistic neutrino covers a distance L in time t (given our $c = 1$ units), and has $E \approx p + \frac{m^2}{2p}$. Consequently, we have that

$$P(\nu_\alpha \to \nu_\beta)(t) = \sum_{i=1}^{N} \sum_{j=1}^{N} U^*_{\alpha i} U_{\alpha j} U_{\beta i} U^*_{\beta j} e^{-i\frac{L}{2p}\Delta m^2_{ij}}, \tag{2.46}$$

where $\Delta m^2_{ij} = m^2_i - m^2_j$. This expression can be more conveniently written as

$$\begin{aligned} P(\nu_\alpha \to \nu_\beta)(t) = \delta_{\alpha\beta} &- 4 \sum_{i>j} Re(U^*_{\alpha i} U_{\beta i} U_{\alpha j} U^*_{\beta j}) \sin^2\left(\frac{\Delta m^2_{ij} L}{4E}\right) \\ &+ 2 \sum_{i>j} Im(U^*_{\alpha i} U_{\beta i} U_{\alpha j} U^*_{\beta j}) \sin^2\left(\frac{\Delta m^2_{ij} L}{2E}\right). \end{aligned} \tag{2.47}$$

From this expression it is straightforward to see that, if neutrinos are massless, oscillations do not occur. Also, it is the sinusoidal nature of this expression that motivates the term "neutrino oscillations." It should be noted that, if CPT conservation holds, then $P(\overline{\nu}_\alpha \to \overline{\nu}_\beta)(t)$ is the same as $P(\nu_\alpha \to \nu_\beta)(t)$ but with $U \to U^*$.

2.2.1.2 Three-Flavor Neutrino Mixing

The expressions derived in the previous section are good for N generations of neutrinos. As shown in Sect. 2.1.3, the best experimental knowledge to date indicates that there are three generations of neutrinos, as contemplated by the Standard Model. The PMNS matrix is commonly parameterized in terms of six free parameters, which are the three mixing angles θ_{12}, θ_{13}, and θ_{23}, as well as three phases δ, α_1 and α_2:

$$\begin{aligned} U = \begin{pmatrix} U_{e1} & U_{e2} & U_{e3} \\ U_{\mu 1} & U_{\mu 2} & U_{\mu 3} \\ U_{\tau 1} & U_{\tau 2} & U_{\tau 3} \end{pmatrix} &= \begin{pmatrix} 1 & 0 & 0 \\ 0 & c_{23} & s_{23} \\ 0 & -s_{23} & c_{23} \end{pmatrix} \begin{pmatrix} c_{13} & 0 & s_{13}e^{-i\delta} \\ 0 & 1 & 0 \\ -s_{13}e^{i\delta} & 0 & c_{13} \end{pmatrix} \begin{pmatrix} c_{12} & s_{12} & 0 \\ -s_{12} & c_{12} & 0 \\ 0 & 0 & 1 \end{pmatrix} \\ &\times \begin{pmatrix} e^{i\alpha_1/2} & 0 & 0 \\ 0 & e^{i\alpha_2/2} & 0 \\ 0 & 0 & 1 \end{pmatrix}, \end{aligned} \tag{2.48}$$

where c_{ij} denotes $\cos\theta_{ij}$ and s_{ij} denotes $\sin\theta_{ij}$. In this way, the matrix is naturally separated into its atmospheric (the "23 sector"), cross-mixing (the "13 and δ" sector), solar (the "12 sector") and Majorana (the "α_1, α_2 sector") components. When seen as one matrix, the PMNS matrix is given by

$$U = \begin{pmatrix} c_{12}c_{13} & s_{12}c_{13} & s_{13}e^{-i\delta} \\ -s_{12}c_{23} - c_{12}s_{23}s_{13}e^{i\delta} & c_{12}c_{23} - s_{12}s_{23}s_{13}e^{i\delta} & s_{23}c_{13} \\ s_{12}s_{23} - c_{12}c_{23}s_{13}e^{i\delta} & -c_{12}s_{23} - s_{12}c_{23}s_{13}e^{i\delta} & c_{23}c_{13} \end{pmatrix}, \tag{2.49}$$

where we have neglected the Majorana phases. The mixing angles determine how much a weak eigenstate couples to a mass eigenstate. The phase δ is nonzero only if neutrino oscillations violate CP symmetry, which is why this phase is also commonly referred to as δ_{CP}. The α_1 and α_2 phases are known as Majorana phases, and are nonzero only if neutrinos are Majorana particles. As we will see later on, the values of these three phases are currently unknown.

From the previous parameterization the oscillation probabilities can be derived using the formalism developed in the previous section. The resulting expressions can be very cumbersome, and typically approximations are used instead. Given that ν_e appearance is the main topic of this thesis, we show as an example the full expression for the $P(\nu_\mu \to \nu_e)$ oscillation probability in vacuum:

$$\begin{aligned} P(\nu_\mu \to \nu_e) = {} & s_{23}^2 \sin^2 2\theta_{13} \sin^2 \frac{\Delta m_{31}^2 L}{4E} + c_{13}^2 c_{23}^2 \sin^2 2\theta_{12} \sin^2 \frac{\Delta m_{21}^2 L}{4E} \\ & + 8c_{13}^2 s_{13} c_{12} s_{12} s_{23} c_{23} \sin \frac{\Delta m_{21}^2 L}{4E} \sin \frac{\Delta m_{31}^2 L}{4E} \cos\left(\frac{\Delta m_{32}^2 L}{4E} + \delta\right) \\ & - 2s_{12}^2 s_{23}^2 \sin^2 2\theta_{13} \sin \frac{\Delta m_{21}^2 L}{4E} \sin \frac{\Delta m_{31}^2 L}{4E} \cos \frac{\Delta m_{32}^2 L}{4E} \\ & + 4c_{13}^2 s_{12}^3 s_{13} s_{23} (s_{23} s_{13} s_{12} - 2c_{12} c_{23} cos\delta) \sin^2 \frac{\Delta m_{21}^2 L}{4E}. \end{aligned} \tag{2.50}$$

When matter effects (described in Sect. 2.2.1) are involved the expressions become prohibitively long. This is why when studying the full ν_e appearance probability in Chap. 4 we use a series approximation to distinguish its most important features. It should also be mentioned that, in many cases, it is not necessary to use the full expression of the probability as the two-flavor approximation is completely adequate to describe the disappearance or the appearance of a particular neutrino flavor. This is shown next.

2.2.1.3 Two-Flavor Approximation

The two-flavor approximation can be used if either:

- the produced flavor eigenstate $|\nu_\alpha\rangle$ couples significantly to only two mass eigenstates,

- one mass splitting Δm^2_{ij} is much larger than the other splitting Δm^2_{jk} and the experiment is sensitive to one of the two (i.e., if $\frac{\Delta m^2 L}{4E} \gtrsim 1$).

For $N = 2$, the PMNS matrix only has one free parameter θ and can thus be parameterized as

$$U = \begin{pmatrix} \cos\theta & \sin\theta \\ -\sin\theta & \cos\theta \end{pmatrix}. \tag{2.51}$$

Using Eq. 2.47 one finds that the probability for ν_β appearance in a ν_α beam greatly reduces to

$$P(\nu_\alpha \rightarrow \nu_\beta) = \sin^2 2\theta \sin^2\left(\frac{\Delta m^2 L}{4E}\right), \tag{2.52}$$

which means that the survival probability of ν_α is given by

$$P(\nu_\alpha \rightarrow \nu_\alpha) = 1 - \sin^2 2\theta \sin^2\left(\frac{\Delta m^2 L}{4E}\right). \tag{2.53}$$

The phase responsible for the oscillations is often rewritten as

$$\frac{\Delta m^2 c^3 L}{4\hbar E} \approx 1.267 \frac{\Delta m^2}{\mathrm{eV}^2} \frac{L}{\mathrm{km}} \frac{\mathrm{GeV}}{E}, \tag{2.54}$$

where c and $\hbar$ have been restored.

An example where the two-flavor approximation works very well is ν_μ disappearance in MINOS:

$$P(\nu_\mu \rightarrow \nu_\mu) \cong 1 - \sin^2 2\theta_{23} \sin^2\left(\frac{\Delta m^2_{32} L}{4E}\right). \tag{2.55}$$

This is because θ_{13} is very small and because $\Delta m^2_{32} \gg \Delta m^2_{21}$. The $P(\nu_\mu \rightarrow \nu_\mu)$ oscillation probability for MINOS' baseline and with the current best values of the oscillation parameters is shown as a function of energy in Fig. 2.6. The first oscillation peak (i.e., the highest energy at which $P(\nu_\mu \rightarrow \nu_\mu)$ reaches a minimum) is located at $\sim$1.5 GeV. Below that value, the frequency of the oscillations increases very rapidly. After the first oscillation peak, the strength of the oscillations diminishes (i.e., $P(\nu_\mu \rightarrow \nu_\mu)$ approaches unity). In order to make a good

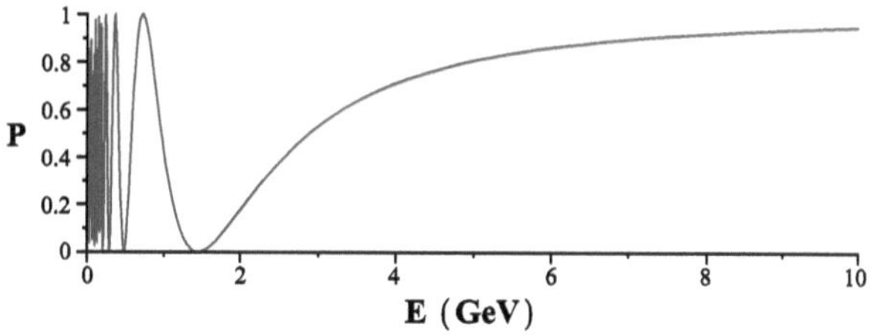

Fig. 2.6 Probability P of ν_μ survival in MINOS as a function of energy, with $\sin^2(2\theta_{23}) = 1$ and $\Delta m^2_{32} = 2.43 \times 10^{-3}\,\mathrm{eV}^2$. Oscillations are strongest at low ($\lesssim$5 GeV) energies

observation of ν_μ disappearance in MINOS it is thus convenient to have as many events as possible around 1.5 GeV.[9] Similarly, the dominant term in $P(\nu_\mu \rightarrow \nu_e)$ is proportional to $\sin^2 \frac{\Delta m^2_{31} L}{4E}$. When searching for ν_e appearance it is thus important to have a high signal selection efficiency around 1.5 GeV.

2.2.1.4 Matter Effects

An additional effect must be taken into account in order to get the exact oscillation probabilities. As neutrinos travel in matter, all three flavors undergo neutral-current interactions with the electrons and the quarks that compose matter. Electron neutrinos however also experience charged-current scattering,

$$\nu_e + e^- \rightarrow \nu_e + e^-, \tag{2.56}$$

while muon and tau neutrinos cannot. This means that the electron neutrino potential has an additional contribution,

$$V^e_m = \pm\sqrt{2} G_F n_e, \tag{2.57}$$

where G_F is Fermi's constant and n_e is the electron density of the medium. In the case of antineutrinos, the potential becomes negative. Given that this potential only affects one flavor of neutrinos, it changes the relative oscillation rates. This effect was first pointed out by Mikheyev, Smirnov and Wolfenstein [50, 51] and is accordingly termed the "MSW effect."

In order to demonstrate the impact on the oscillation probabilities, we consider the two-flavor approximation. If the mass eigenstates are $|\nu_1\rangle$ and $|\nu_2\rangle$, the evolution equation for these states can be written as

$$i\frac{d}{dt}\begin{pmatrix} \nu_1 \\ \nu_2 \end{pmatrix} = H \begin{pmatrix} \nu_1(t) \\ \nu_2(t) \end{pmatrix}, \tag{2.58}$$

where

$$H = \begin{pmatrix} E_1 & 0 \\ 0 & E_2 \end{pmatrix} \approx p + \begin{pmatrix} m_1^2/2E & 0 \\ 0 & m_2^2/2E \end{pmatrix}. \tag{2.59}$$

We can rewrite Eq. 2.58 in terms of the weak eigenstates,

$$i\frac{d}{dt}\begin{pmatrix} \nu_\alpha \\ \nu_\beta \end{pmatrix} = H' \begin{pmatrix} \nu_\alpha(t) \\ \nu_\beta(t) \end{pmatrix}, \tag{2.60}$$

---

[9] As shown in Fig. 3.4, the MINOS neutrino spectrum in the low-energy configuration peaks at $\sim$3 GeV.

where the PMNS matrix of Eq. 2.51 is used to obtain $H' = UHU^{\dagger}$. Neglecting the terms that are proportional to the unit matrix, the modified Hamiltonian H' thus takes the form:

$$H' = \frac{\Delta m^2}{2E} \begin{pmatrix} \sin^2\theta + \frac{2E}{\Delta m^2} V_m^e & -\sin\theta\cos\theta \\ -\sin\theta\cos\theta & \cos^2\theta \end{pmatrix}. \tag{2.61}$$

The modified Hamiltonian can be diagonalized by introducing a new mixing angle θ_m such that

$$H' = \frac{\Delta m^2}{2E} U(\theta_m) \begin{pmatrix} \lambda_1 & 0 \\ 0 & \lambda_2 \end{pmatrix} U^{\dagger}(\theta_m) = \frac{\Delta m^2}{2E} U(\theta_m) \begin{pmatrix} 0 & 0 \\ 0 & \lambda_2 - \lambda_1 \end{pmatrix} U^{\dagger}(\theta_m) + C, \tag{2.62}$$

where C is a term proportional to the unit matrix. Comparing Eq. 2.62 with Eq. 2.61, we find:

$$\lambda_{1,2} = \frac{1}{2}\left(1 + \frac{\Delta m^2}{2E} V_m^e \pm \sqrt{\sin^2 2\theta + \left(\cos 2\theta - \frac{2E}{\Delta m^2} V_m^e\right)^2}\right), \tag{2.63}$$

$$\cos 2\theta_m = \frac{\cos 2\theta - \frac{\Delta m^2}{2E} V_m^e}{\sqrt{\sin^2 2\theta + \left(\cos 2\theta - \frac{2E}{\Delta m^2} V_m^e\right)^2}}, \tag{2.64}$$

$$\sin 2\theta_m = \frac{\sin 2\theta}{\sqrt{\sin^2 2\theta + \left(\cos 2\theta - \frac{2E}{\Delta m^2} V_m^e\right)^2}}. \tag{2.65}$$

We thus recover the same situation as in vacuum oscillations, but with

$$\sin^2 2\theta_m \to \frac{\sin^2 2\theta}{f_{\mathrm{MSW}}^2}, \quad \Delta m^2 \to \Delta m^2 f_{\mathrm{MSW}}, \tag{2.66}$$

where

$$f_{\mathrm{MSW}} = \sqrt{\sin^2 2\theta + \left(\cos 2\theta - \frac{2E}{\Delta m^2} V_m^e\right)^2}. \tag{2.67}$$

It is interesting to note that the oscillations can be maximal ($\sin 2\theta_m = 1$) even for small vacuum mixing angles. This occurs for an electron density given by

$$n_e = \frac{\Delta m^2 \cos 2\theta}{2\sqrt{2} G_F E} \tag{2.68}$$

Matter effects can thus significantly affect the effective mass splittings and mixing angles. Probably the most significant example of this is the solar neutrino problem, as shown in the next section. The effect is different for neutrinos and antineutrinos, and is sensitive to the sign of Δm^2. Given that most ν_μ neutrinos that disappear transition (presumably) to ν_τ, the MSW effect is not a concern for ν_μ disappearance analyses, but it is taken into account in the ν_e appearance analysis described in this thesis. Chapter 4 includes a description of the MSW effect's impact on the ν_e appearance probability in MINOS.

2.2.2 Precision Measurements of Neutrino Oscillations

The field of neutrino physics has entered an era of precision measurements, particularly in the solar and atmospheric sectors. Prior to the publication of the SNO results, the results from the Homestake, Kamiokande, GALLEX, and SAGE solar experiments allowed two distinct regions of Δm^2_{21} vs. θ_{12} parameter space, respectively named the Large Mixing Angle (LMA) and the Small Mixing Angle (SMA) solutions. The SNO experiment is more sensitive to higher energy neutrinos, where matter effects are stronger. The SNO results in 2002 [52] and 2004 [53] conclusively showed that the average survival probability of solar neutrinos from the sun is only $\sim$34%, thus ruling out the SMA solution and proving the occurrence of MSW effects in the sun. Furthermore, from the sign of the observed MSW effects, it was established that the second mass eigenstate is more massive than the first one (i.e., $m_2 > m_1$).

The Kamioka Liquid Scintillator Antineutrino Detector (KamLAND) experiment, which looks at the disappearance of electron antineutrinos from terrestrial sources and in a different kinematic regime, confirmed the results of all the previous solar experiments. The KamLAND experiment is situated in Japan, where it is surrounded by 53 commercial nuclear reactors. The experiment consists of an 18 m diameter stainless steel spherical vessel with 1,879 PMTs mounted on the inner surface. Inside the sphere is a 13 m diameter nylon balloon filled with liquid scintillator. Reactor antineutrinos are detected via the inverse β-decay reaction (of Eq. 2.1). The light from the positron's kinetic energy, the e^+e^- annihilation, and the recoiling neutron capture on hydrogen, are used to identify events. The energy spectrum of KamLAND antineutrinos, shown on the top of Fig. 2.7, constitutes one of the clearest signatures of oscillation-driven neutrino disappearance existing to date [54]. The bottom of Fig. 2.7 shows the region of parameter space allowed by the solar experiments and by KamLAND. The solar oscillation parameters are currently known to better than 3.5% precision [15]:

$$\sin^2 2\theta_{12} = 0.87 \pm 0.03, \tag{2.69}$$

$$\Delta m^2_{21} = (7.59 \pm 0.20) \times 10^{-5}\,\text{eV}^2. \tag{2.70}$$

Furthermore, by obtaining results that are consistent with those of the solar experiments (which measure the disappearance of ν_e neutrinos), KamLAND has

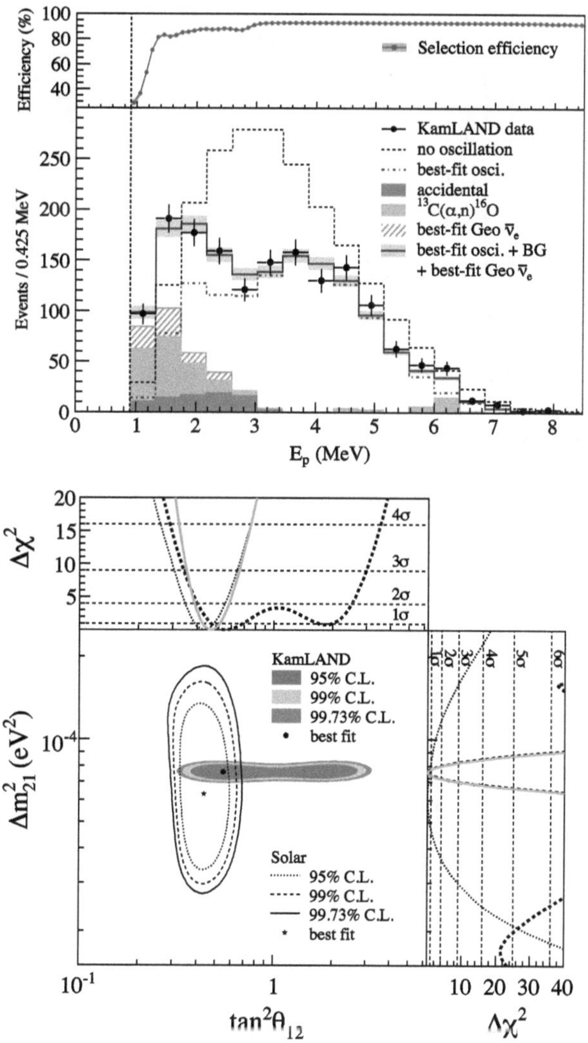

Fig. 2.7 Prompt event energy spectrum of $\overline{\nu}_e$ candidate events (top) and allowed region for neutrino oscillation parameters from KamLAND and solar neutrino experiments (bottom). All histograms corresponding to reactor spectra and expected backgrounds incorporate the energy-dependent selection efficiency shown on the top. For the bottom plot, the side panels show the $\Delta\chi^2$-profiles for KamLAND (*dashed*) and solar experiments (*dotted*) individually, as well as their combination (solid). Images obtained from [54]

also established that $|\Delta m^2_{21} - \Delta\overline{m}^2_{21}| \lesssim 1.3 \times 10^{-3}\,\text{eV}^2$ [55]. There is thus no evidence of CPT violation in the solar sector at this point.

The atmospheric sector has also been extensively studied by different experiments, and Super-Kamiokande and MINOS are the ones that currently dominate. MINOS is the long baseline neutrino oscillation experiment which is fully described in Chap. 3 and whose data constitute the basis for this thesis. MINOS' primary goal is to make a precise measurement of ν_μ disappearance by comparing the unoscillated energy spectrum at the Near Detector to the oscillated energy spectrum at the Far Detector. The latest MINOS results [56] on ν_μ disappearance corresponding to 3.36×10^{20} protons-on-target (POT) are shown in Fig. 2.8, together with the Super-Kamiokande limits from [57, 58]. Both experiments obtain

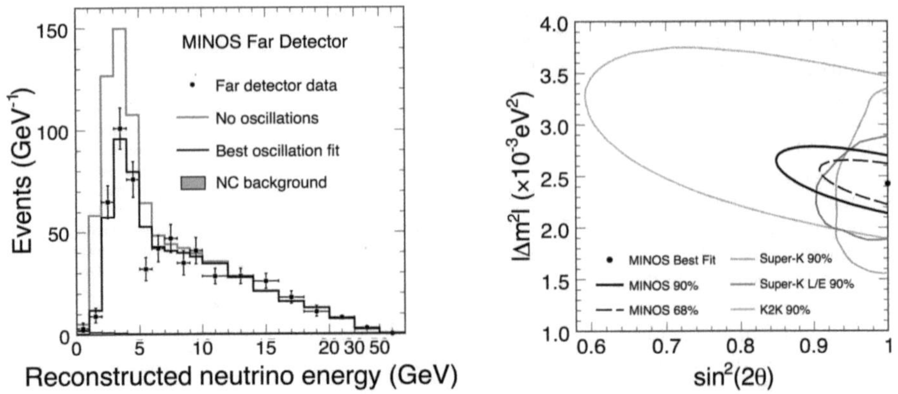

Fig. 2.8 Reconstructed energy spectrum of ν_μ CC-like events (*left*) and allowed region for neutrino oscillation parameters from MINOS and Super-Kamiokande (*right*)

consistent results and are complementary. MINOS is more sensitive to Δm^2_{32} while Super-Kamiokande is more sensitive to $\sin^2 2\theta_{23}$. The current limits on the atmospheric neutrino parameters are given by

$$\Delta m^2_{32} = (2.43 \pm 0.13) \times 10^{-3}\,\mathrm{eV}^2, \tag{2.71}$$

$$\sin^2 2\theta_{23} > 0.92\ (90\%\ \mathrm{C.L.}). \tag{2.72}$$

Moreover, the disappearance signal seen by MINOS can be used to test alternative disappearance models. The latest MINOS results indicate that neutrino decay [59] and neutrino decoherence [60] are ruled out at the 3.7σ and 5.7σ levels respectively.

Neutrino oscillations are thus a firmly established phenomenon, having been observed by many experiments operating in different regimes. Because of their implication of massive neutrinos, they also constitute the first confirmed evidence for physics beyond the Standard Model. Consequently, it is of the utmost importance to continue to investigate neutrinos as they may point the way to a more complete theory. As described in the next section, there are many open questions in the field of neutrino physics that require an answer. This thesis addresses two in particular.

2.3 Open Questions in Neutrino Physics

2.3.1 General Questions

The discovery that neutrinos have mass introduced a number of questions that remain without an answer to this day:

What is the rest mass of neutrinos? Neutrino oscillation measurements are sensitive only to the difference of squared masses of the mass eigenstates. Given that Δm^2_{32} and Δm^2_{21} are nonzero, we know that at least two of the mass eigenstates have mass. We do not know however the value of the mass of the lowest mass eigenstate.

The tightest limits on the neutrino mass existing to date come from cosmology. Neutrinos are, after all, the second most abundant particle in the universe and thus have a strong impact on many different aspects of the universe's evolution. The advent of precision data from the cosmic microwave background has placed the strongest bounds on the neutrino mass. The latest Wilkinson Microwave Anisotropy Probe (WMAP) data, combined with the data from Supernovas and Baryon Acoustic Oscillations, restrict the total mass of neutrinos to $\Sigma m_\nu < 0.67\,\text{eV}$ (95% C.L.) [61]. Many other estimates exist which also combine the data from, for instance, galaxy surveys and the Lyman-alpha forest [62]. The most recent estimates place an upper limit on the total neutrino mass of $\lesssim 1\,\text{GeV}$ (an extensive review is available in Chap. 18 of [63]).

Moreover, similar bounds on the neutrino mass exist from experiments that measure the endpoint of the β decay spectrum. These experiments use tritium, which has one of the least energetic β decays, and are designed to measure the numbers of electrons emitted with energies close to the total energy (18.6 keV in the case of tritium). The final results from the Mainz experiment in Germany [64] reported an upper limit on the electron neutrino mass of 2.3 eV (95% C.L.). A similar experiment, the Karlsruhe Tritium Neutrino Experiment (Katrin), is expected to start operating in the summer of 2012, and is sensitive to masses as low as 0.2 eV (90% C.L.) [65].

If neutrino masses are smaller than ~0.2 eV, they could still be measured through the observation of neutrinoless double-beta decay. In order for this process to occur however it is necessary for neutrinos to be Majorana particles. This question is addressed next.

Are neutrinos their own anti-particles? Massive neutrinos can be incorporated in the Standard Model Lagrangian in the form of Dirac and/or Majorana mass terms. It is possible to add a Dirac mass term for neutrinos (see Eq. 2.34) if we assume that right-handed neutrinos exist. The upper limits on neutrino mass discussed above however imply that the Yukawa coupling leading to neutrino mass upon electroweak symmetry breaking must be at least six orders of magnitude smaller than the Yukawa couplings of the other fermions.

One of the ways to avoid this fine-tuning is with the so-called seesaw mechanism. The most general mass term we can write for neutrinos is a combination of Dirac and Majorana terms:

$$\mathcal{L} = -\frac{1}{2}\begin{pmatrix}\overline{\nu}_L & \overline{\nu}^C_L\end{pmatrix}\begin{pmatrix} m_L & m_D \\ m_D & m_R\end{pmatrix}\begin{pmatrix}\nu^C_R \\ \nu_R\end{pmatrix} + h.c. \tag{2.73}$$

The eigenvalues of the mass matrix are

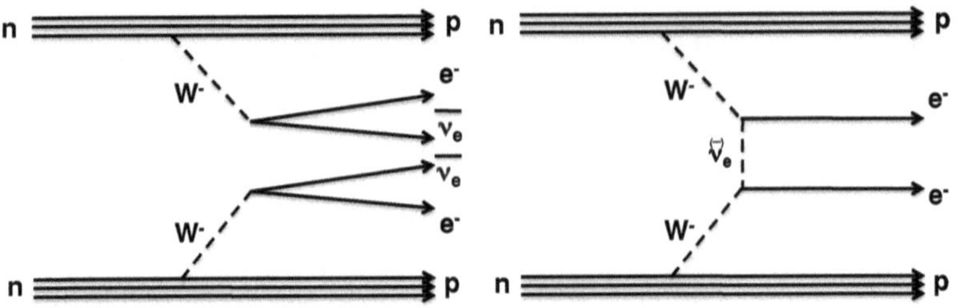

Fig. 2.9 Feynman diagrams for double beta decay (*left*) and neutrinoless double beta decay (*right*). Neutrinoless double beta decay can only occur if neutrinos are Majorana particles

$$m_{+,-} = \frac{1}{2}\left[(m_L + m_R) \pm \sqrt{(m_L - m_R)^2 + 4m_D^2}\right]. \tag{2.74}$$

If m_D is approximately the mass of the other fermions, then we have that $m_+ \approx m_D^2/m_R$ and that $m_- \approx m_R$. Therefore, through the postulation of a very heavy neutrino field (that can be sterile or coupled through new physics) the large difference in the masses between the neutrinos and the other leptons can be naturally explained.

There are literally hundreds of versions of seesaw models, as well as many others.[10] In any case, the motivation exists for neutrinos to be Majorana particles, which means that they would be their own anti-particles. If that is the case, then neutrinoless double beta decay ($0\nu\beta\beta$) could occur, as illustrated by Fig. 2.9. The signal for $0\nu\beta\beta$ decay is a monoenergetic peak at the endpoint of the double β-decay spectrum. Through a measurement of the $0\nu\beta\beta$-decay rate the neutrino mass can be obtained. Current $0\nu\beta\beta$-decay experiments can reach as low as $\sim$0.03 eV [66].

What is the right neutrino mass hierarchy? As previously explained, the observation of MSW effects in the sun established that $m_2 > m_1$. Our knowledge of the atmospheric sector however comes from the observation of atmospheric and beam ν_μ disappearance. As shown in Sect. 2.2.1, the survival of ν_μ neutrinos is given to a very good approximation by

$$P(\nu_\alpha \to \nu_\alpha) \cong 1 - \sin^2 2\theta_{23} \sin^2\left(\frac{\Delta m_{32}^2 L}{4E}\right), \tag{2.75}$$

which is insensitive to the sign of Δm_{32}^2. The subleading terms are just too small to make any visible contribution in our current experiments. Consequently, we still do not know if the mass of the third mass eigenstate m_3 is higher (standard hierarchy) or lower (inverted hierarchy) than m_2. Both mass hierarchies are illustrated in Fig. 2.10.

Are there more than three neutrinos? All neutrino experiments to date have yielded results that are consistent with three-flavor neutrino oscillations, with one

[10] For instance, there are models where neutrino masses are generated through radiative corrections.

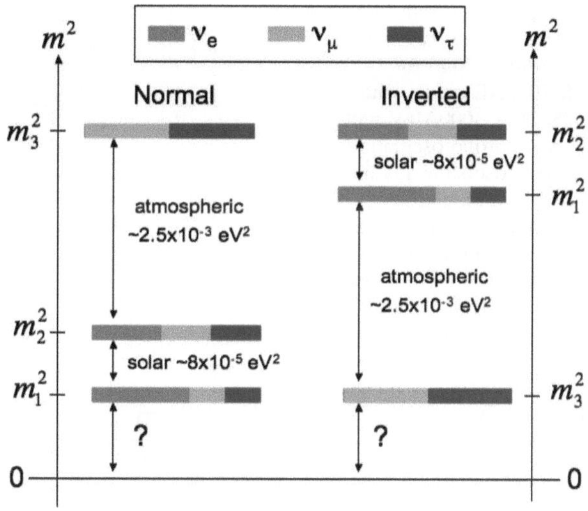

Fig. 2.10 Three-flavor neutrino mass spectrum in the normal and inverted mass hierarchies cases

exception. The LSND experiment, which ran until 1998 at Los Alamos National Laboratory, saw a 3.8σ excess of $\overline{\nu}_e$ events in a $\overline{\nu}_\mu$ beam. Due to its short baseline of 30 m, the LSND observation requires a mass splitting $\Delta m^2 \gtrsim 0.02\,\text{eV}^2$ when interpreted as an oscillation signal.

Given that the LSND mass splitting is incompatible with solar or atmospheric oscillations, one possibility is that neutrinos oscillate to a sterile neutrino(s). The MiniBooNE experiment at Fermilab has recently released results [67] that mostly exclude the oscillation interpretation of LSND at 90% C.L. with neutrinos, as illustrated in Fig. 2.11. Nevertheless, given that the LSND result was obtained with antineutrinos, sterile neutrinos can still account for all of the observations if CP or CPT is violated. For instance, in [68] a scenario with 2 sterile neutrinos and CP violation is proposed. Furthermore, as explained in [69], a better fit to the global data is obtained if a CPT violating mass spectrum is used in combination with sterile neutrino(s). In that case the neutrino mass splittings could be different from the antineutrino ones, giving more degrees of freedom. Therefore, the resolution of the LSND anomaly necessitates a verdict concerning the possibility of CP and/or CPT violation in the neutrino sector. This is a question that we address in this thesis.

Before moving on, it should be mentioned that MINOS is sensitive to oscillations to sterile neutrinos by looking for neutral-current event disappearance. No evidence of oscillations to sterile neutrinos has been found in MINOS to date [70].

2.3.2 Questions Addressed by This Thesis

Do neutrinos obey CPT and CP conservation? The motivation to search for CP and CPT violation in the neutrino sector goes beyond resolving the LSND

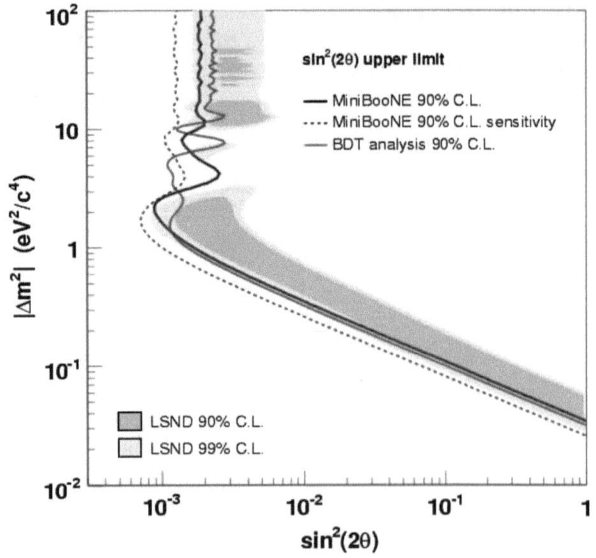

Fig. 2.11 Sensitivity (*dashed black line*) and limit (*solid black line*) for MiniBooNE events with $475 < E_\nu < 3000\,\text{MeV}$ within a two neutrino oscillation model. Also shown is the limit from the boosted decision tree analysis (*thin solid curve*) for events with $300 < E_\nu < 3000\,\text{MeV}$. The MiniBooNE limits mostly exclude the allowed regions derived from the results of the LSND experiment. Image obtained from [67]

anomaly. The observation of leptonic CP and/or CPT violation would be a crucial step forward in our understanding of the most fundamental properties of nature. Also, such observation could hold the key to some of the greatest unanswered questions of our day, such as the predominance of matter over anti-matter in the universe. It is expected that, after the Big Bang, the universe was filled with equal amounts of matter and anti-matter, which annihilated as the universe cooled. However, for unknown reasons, roughly one in every 10 billion particles of matter survived. With Majorana neutrinos it is possible to explain the excess of matter over anti-matter with leptogenesis, i.e. an asymmetry between lepton and anti-lepton creation in the early universe [71]. In the hot big bang, heavy right-handed neutrinos could have been created. Being their own antiparticle, these neutrinos could have then decayed into left-handed neutrinos or right-handed antineutrinos together with Higgs bosons, which would have undergone further decays into heavy quarks. Even small differences in the decay probabilities to matter and anti-matter could explain the current excess of matter in the universe.

An essential ingredient for leptogenesis is leptonic CP violation [72]. CP violation has been extensively documented in the quark sector (through neutral kaon decays for instance [73]), and it is thus expected to occur in the leptonic sector as well.[11] As can be seen in the PMNS matrix parameterization of Eq. 2.48 however, the CP-violating phase δ is directly coupled to $\sin\theta_{13}$. This means that no CP-violation will occur in the neutrino sector if θ_{13} is zero. Consequently, it is necessary to first determine the value of θ_{13}.

[11] The amount of CP violation observed in the quark sector is insufficient to explain the asymmetry of matter and anti-matter in the universe. Hence, we look for larger effects in the leptonic sector.

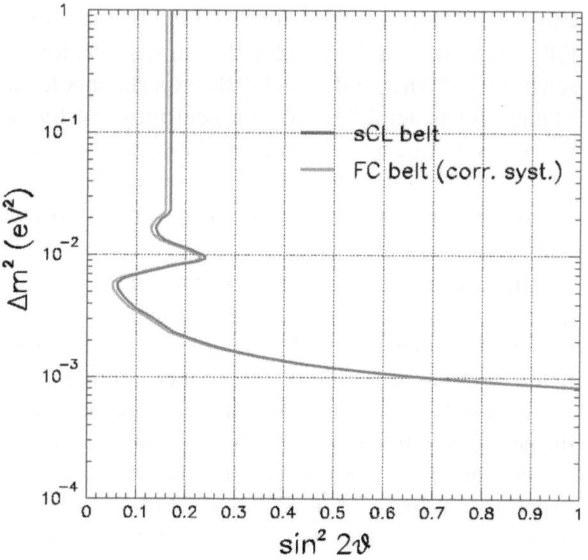

Fig. 2.12 Exclusion plot at 90% C.L. in the $\sin^2 2\theta_{13}$ vs. Δm^2_{31} plane obtained from the CHOOZ experiment. The standard C.L. and the Feldman–Cousins based C.L. contour are shown. Image obtained from [74]

CPT conservation on the other hand can be tested directly in MINOS through the study of antineutrino oscillations. As shown in Chap. 10, MINOS has a unique opportunity to perform a precise test of CPT conservation in the atmospheric neutrino sector for the first time.

What is the value of θ_{13}? θ_{13} is the only mixing angle in the 3×3 PMNS neutrino matrix that remains unknown. The angle is known to be small, and the most stringent upper limit on it is the one set by the CHOOZ [74] experiment in France. The CHOOZ experiment was a liquid scintillator detector that looked for reactor $\overline{\nu}_e$ disappearance at a distance of ~ 1 km from the source. At that baseline, the $\overline{\nu}_e$ disappearance probability is approximately given by

$$P(\overline{\nu}_e \to \overline{\nu}_e) \cong \sin^2 2\theta_{13} \sin^2 \theta_{23} \sin^2\left(\frac{\Delta m^2_{31} L}{4E}\right), \tag{2.76}$$

where we assume CPT conservation. The CHOOZ experiment saw no $\overline{\nu}_e$ disappearance and thus set a limit on the θ_{13} angle as a function of the mass splitting Δm^2_{31}. This limit is shown in Fig. 2.12, and for a value[12] of $\Delta m^2_{31} \approx 2.43 \times 10^{-3}\,\text{eV}^2$ it corresponds to

$$\sin^2(2\theta_{13}) < 0.15 \quad (90\%\,\text{C.L.}). \tag{2.77}$$

An aggressive experimental program has been set in motion to measure θ_{13}. This program includes the the two reactor experiments Double-CHOOZ [75] and

[12] Please note that $\Delta m^2_{31} = \Delta m^2_{32} + \Delta m^2_{21} \approx \Delta m^2_{32}$, which is why we use the value of the atmospheric mass splitting.

Daya Bay [76], as well as the two long baseline experiments NOνA [77] and T2K [78]. Daya Bay, NOνA and T2K can reach below $\sin^2 2\theta_{13} \sim 0.01$ (90% C.L.) in sensitivity.[13] The NOνA and T2K experiments have an extra advantage however. Matter effects act differently for neutrinos and for antineutrinos, as well as for the two mass hierarchies. Therefore, if a signal is seen, the comparison of $P(\nu_\mu \rightarrow \nu_e)$ to $P(\overline{\nu}_\mu \rightarrow \overline{\nu}_e)$ may allow for the disentanglement of CP violation from the mass hierarchy of the neutrino sector. This can only occur if θ_{13} is large enough though ($\sin^2(2\theta_{13}) \gtrsim 0.02$).

The determination of the θ_{13} mixing angle is thus one of the highest priorities in the field of neutrino physics. Due to its link to CP violation, the value of θ_{13} has profound implications in the nature of particle physics which, in turn, affect other fields such as cosmology. Moreover, through θ_{13}-driven oscillations, a good chance exists for the mass hierarchy of the neutrino sector to be addressed. Before the next generation experiments come online, the MINOS experiment has a chance to make the first measurement of a nonzero θ_{13} if this angle lies close to the current experimental limit set by CHOOZ. How to make the most of the MINOS data set in order to maximize the sensitivity to θ_{13} is the main topic of this thesis.

References

1. A.H. Becquerel, On the rays emitted by phosphorescence, Compt. Rend. Math. **122**, 420–421 (1896).
2. J. Chadwick, Distribution in intensity in the magnetic spectrum of the β-rays of radium, Ver. Dtsch. Physik. Ges. **16**, 383–391 (1914)
3. C.D. Ellis, W.A. Wooster, The average energy of disintegration of radium E. Proc. R. Soc. London Ser. A **117**(776), 109–123 (1927)
4. W. Pauli, Letter to the Physical Society of Tübingen (1930). The letter is reproduced in *Are There Really Neutrinos?: An Evidential Story*, ed. by Allan Franklin (Westview Press, Boulder, 2001)
5. J. Chadwick, Possible existence of a neutron, Nature **129**, 312 (1932)
6. E. Fermi, An attempt of a theory of beta radiation. 1, Z. Phys. **88**, 161–177 (1934)
7. H. Bethe, R. Peierls, The 'neutrino', Nature **133**, 532 (1934)
8. C. Powell, P. Fowler, D. Perkins, The study of elementary particles by the photographic method, Nature **159**, 694 (1947)
9. C.L. Cowan Jr., F. Reines et al., Detection of the free neutrino: a confirmation, Science **124**(3212), 103–104 (1956)
10. R. Davis, D. Harmer, Attempt to observe the $Cl^{37}(\overline{\nu}, e^-)A^{37}$ reaction induced by reactor antineutrinos, Bull. Am. Phys. Soc. **4**, 217 (1959)
11. E.J. Konopinski, H.M. Mahmoud, The universal Fermi interaction, Phys. Rev. **92**, 1045–1049 (1953)
12. G. Danby et al., Observation of high-energy neutrino reactions and the existence of two kinds of neutrinos, Phys. Rev. Lett. **9**(1), 36–44 (1962)

[13] The Double CHOOZ experiment can only go down to $\sin^2 2\theta_{13} \sim 0.03$, although it will be the first of the four experiments to come online.

13. M.L. Perl et al., Evidence for anomalous lepton production in $e+-e-$ annihilation, Phys. Rev. Lett. **35**(22), 1489–1492 (1975)
14. K. Kodama et al., Observation of tau neutrino interactions, Phys. Lett. B **504**, 218–224 (2001)
15. C. Amsler et al. (Particle Data Group), Review of particle physics, Phys. Lett. **B**(667), 1 (2008)
16. J.R. Ellis, Beyond the standard model with the LHC, Nature **448**, 297–301 (2007)
17. S.L. Glashow, Partial symmetries of weak interactions, Nucl. Phys. **22**, 579–588 (1961)
18. A. Salam, J.C. Ward, Electromagnetic and weak interactions, Phys. Lett. **13**, 168–171 (1964)
19. S. Weinberg, A model of leptons, Phys. Rev. Lett. **19**, 1264–1266 (1967)
20. C.S. Wu et al., Experimental test of parity conservation in beta decay, Phys. Rev. **105**(4), 1413–1415 (1957)
21. M. Goldhaber, L. Grodzins, A.W. Sunyar, Helicity of neutrinos, Phys. Rev. **109**(3), 1015–1017 (1958)
22. M.E. Peskin, D.V. Schroeder, *An Introduction To Quantum Field Theory (Frontiers in Physics)*, (Westview Press, Boulder, 1995)
23. F. Englert, R. Brout, Broken symmetry and the mass of gauge vector mesons, Phys. Rev. Lett. **13**(9), 321–323 (1964)
24. P.W. Higgs, Broken symmetries and the masses of gauge bosons, Phys. Rev. Lett. **13**(16), 508–509 (1964)
25. G.S. Guralnik, C.R. Hagen, T.W.B. Kibble, Global conservation laws and massless particles, Phys. Rev. Lett. **13**(20), 585–587 (1964)
26. N. Cabibbo, Unitary symmetry and leptonic decays, Phys. Rev. Lett. **10**(12), 531–533 (1963)
27. M. Kobayashi, T. Maskawa, *CP*-Violation in the renormalizable theory of weak interaction, Progr. Theor. Phys. **49**(2), 652–657 (1973)
28. D. Griffiths, *Introduction to Elementary Particles*, 2nd edn. (Wiley-VCH, Weinheim, 2008)
29. R.D. Field, R.P. Feynman, A parametrization of the properties of quark jets, Nucl. Phys. **B136**, 1 (1978)
30. T.D. Lee, C.N. Yang, Parity nonconservation and a two-component theory of the neutrino, Phys. Rev. **105**, 1671–1675 (1957)
31. F.J. Hasert et al., Observation of neutrino-like interactions without muon or electron in the Gargamelle neutrino experiment, Phys. Lett. **B46**, 138–140 (1973)
32. S. van der Meer, Stochastic cooling and the accumulation of antiprotons, Rev. Mod. Phys. **57**(3), 689–697 (1985)
33. G. Arnison et al., Experimental observation of isolated large transverse energy electrons with associated missing energy at $\sqrt{s} = 540$ GeV, Phys. Lett. B **122**, 103–116 (1983)
34. M. Banner et al., Observation of single isolated electrons of high transverse momentum in events with missing transverse energy at the CERN pp collider, Phys. Lett. B **122**, 476–485 (1983)
35. The ALEPH Collaboration et al, Precision electroweak measurements on the *Z* resonance, hep-ex/0509008 (2006)
36. T. Wyatt, High-energy colliders and the rise of the standard model, Nature **448**, 274–280 (2007)
37. B.T. Cleveland et al., Measurement of the solar electron neutrino flux with the homestake chlorine detector, Astrophys. J. **496**(1), 505–526 (1998)
38. J.N. Bahcall, Solar neutrinos. I: Theoretical, Phys. Rev. Lett. **12**, 300–302 (1964)
39. J.N. Abdurashitov et al., The Russian–American Gallium Experiment (SAGE) Cr neutrino source measurement, Phys. Rev. Lett. **77**(23), 4708–4711 (1996)
40. W. Hampel et al., GALLEX solar neutrino observations: results for GALLEX IV, Phys. Lett. **B447**, 127–133 (1999)
41. K.S. Hirata et al., Real time, directional measurement of ^{8}B solar neutrinos in the Kamiokande-II detector, Phys. Rev. **D44**, 2241–2260 (1991)

42. Y. Fukuda et al., Evidence for oscillation of atmospheric neutrinos, Phys. Rev. Lett. **81**(8), 1562–1567 (1998)
43. T.J. Haines et al., Calculation of atmospheric neutrino-induced backgrounds in a nucleon-decay search, Phys. Rev. Lett. **57**(16), 1986–1989 (1986)
44. S.P. Ahlen et al., Atmospheric neutrino flux measurement using upgoing muons, Phys. Lett. **B357**, 481–486 (1995)
45. W.W.M. Allison et al., Measurement of the atmospheric neutrino flavour composition in Soudan-2, Phys. Lett. **B391**, 491–500 (1997)
46. Q.R. Ahmad et al., Measurement of the rate of $\nu_e + d \rightarrow p + p + e^-$ interactions produced by ^{8}B solar neutrinos at the Sudbury Neutrino Observatory, Phys. Rev. Lett. **87**(7), 071301 (2001)
47. B. Aharmim et al., Measurement of the ν_e and total B-8 solar neutrino fluxes with the Sudbury Neutrino Observatory phase I data set, Phys. Rev. **C75**, 045502 (2007)
48. B. Pontecorvo, Mesonium and antimesonium, Sov. Phys. JETP **6**, 429 (1957)
49. Z. Maki, M. Nakagawa, S. Sakata, Remarks on the unified model of elementary particles, Progr. Theor. Phys. **28**(5), 870–880 (1962)
50. L. Wolfenstein, Neutrino oscillations in matter, Phys. Rev. D **17**(9), 2369–2374 (1978)
51. S.P. Mikheyev, A.Y. Smirnov, Resonance enhancement of oscillations in matter and solar neutrino spectroscopy, Sov. J. Nucl. Phys. **42**, 913–917 (1985)
52. Q.R. Ahmad et al., Measurement of day and night neutrino energy spectra at SNO and constraints on neutrino mixing parameters, Phys. Rev. Lett. **89**(1), 011302 (2002)
53. S.N. Ahmed et al., Measurement of the total active ^{8}B solar neutrino flux at the Sudbury Neutrino Observatory with enhanced neutral current sensitivity, Phys. Rev. Lett. **92**(18), 181301 (2004)
54. S. Abe et al. (the KamLAND Collaboration), Precision measurement of neutrino oscillation parameters with KamLAND, Phys. Rev. Lett. **100**(22), 221803 (2008)
55. H. Murayama, CPT tests: kaon vs neutrinos, Phys. Lett. **B597**, 73–77 (2004)
56. P. Adamson et al., Measurement of neutrino oscillations with the MINOS detectors in the NuMI beam, Phys. Rev. Lett. **101**, 131802 (2008)
57. Y. Ashie et al., Evidence for an oscillatory signature in atmospheric neutrino oscillations, Phys. Rev. Lett. **93**(10), 101801 (2004)
58. Y. Ashie et al., Measurement of atmospheric neutrino oscillation parameters by Super-Kamiokande I, Phys. Rev. D **71**(11), 112005 (2005)
59. V. Barger et al., Neutrino decay as an explanation of atmospheric neutrino observations, Phys. Rev. Lett. **82**(13), 2640–2643 (1999)
60. G.L. Fogli et al., Status of atmospheric neutrino $\nu_\mu \rightarrow \nu_\tau$ oscillations and decoherence after the first K2K spectral data, Phys. Rev. D **67**(9), 093006 (2003)
61. E. Komatsu et al., Five-year Wilkinson microwave anisotropy probe (WMAP) observations: cosmological interpretation, Astrophys. J. Suppl. **180**, 330–376 (2009)
62. S. Hannestad, Neutrino mass bounds from cosmology, Nucl. Phys. Proc. Suppl. **145**, 313–318 (2005)
63. R.N. Mohapatra, P.B. Pal, *Massive Neutrinos in Physics and Astrophysics*, 3rd edn. World Scientific Lecture Notes in Physics, vol. 72 (World Scientific, Singapore, 2004) pp. 1–451
64. C. Kraus et al., Final results from phase II of the Mainz Neutrino Mass Search in tritium β decay, Eur. Phys. J. **C40**, 447–468 (2005)
65. A. Osipowicz et al., KATRIN: a next generation tritium beta decay experiment with sub-eV sensitivity for the electron neutrino mass FZKA report 6691 (2001)
66. F.T. Avignone III, S.R. Elliott, J. Engel, Double beta decay, Majorana neutrinos, and neutrino mass, Rev. Mod. Phys. **80**(2), 481 (2008)
67 A.A. Aguilar-Arevalo et al., A search for electron neutrino appearance at the $\Delta m^2 \sim 1\text{eV}^2$ scale, Phys. Rev. Lett. **98**, 231801 (2007)
68. G. Karagiorgi et al., Leptonic CP violation studies at MiniBooNE in the (3+2) sterile neutrino oscillation hypothesis, Phys. Rev. **D75**, 013011 (2007)

69. G. Barenboim, J.D. Lykken, Status of CPT-violating neutrinos, hep-ph/0908.2993 (2009)
70. P. Adamson et al., Search for active neutrino disappearance using neutral-current interactions in the MINOS long-baseline experiment, Phys. Rev. Lett. **101**, 221804 (2008)
71. E. Ma, Leptogenesis: Neutrinos and new lepton flavor violation at the TeV energy scale, hep-ph/0010139 (2000)
72. A.D. Sakharov, Violation of CP invariance, C asymmetry, and baryon asymmetry of the universe, Sov. Phys. JETP Lett. (5), 24–27 (1967)
73. J.H. Christenson et al., Evidence for the 2π Decay of the K_2^0 Meson, Phys. Rev. Lett. **13**, 138–140 (1964)
74. M. Apollonio et al. (the CHOOZ Collaboration), Search for neutrino oscillations on a long base-line at the CHOOZ nuclear power station, Eur. Phys. J. **C**(27), 331–374 (2003)
75. F. Ardellier et al., Double Chooz: a search for the neutrino mixing angle θ_{13}, hep-ex/0606025 (2006)
76. X. Guo et al., A precision measurement of the neutrino mixing angle θ_{13} using reactor antineutrinos at Daya Bay, hep-ex/0701029 (2007)
77. D.S. Ayres et al., NOvA proposal to build a 30-kiloton off-axis detector to study neutrino oscillations in the Fermilab NuMI beamline, hep-ex/0503053 (2004)
78. Y. Itow et al., The JHF-Kamioka neutrino project, hep-ex/0106019 (2001)

Chapter 3
The MINOS Experiment

Several of the open questions discussed at the end of the last chapter can be addressed in the MINOS experiment. MINOS is a long-baseline neutrino oscillation experiment that samples a beam consisting primarily of muon neutrinos in two locations. An overview of MINOS is given first, followed by a description of its main components: the beam and the detectors. A discussion of the MINOS data and their calibration is provided at the end of this chapter.

3.1 An Overview of the Experiment

The Main Injector Neutrino Oscillation Search (MINOS) experiment is a long baseline neutrino oscillation experiment. As such, it consists of three basic components: a beam of neutrinos, a Near Detector and a Far Detector.

The neutrino beam is produced by 120 GeV protons from Fermilab's Main Injector [1]. The Near Detector is located ~ 1 km downstream of the neutrino production target, also in Fermilab. Due to its proximity to the beam source, the Near Detector is exposed to a very high flux of neutrinos. This allows it to measure the beam's flavor composition and energy spectrum to a very high statistical precision, before oscillations take place. The beam is then sampled a second time, after any oscillations have occurred, by the Far Detector, located 735 km away from the Near Detector in the Soudan Underground Laboratory in northern Minnesota. This is illustrated in Fig. 3.1.

The two-detector concept allows for a strong reduction of systematic errors. The unoscillated energy spectrum of neutrinos is directly measured in the Near Detector and then compared with the oscillated one measured at the Far Detector. In this way, uncertainties in the cross-sections, in the flux and in the neutrino event simulation, among others, cancel to first order, effectively reducing the dependence on the simulation.

J. P. Ochoa Ricoux, *A Search for Muon Neutrino to Electron Neutrino Oscillations in the MINOS Experiment*, Springer Theses, DOI: 10.1007/978-1-4419-7949-0_3,

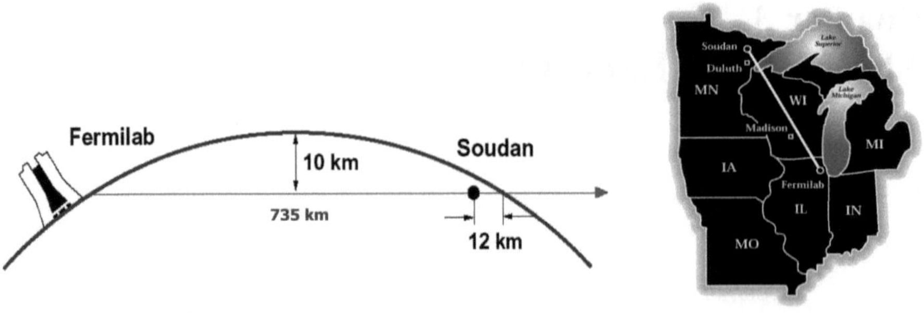

Fig. 3.1 Schematic of the MINOS experiment, as viewed from the side (*left*) and from above (*right*). A beam of neutrinos produced at Fermilab travels underground from Illinois to Minnesota. The beam is sampled both at Fermilab and at the Soudan Underground Laboratory, 735 km from the neutrino source

In the following sections each of the components of the MINOS experiment is described in detail.

3.2 The NuMI Neutrino Beam

3.2.1 Basic Principle

The MINOS experiment uses the neutrino beam produced by the Neutrinos at the Main Injector (NuMI) facility at Fermilab. The NuMI beam was designed to provide a high-intensity flux of muon neutrinos to the MINOS experiment. It will also be used by next-generation experiments such as NOνA [2] and MINERνA [3].

The main components of the NuMI beamline are illustrated in Fig. 3.2. A 120 GeV proton beam is extracted from Fermilab's Main Injector storage ring and allowed to strike a graphite target, producing a spray of secondary particles. The positively charged secondary mesons from the target are focused by a system of two toroidal magnets called "horns" into a 675 m long decay pipe, where they decay primarily into muons and muon neutrinos:

$$\pi^+ \rightarrow \mu^+ + \nu_\mu,$$
$$K^+ \rightarrow \mu^+ + \nu_\mu.$$

At the end of the decay pipe a beam absorber followed by approximately 240 m of unexcavated rock stop the remnant hadrons and charged leptons, leaving only neutrinos.

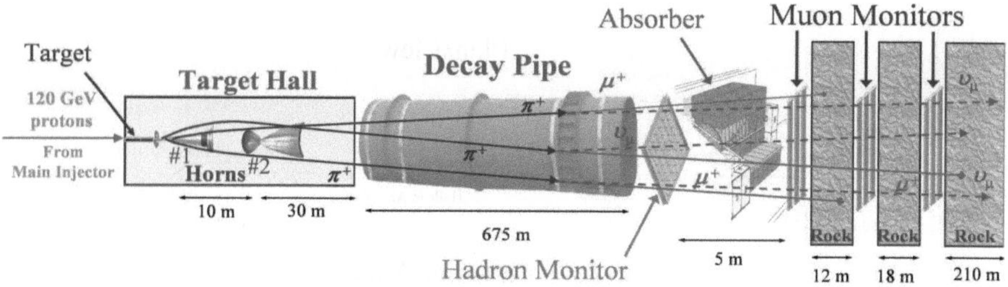

Fig. 3.2 Major components of the NuMI beamline. 120 GeV protons from Fermilab's Main Injector impinge on a graphite target, producing a spray of secondary particles. The positively charged particles from the target are focused by two parabolic horns into a 675 m evacuated decay pipe, followed by a hadron absorber and 240 m of un-excavated rock

3.2.2 Description of the Main Components

We now describe each of the main components of NuMI in more detail. The Main Injector (MI) receives multiple batches of protons from the 8 GeV Booster accelerator. Protons are uniformly spread across batches. In most MI cycles seven batches are accelerated, of which two are devoted to anti-proton production for the Tevatron accelerator program [4]. The other five are bent downward by 58 mrad towards the Far Detector and transported 350 m to the NuMI target, as illustrated by Fig. 3.3. When there is no anti-proton production NuMI receives all the batches from the Main Injector and operates in the so-called NuMI-only mode [5]. The NuMI spills typically contain 2.5×10^{13} protons, have a cycle time of 2.2–2.4 s, and last for 8.6–11 μs.

The data used in this thesis were taken in the configuration just described. Nevertheless, it should be pointed out that much work has recently gone into increasing the proton intensity for both NuMI and anti-proton production. This has been achieved with a technique called "slip stacking," where two booster batches are stacked end to end but with slightly different momenta. The two batches "slip" relative to each other until they finally overlap [7]. Starting in 2008, the MI has been operating in "multi-batch slip stacking" mode, where five batches are initially loaded and six more batches are loaded and "slipped" into the spaces midway between the first five. Of these, two are dedicated for anti-proton production and nine for NuMI, increasing the NuMI intensity to approximately 3.1×10^{13} protons per spill. In "NuMI-only mode," the intensity reaches as high as 3.7×10^{13} protons per spill. This increase in proton intensity will be very beneficial for future analyses.

Before hitting the target the proton beam goes through a 1.5 m long collimating graphite baffle that protects the downstream components of the beamline. In order to relieve thermal stresses induced during the beam spill, the $6.4 \times 15 \times 940$ mm target is segmented longitudinally into 47 segments, or fins. The beam spot size at the target is 1.2–1.5 mm. The target is water cooled and electrically isolated. The

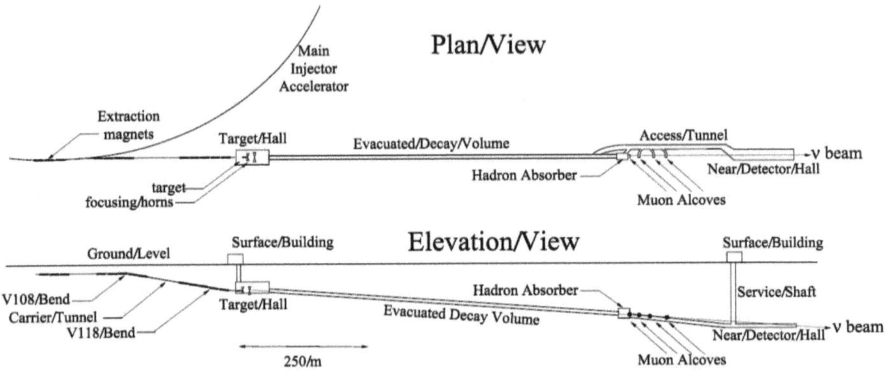

Fig. 3.3 Plan and elevation views of the NuMI facility. A proton beam is directed onto the target at an angle of 58 mrad with respect to the surface. The secondary pions and kaons from the target are focused into an evacuated decay volume via magnetic horns. Ionization chambers at the end of the beamline measure the secondary hadron beam and tertiary muon beam. Image obtained from [6]

total target length represents approximately 1.9 hadronic radiation lengths, which means that most of the protons interact in it to produce pions and kaons.

The two magnetic horns [8] that focus the particles produced in the target are pulsed with a 200 kA current, yielding a maximum 30 kG toroidal field. The magnetic field selects particles produced at the target by charge-sign and momentum. The parabolic-shaped inner conductors in the horns allow them to act as a focusing lens, where the focal length for point-to-parallel focusing is proportional to the particle's momentum. In their standard configuration, the horns focus positively charged particles and defocus negatively charged ones. It is possible to invert this situation by reversing the current in the horns, as described in Sect. 10.2.3. In that case a beam composed mostly of muon antineutrinos is obtained.

The particles from the target are focused by the horns into a 675 m long, 2 m diameter steel pipe evacuated to 0.5 Torr in order to reduce meson absorption and scattering.[1] The length of the decay pipe corresponds approximately to one decay length of a 10 GeV pion, thus providing most of the low-energy pions and kaons with enough time to decay. The decay pipe is surrounded by 2.5–3.5 m of concrete shielding and by water cooling lines that remove the 150 kW of beam heating deposited in the steel and the concrete. At the end of the decay pipe is a beam absorber consisting of a $1.2 \times 1.2 \times 2.4\ \mathrm{m}^3$ aluminum core which absorbs 65 kW of beam power. The core is surrounded by a 1 m layer of steel blocks followed by a 1.5 m layer of concrete.

---

[1] In December 2007 the decay pipe was filled with helium gas at atmospheric pressure to avoid a possible failure of the decay pipe window. The data used for this thesis, however, were obtained with the evacuated beam pipe.

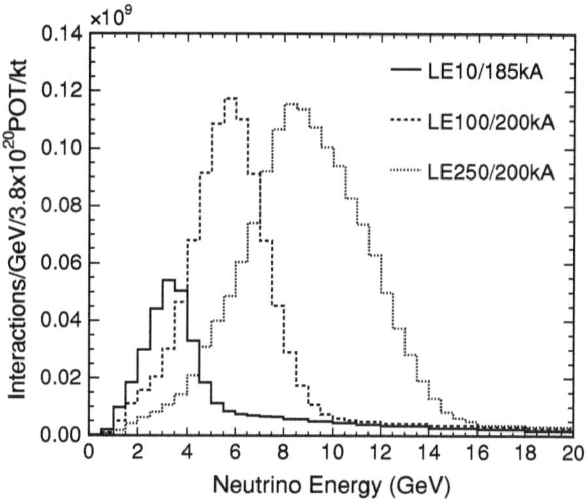

Fig. 3.4 Simulated rate of ν_μ CC interactions in the MINOS Near Detector. Three spectra are shown, corresponding to the low-energy (LE), pseudo-medium energy (pME) and pseudo-high energy (pHE) beam configurations. For these configurations the separation between the target and the horns is 10, 100 and 250 cm, and the horn current is 185, 200 and 200 kA respectively. Image obtained from [6]

At the end of the beamline the remnant hadrons and the tertiary muons from pion and kaon decays are monitored with ionization chambers [9]. These chambers are labeled "Hadron Monitor" and "Muon Monitors" in Fig. 3.2. More details about all the other monitoring devices, as well as about NuMI in general, can be found in [10].

3.2.3 Beam Configuration

The target assembly is mounted on a rail-drive system which allows for the horn-target separation to vary, with a maximum of 2.5 m. By varying the relative position between the horns and the target the optimal momentum focus for pions changes, thus modifying the neutrino energy of the beam. Figure 3.4 shows the ν_μ charged-current (CC) spectra in the low-energy (LE), pseudo[2]-medium energy (pME) and pseudo-high energy (pHE) beam configurations, which correspond to horn-target separations of 10, 100 and 250 cm respectively. The spectra peak at 3.3, 5.6 and 8.6 GeV respectively.

[2] By varying the separation between the horns it is also possible to produce a higher energy beam. Such a reconfiguration of the NuMI target hall, while part of the original plan, is very time consuming and unlikely to occur in all the lifetime of the experiment. In order to distinguish this situation with the one where only the target is moved relative to the horns, the prefix "pseudo" is commonly used to denote the latter. In this thesis however the terms ME and pME, as well as HE and pHE, are used interchangeably and refer to the same beam configuration.

As explained in Sect. 2.2.1, when studying neutrino oscillations in MINOS it is beneficial to concentrate as many events as possible around the first oscillation peak, which is located approximately at $\sim$1.5 GeV. Because of this, the LE configuration is the standard configuration for NuMI running,[3] although small amounts of pME and pHE data have also been collected. These higher-energy data are useful when discriminating against other disappearance models [13], as well as for studies such as the one presented in the Appendix C.

3.2.4 *Composition*

The branching ratios for the decays of π^+ and K^+ mesons into muon neutrinos are very high. In particular, a π^+ decays 99.988% of the time into a $\mu^+\nu_\mu$ pair [14]. Moreover, more than 80% of the beam is produced by π^+ decay. Because of this approximately 92.9% of the NuMI beam in the LE configuration at the Near Detector consists of muon neutrinos. There is also a small fraction of negatively charged mesons π^- and K^- that go through the center of the horns undeflected and which decay into muon antineutrinos. These muon antineutrinos make up approximately 5.8% of the beam in the Near Detector and are described in more detail in the second half of Chap. 10. Also, there is a small ν_e component which constitutes 1.3% of the beam and which originates primarily from secondary muon decay $\mu^+ \rightarrow e^+ + \overline{\nu}_\mu + \nu_e$. The beam ν_e component is described in more detail in Sect. 6.2, as it constitutes an irreducible background to the ν_e appearance search.

3.3 The MINOS Detectors

3.3.1 *Detector Technology*

As previously mentioned, the MINOS experiment samples the NuMI beam before and after oscillations have occurred in order to reduce systematic errors. To take full advantage of this asset it is necessary for the two detectors to be as similar as possible.

It is not optimal, at fixed cost, for the two detectors to be exactly identical in size and in readout. This is simply because, even with approximately 1/5 of the Far Detector's mass, the Near Detector sees an event rate that is on the order of 10^4 times larger than in the Far Detector. This yields more than sufficient data for beam

[3] The NuMI beam was designed to be used at a higher energy configuration. In 1998, the value of the atmospheric mass splitting reported by Super-Kamiokande was of $(4 - 6) \times 10^{-3}\text{eV}^2$ at 90% C.L. [11], which puts the first oscillation peak at approximately 2.5–3.5 GeV. Previous experiments such as Soudan 2 [12] had yielded even higher values for Δm^2_{32}.

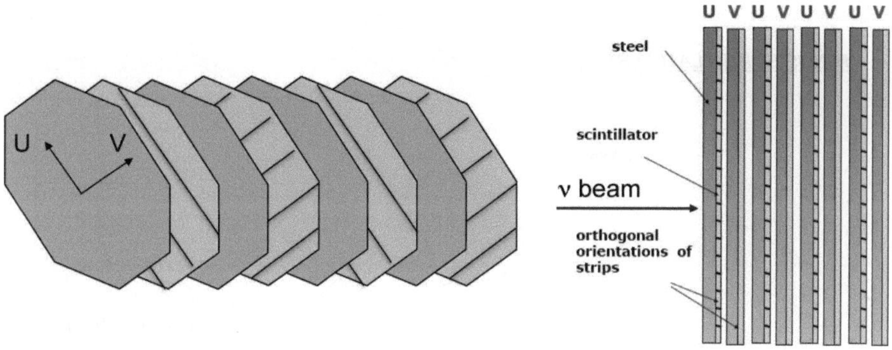

Fig. 3.5 Schematic of the MINOS steel-scintillator sampling calorimeters as viewed from an angle (*left*) and from the side (*right*). The detectors consist of planes of scintillator and steel. Alternating scintillator planes have strips perpendicular to one another. Even though the planes shown here have the octagonal shape of the Far Detector planes, the principle remains the same for the Near Detector

characterization.[4] Still, it is possible for the two detectors to be identical in terms of the environment where the neutrinos interact. In other words, their basic detection technology can be the same.

3.3.1.1 Basic Principle

Both MINOS detectors are steel and scintillator sampling calorimeters, which means that they consist of alternating layers of scintillator and steel. Each detector plane consists of a 2.54 cm thick sheet of steel on which is mounted a 1 cm sheet of solid polysterene scintillator. A 2.41 cm air gap is left between two successive planes for mechanical reasons. The steel provides the bulk 95% of the mass as a relatively inexpensive target for neutrino interactions, while the scintillator constitutes the active detector element needed to provide tracking and energy deposit information.

The scintillator is segmented into 4.1 cm wide parallel strips. Alternating scintillator planes have strips perpendicular to one another allowing reconstruction of events in three dimensions, as illustrated in Fig. 3.5. The two strip orientations are called "U" and "V", since neither is horizontal or vertical, the traditional x and y orientations.

[4] It should be noted that further reducing the size would compromise containment and reduce the effectiveness of the Near Detector.

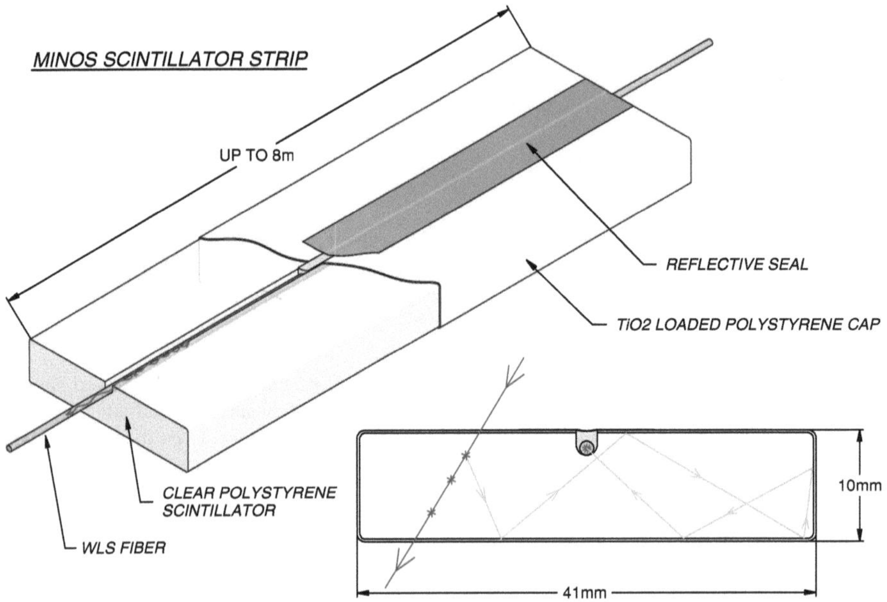

Fig. 3.6 Cutaway drawing of a single scintillator strip. Light produced by an ionizing particle is multiply reflected inside the strip until absorbed by the WLS fiber, at which point it is wavelength shifted and reemitted isotropically. The resulting wave shifted photons whose directions fall within the total internal reflection cones are transported along the fiber to the edges of the detector, subsequently being routed to the photodetectors. Image obtained from [16]

3.3.1.2 Light Collection

The strips are made of polystyrene doped with the fluors PPO (1% by weight) and POPOP[5] (0.03% by weight)[15]. Each scintillator strip is extruded with a 2.0 mm wide, 2.5 mm deep groove driven along it. A 1.2 mm diameter wavelength-shifting fiber (WLS) is glued into the groove with optical epoxy in order to collect the light produced by the interaction of charged particles with the scintillator, as illustrated by Fig. 3.6. A reflective seal of aluminized Mylar tape is placed over the groove after the WLS fiber has been glued. The strips are up to 8 m long, and are coextruded with a 0.25 mm thick layer of titanium dioxide TiO_2, which acts as a reflector that traps the light until it is absorbed by the WLS fibre.

The WLS fibre is a double-clad polystyrene fibre doped with Y11 fluor. It absorbs the blue ($\lambda \sim 420$ nm) photons from the scintillator and emits them isotropically in the green ($\lambda \sim 530$ nm). Those photons whose directions fall within the total internal reflection cones are transported along the fiber. Groups of strips are encased within aluminum sheets into light-tight modules of several different

[5] PPO stands for (2,5-diphenyloxazole) and POPOP for (1,4-bis(5-phenyloxazol-2-yl)benzene).

sizes and shapes. At the end of each module, the WLS fibers are routed into an optical connector. Light is then carried by a cable made with clear polystyrene fiber to a photomultiplier (PMT) box.

3.3.1.3 Magnetic Field

The steel plates in both detectors are magnetized, which allows for momentum measurement and charge-sign identification.

Water-cooled coils running through the middle of each detector supermodule[6] and returning outside the steel are used to produce average magnetic fields of 1.28 T and of 1.42 T in the Near and Far detectors respectively. In the Near Detector, the coil is offset by 55.8 cm from the plane center in such a way for the beam spot to be midway between the coil hole and the left vertical edge of the plane. The Far Detector uses two separate 190-turn coils of stranded copper wire with Teflon insulation, with each coil dissipating 20 kW. The Near Detector uses a single 8-turn coil made from cold conformed aluminum, dissipating 47 kW.

In their forward (and standard) polarity the magnetic fields of both detectors are such that negatively charged particles, such as the μ^-'s resulting from ν_μ CC interactions, bend toward the coil. By the same token, positively charged particles such as μ^+'s from $\bar{\nu}_\mu$ CC interactions bend toward the outside of the detector. The ability to distinguish between neutrinos and antineutrinos on an event-by-event basis is a unique characteristic of MINOS, relative to other neutrino experiments. Chap. 10 discusses the possibilities that open up as a result.

3.3.2 The Far Detector

The MINOS Far Detector is located at the Soudan Underground Laboratory in northern Minnesota at a depth of 2070 meters-water-equivalent (705 m). It is placed 735 km from the NuMI target, and has a mass of 5,400 metric tons (or 5.4 kt).

The Far Detector consists of 486 steel planes arranged in two supermodules separated by a 1.1 m gap. Each steel plane is an 8 m wide octagon. With the exception of the first and last planes, they are all covered by 192 scintillator strips that run at 45° from the vertical. The strips are arranged into eight separate modules as shown in Fig. 3.7.

Each strip end is read out from both ends by 16-anode Hamamatsu PMTs (referred to as M16). In order to reduce the number of PMTs needed, eight fibers are directed into a single PMT pixel. This is commonly referred to as "multiplexing" or, more accurately, "optical summing." The strip-to-pixel patterns are

[6] As seen below, the Far Detector is divided into two supermodules.

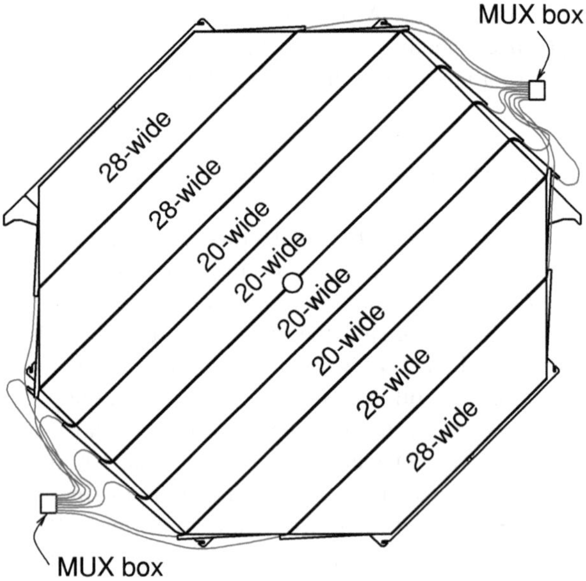

Fig. 3.7 Layout of *U* modules on a Far Detector plane. The *V* configuration is equivalent but rotated by 90 °. Eight fibers from different modules are directed into a single PMT pixel. Strips are read out from their two ends. The MUX boxes are the enclosures that contain the PMTs. Image obtained from [16]

different on the two ends of the strips, which allows for the identification of the strip that was hit during reconstruction.

Only a couple of NuMI neutrinos interact in the Far Detector per day. Moreover, the cosmic event rate is on the order of 0.5 Hz. The signal rate is thus dominated by the detector noise, which originates from spontaneous emission in the WLS fibers and from PMT dark noise, at a rate of approximately $\sim$4 kHz per 16 channel PMT. This low rate allows for a tolerable level of deadtime, and was one of the main factors in the electronics design. Each PMT is read out by a Viking VA chip, which is a multi-channel application specific integrated circuit produced by the Norwegian company Ideas ASA. Three VA chips are mounted on a VA front-end board (VFB), which provides power and biasing controls for the VA chips. The VFB also contains a discriminator chip ASDLite, which compares the PMT dynode signal with a common programmable threshold commonly set at $\sim$0.25 photoelectrons (PEs). The analog output from the VA chip is multiplexed into an analog to digital (ADC) converter by the VA Mezzanine Modules (VMMs), which control two VFBs each. When the threshold is reached, the ASDLite signals the VA readout controller (VARC), which houses 6 VMMs and thus controls 36 PMTs with 16 channels each. The digitization of a VA chip requires approximately 5 μs, during which time the given VA chip does not record any new signals. In order to reduce the electronics deadtime, a coincidence of two or more signals within 400 ns from ADSLite chips in the same VARC are required in order to initiate digitization and read out. This is referred to as the "two out of thirty-six" condition. The VARCs also handle the pedestal subtraction and the sparsification. More details about the Far Detector front-end electronics can be found in [17].

A data acquisition system (DAQ) continuously reads the front-end electronics. There are higher-level trigger conditions implemented in the DAQ software in order to decide whether or not to save the digitized data to file. For instance, the two detectors are synchronized with GPS clocks. When a beam spill occurs, a signal from the beam is sent to the Near Detector where it triggers the readout and where it is timestamped by the timing system. This timestamp is then sent over the network to the Far Detector, where it causes the continuous readout of the data within a 100 μs window around the time of the spill. It should be noted that the DAQ has enough data buffering to wait several seconds for the GPS timestamp of the spill to arrive. There are other triggers also implemented in the DAQ system. There is for example a "plane trigger" which activates the readout when four out of five consecutive planes have a dynode trigger. The plane trigger allows MINOS to record atmospheric neutrino interactions that occur outside the beam spill window. Also, fake spill triggers are generated to monitor backgrounds. Finally, there are also triggers used for debugging and for calibration.

A picture of the Far Detector is shown in Fig. 3.8. As seen there, the Far Detector is surrounded on the top and on the sides by a cosmic ray veto shield. The veto shield is built from the same 8 m long modules each containing 20 strips that are used in the main body of the detector, but with the orientation of the strips aligned along the beam direction. Consequently, it is read out in the same fashion as just described. By tagging cosmic rays with high efficiency and thus greatly reducing the cosmic ray background, the veto shield is an essential tool for the different atmospheric neutrino analyses in MINOS, such as the analysis detailed in [18]. Part of our initial work in MINOS consisted of tuning and calibrating the

Fig. 3.8 The MINOS Far Detector viewed from the front. The cosmic ray veto shield is visible on the top and on the sides of the detector

MINOS veto shield, which required the implementation of new reconstruction software tools. This work is presented in Appendix A.

3.3.3 The Near Detector

The Near Detector is located in the Fermilab complex, approximately 1 km from the NuMI target, at a depth of 225 m water equivalent (100 m). Due to its proximity to the NuMI beam source, the Near Detector sees a flux of neutrinos that is on the order of 10^6 larger than the one seen by the Far Detector. Consequently, it need not be as massive.

The Near Detector has a total mass of 980 metric tons. It consists of 282 planes arranged in a single magnetized module. The steel plates have the shape of a "squashed octagon," with dimensions 4.8 m wide and 3.8 m high. There are two different types of Near Detector planes: partially instrumented and fully instrumented. Both of them are illustrated in Fig. 3.9. Fully instrumented planes have 96 strips, and they are attached to one out of every five steel plates in the detector. Partially instrumented planes have 65 strips, and they are attached to four out of five steel plates, but only in the upstream portion of the detector (until plane number 120). This means that, of the 282 Near Detector planes, only 153 are equipped with scintillator. The upstream 121 planes, which are all instrumented, form the calorimeter region. The calorimeter is used to define the interaction

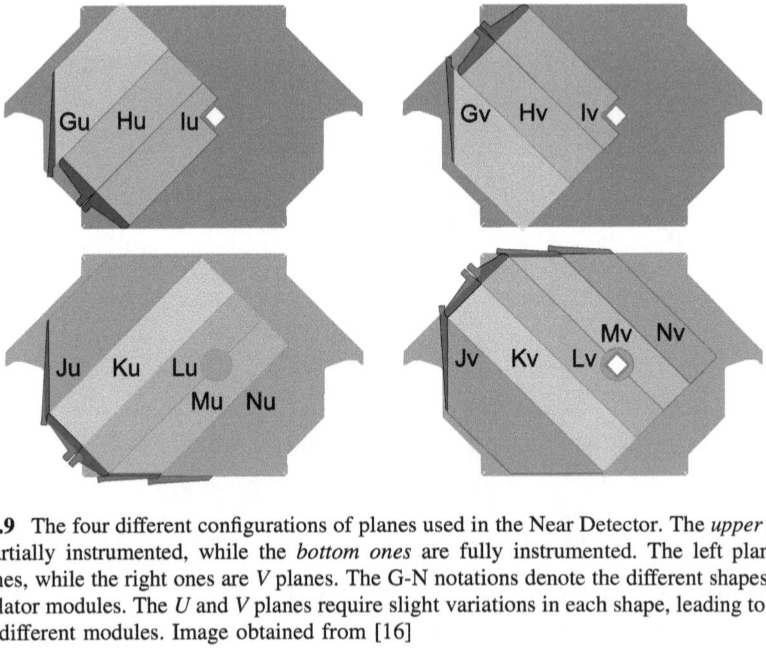

Fig. 3.9 The four different configurations of planes used in the Near Detector. The *upper planes* are partially instrumented, while the *bottom ones* are fully instrumented. The left planes are *U* planes, while the right ones are *V* planes. The G-N notations denote the different shapes of the scintillator modules. The *U* and *V* planes require slight variations in each shape, leading to a total of 16 different modules. Image obtained from [16]

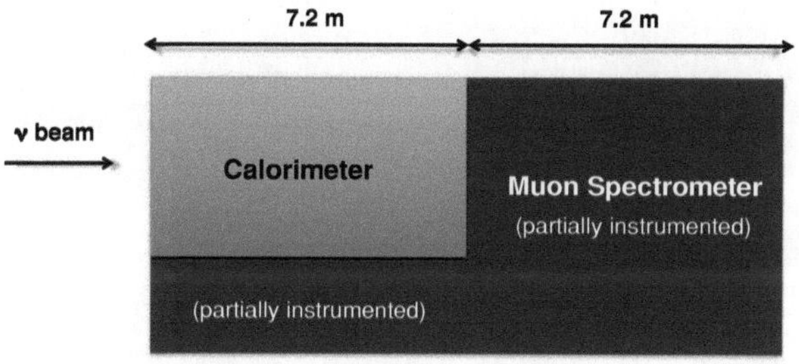

Fig. 3.10 Schematic of the different sections of the Near Detector, as viewed from above. Throughout the entire detector, one of every five planes is fully instrumented. For the first 120 planes the other four are also partially instrumented. Thus the upstream section of the detector is the calorimeter, where every plane is instrumented. In the muon spectrometer only one out of every five planes is instrumented

vertex, to find the upstream portion of muon tracks, and to measure the energy of the neutrino induced hadronic and electromagnetic shower. The downstream 161 planes, where only one in every five is instrumented, are used as a muon spectrometer. Figure 3.10 illustrates the different regions of the Near Detector.

In contrast to what is done in the Far Detector, Near Detector strips are only read out from one end. The other end is glued into a metal guide, cut with a hot knife and covered with two layers of a reflective aluminized Mylar tape to increase the light yield. Given that Near Detector strips are shorter ($\sim$2.8 m on average versus 8 m in the Far Detector), enough photons reach the PMTs with this arrangement. This, along with the other features of the Near Detector that have been discussed so far, is illustrated in Fig. 3.11.

Near Detector strips are read out by 64-anode Hamamatsu PMTs (referred to as M64). Also in contrast to the Far Detector situation, there is no multiplexing in the calorimeter region. In the spectrometer region, four strips are electrically summed into one channel. The ambiguity is resolved in the reconstruction of each event, by extrapolating the well-measured muon track from the calorimeter. The muon spectrometer region is not directly used in the ν_e appearance search described in this thesis.

At the typical beam intensity of 2.2×10^{13} protons-on-target (POT) per spill, an average of 16 neutrino interactions occur in the Near Detector for every NuMI spill in the LE configuration. Of these, about half occur in the calorimeter region and may be fully reconstructed. Consequently, the Near Detector needs fast electronics where there is no deadtime for the duration of a spill. This is achieved through the use of the Charge Integrator and Encoder (QIE) chip, which is a custom integrated circuit developed at Fermilab.[7] The QIE sits on a circuit board called a MINOS Electronics for Neutrinos (MENU), alongside a commercial flash analog-to-digital converted (FADC) and a data buffer. Each PMT pixel is attached

[7] The QIE is also used in experiments such as KTEV and CDF.

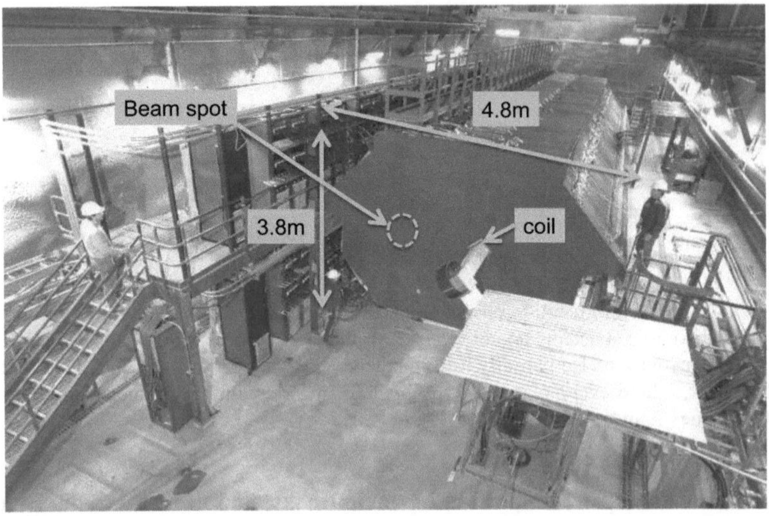

Fig. 3.11 The MINOS Near Detector. The coil hole is located approximately 1.5 m away from the beam center. The crates on the left house all the readout electronics. The detector is only read out from one end

to a QIE, which integrates and processes its charge at the 53 MHz frequency of the Main Injector (i.e., every 18.8 ns). The QIE splits the signal current into eight binary-weighted ranges and integrates each range onto a capacitor, in order to identify the signal range which is output to the FADC. Each QIE is equipped with four independent copies of the circuitry, which allows for continuous operation without deadtime. 16 MENUs are grouped into a MINOS Near Detector Electronics Readout (MINDER) board which is responsible for the timestamping. The system can also acquire data from cosmic rays, by relying on dynode signals from the PMTs to trigger the digitization and read out of the associated channels. In that case, continuous digitization of a period of 150 ns is initiated independently for each PMT. More details about the Near Detector front-end electronics can be found in [19].

3.4 MINOS Data

3.4.1 MINOS Beam Data

MINOS has collected 7.7×10^{20} POT to date, which corresponds to approximately 15 million neutrino interactions in the Near Detector. This constitutes the largest library of neutrino interactions ever recorded.

Figure 3.12 shows the POT exposure that has been accumulated per week, and in total, since the beginning of operations in May 2005. The data are divided into

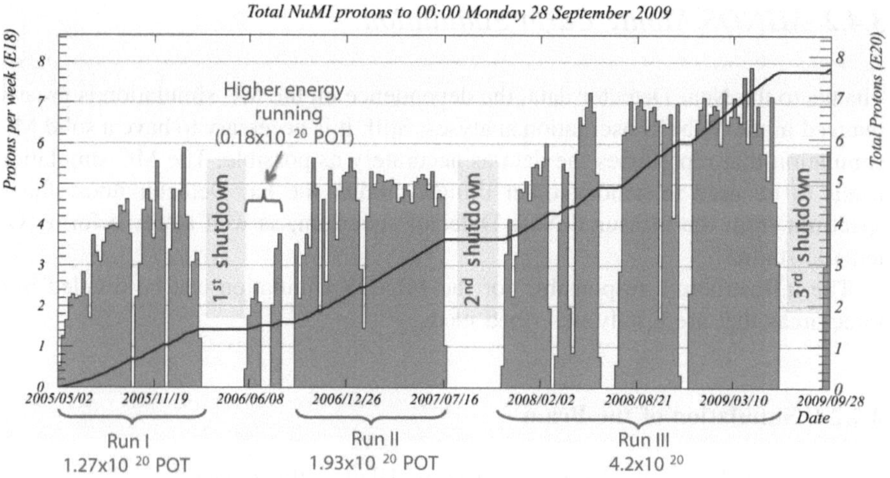

Fig. 3.12 Total number of protons on the NuMI target by September 28, 2008. The number of POTs accumulated every week is depicted by the green histogram, and the total number of POTs accumulated as a function of time is shown by the blue curve

different run periods delineated by the accelerator complex shutdowns. The data used in this thesis corresponds to Runs I and II, i.e., the data collected until the summer of 2007. After data quality cuts these data amount to 3.14×10^{20} POT. No Run III data is used in this thesis since at the time of completion of the analysis the Run III period was not yet finalized. Moreover, starting in Run III the decay pipe was filled with helium in order to prevent the collapse of the decay pipe window. Also, a small systematic decrease in the neutrino flux was observed as Run III progressed, suggesting a degradation of the target [20]. The differences in Run III data thus need to be fully understood before any analyses are performed.

Most of the data taking has been carried out in the LE beam configuration, where the separation between the target and the horns is 10 cm. Some horn-off data has also been taken in the different run periods, which amounts to approximately 1.2×10^{19} POT. In addition, before the start of the Run II period, data in higher-energy configurations were taken. Those data were acquired mostly in the pHE beam configuration, where the separation between the target and the first horn is 250 cm.

Besides a few glitches along the way, data taking has been smooth and very successful. Also, the prospects are very good as the proton intensity clearly increased in Run III as a result of the multi-batch slip stacking regime described in Sect. 3.2.2. Run IV had just begun at the time of this writing. 1.0×10^{19} POT were first accumulated in order to validate the new target. The horn current has now been reversed in order to run in antineutrino mode until March of 2010. The details and the physics motivation for reverse horn current running are addressed at the end of Chap. 10. The experiment is expected to continue until 2010, at which point approximately 10^{21} POT will have been collected in total.

3.4.2 MINOS Monte-Carlo Simulation

Thanks to the Near Detector data, the dependence on the MC simulation is greatly reduced in the MINOS oscillation analyses. Still, it is necessary to have a solid MC simulation that reproduces the data as accurately as possible. The MC simulation needs to be used to, among other things, predict the Far Detector unoscillated spectrum from the measured Near Detector spectrum, as well as to perform systematic studies.

The infrastructure responsible for the MINOS simulation can be divided into three areas that are briefly described next.

3.4.2.1 Simulation of the Beam

The neutrino flux produced by NuMI is modeled in three stages:

1. A simulation of the hadrons produced by the interaction of the 120 GeV protons in the target.
2. The propagation of those hadrons and their progeny through the beamline.
3. The calculation of the probability for a produced neutrino to go through the Near and/or Far detectors.

Stage 1 is modeled in FLUKA05 [21], while stage 2 is modeled in GEANT3 [22]. The simulation takes into effect the effects of the magnetic focusing horns, the surrounding shielding, the decay pipe and the beam absorber. Neutrinos are forced to go through either the Near or Far detectors, and are weighted by the probability of the particular meson decay. The probability is obtained from the kinematics of two-body decays of relativistic mesons.

When the ν_μ CC flux measured in the Near Detector is compared to the simulation's prediction, differences on the order of 30% are found. As seen in Fig. 3.13 however, the discrepancy between the data and the simulation moves with the focusing peak, suggesting a problem with the beam model rather than with cross sections and/or detector modeling. This is not unexpected given the large uncertainties in the production of secondary mesons at the NuMI target, a process which is not strongly constrained by external data.

Given this situation, the Near Detector data is used to constrain the beam model [23]. This is done by parameterizing the longitudinal and transverse momentum distributions of secondary hadrons produced at the target and fitting them to the data obtained in seven different beam configurations.[8] Both the neutrino and antineutrino spectra are fit simultaneously. Other beam model uncertainties are

[8] In addition to the data in the LE, pME and pHE configurations, there exist small amounts of data with the same horn-target separation as in the LE configuration but with horn currents of 170 kA and of 200 kA, as opposed to the standard 185 kA. There is also some data taken in a configuration that is intermediate between pME and pHE, as well as horn-off data.

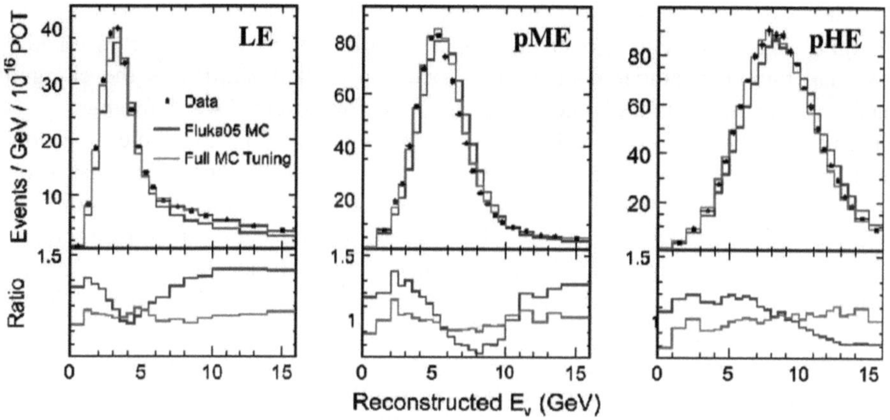

Fig. 3.13 Near Detector spectra of ν_μ CC events in the data compared to the predictions from the original simulation and from the tuned simulation, in the low-energy (LE), pseudo-medium energy (pME) and pseudo-high energy (pHE) beam configurations. The corresponding data to MC ratios are shown in the bottom panels, before and after tuning. In all cases the tuned simulation agrees very well with the data

included in the fit, such as baffle scraping, horn current miscalibration and target misalignment, although the fit's power comes predominantly from hadron production as expected. The output of the fit is used to tune the simulation, and the results can be seen in Fig. 3.13. The data and the tuned MC agree very well. The tuned simulation is the one used throughout this thesis, unless indicated.

3.4.2.2 Simulation of the Neutrino Interactions

Neutrino interactions are modeled with the NEUGEN program [24, 25]. NEUGEN simulates both quasi-elastic and inelastic neutrino scattering in the range of 100 MeV to 100 GeV, and was developed mostly by MINOS collaborators. NEUGEN was first used in the Soudan 2 experiment.

Of particular interest to this thesis is the simulation of hadronic showers, which constitute the main background to the ν_e CC appearance analysis. Hadronization in NEUGEN is handled by the AGKY model [26]. AGKY uses the PYTHIA/JETSET [27] model to simulate hadronic showers at high hadronic invariant masses W but incorporates a phenomenological description of low invariant mass hadronization. The reason for this is that the PYTHIA/JETSET model deteriorates near the pion production threshold. The phenomenological model implemented in AGKY is based on Koba-Nielsen-Olesen (KNO) scaling [28], although it incorporates several improvements. The transition from the KNO-based model to the PYTHIA/JETSET model takes place gradually at an intermediate region ($2.3\,\text{GeV} < W < 3.0\,\text{GeV}$), ensuring the continuity of all the observables as a function of the invariant mass.

The AGKY model was also developed by MINOS collaborators, and represents a significant improvement over previous hadronic models. The model is tuned to external bubble chamber data, with a focus on such quantities as the charged/ neutral pion multiplicities and dispersion, the forward/backward fragments, the fragmentation functions and the transverse momentum. The available data however generally covers an energy region that lies outside the one of interest to our analysis. This means that, for several aspects of the AGKY model, its predictions are basically extrapolations from other energy regions. Consequently, there are many uncertainties in the model which can strongly impact the particle distributions in the hadronic showers involved in the ν_e appearance search. These uncertainties are addressed in Chap. 8.

After hadronization, the particles produced in neutrino scattering can still interact with other particles before exiting the target nucleus. These interactions are handled by the INTRANUKE package [29] which is integrated in NEUGEN. INTRANUKE incorporates pion elastic and inelastic scattering, as well as single charge exchange and absorption. The calculations are gauged by a comparison of exclusive final states in neutrino bubble chamber experiments with deuterium and neon targets. The uncertainties in INTRANUKE are also addressed in Chap. 8.

3.4.2.3 Simulation of the Detectors

Neutrinos from the flux simulation are randomly sampled and traced through the Near and Far Detector halls, in such a way that events are generated both inside the detectors as well as in the support structure and rock. The simulation includes a detailed geometric model of the detector. The position of individual strips is known to approximately 1 mm from cosmic ray tracks. Also, the magnetic field is modeled via finite element analysis driven by bench measurements of the B-H curve of steel samples.

Once a neutrino is generated, the particles produced by NEUGEN are passed to the detector model, which is based on GEANT3 [30]. The detector simulation then generates raw energy depositions in the scintillator strips. At that point, photons are generated in the scintillator and propagated to the PMT cathode by a C++ package called "PhotonTransport." PhotonTransport simulates the capture and the transmission of photons in the WLS fibers, as well as the conversion of photons to photoelectrons in the PMTs. Features such as PMT non-linearity, noise and triggers are incorporated.

Another feature that is incorporated in the simulation and which is of particular interest to this thesis is the crosstalk simulation. Crosstalk is the phenomenon by which a small amount of light leaks to PMT pixels adjacent to the one being illuminated. Crosstalk can be divided into two main categories: optical crosstalk and charge crosstalk. Optical crosstalk occurs when photons emerging from the end of a fiber spread due to optical reflections and refractions, or when photoelectrons close to the edge of a pixel strike a neighboring dynode chain. Charge crosstalk occurs when some charge in the multiplication process spreads to

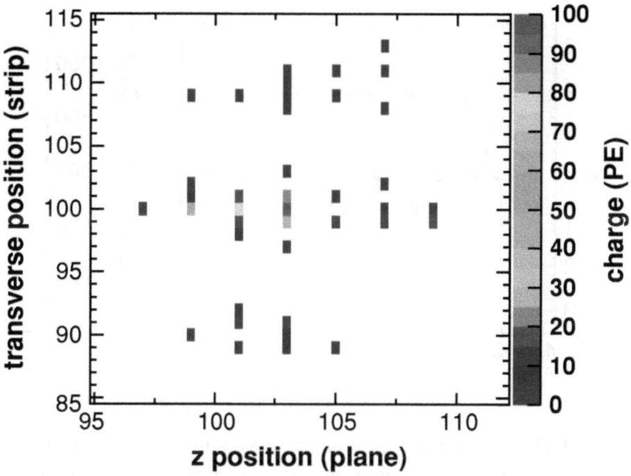

Fig. 3.14 A simulated ν_e CC event. The crosstalk hits form a cluster approximately ten strips below and above the shower hits

neighboring channels or when charge is induced through the capacitive coupling on the anodes.

Based on test-stand measurements [31, 32], the PMTs at both detectors have been found to leak less than 4% of their light onto neighboring pixels. Nevertheless, cross-talk can still have a strong impact in the topology of showers, as illustrated by Fig. 3.14. Crosstalk is implemented in the simulation based on the crosstalk fractions to nearest-neighbor, diagonal-neighbor and non-neighbor pixels as measured at the test stands. As will be seen in Chap. 8, the performance of the current crosstalk model is not good enough for the ν_e appearance search. In particular, the model fails to account for an excess of low pulse-height hits observed in the data whose exact origin remains unknown. The solution is to remove most crosstalk hits by applying a pulse-height cut to the selection.

3.4.3 MINOS Reconstruction

Before being analyzed, both the data and the simulation are processed through the MINOS event reconstruction. The reconstruction is a C++ based framework whose goal is to estimate the visible energy of the different neutrino interactions and to provide a distinct set of quantities that describe each event. The input to the reconstruction is the digitized readout recorded during a beam spill or during a cosmic ray event. This information is referred to as a "snarl."

A snarl can contain multiple events, especially if it is a Near Detector snarl. The first step in the reconstruction is to divide the activity in the detector into one or more events. Figure 3.15 shows the example of a beam spill as observed in the Near Detector. Hits from a single interaction are identified using timing and spatial information. In the Far Detector the rate is much lower, and there is rarely more than one event per beam spill.

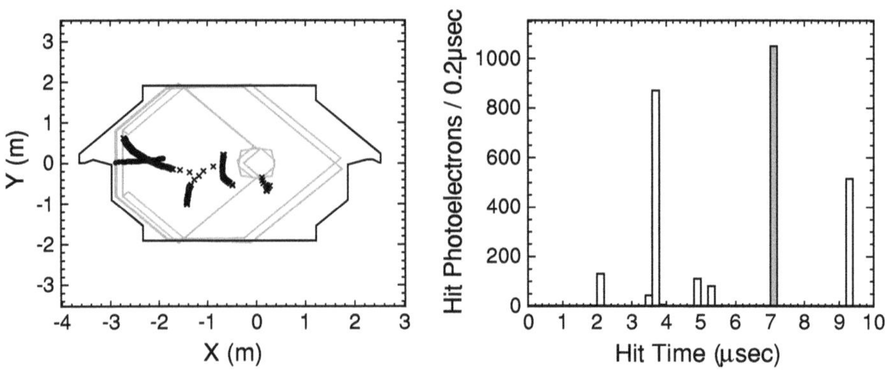

Fig. 3.15 One beam spill as observed in the Near Detector. For clarity a spill containing a smaller than average number of neutrino interactions was chosen. The left plot shows the horizontal and vertical position of track hits. The right plot shows the detector signal as a function of time, where the bin width is about ten times larger than the detector's timing resolution. Image obtained from [6]

A track-finding algorithm is then applied to each event. The algorithm operates by finding small track-like segments and then, when possible, joining them to form a "seed track." The seed track is then iteratively passed through a Kalman filter, which relies on the information from muon propagation in order to account for the effects of noise and of multiple scattering [33]. The track's momentum is obtained from the fitted curvature, which also provides information concerning the particle's charge sign. If the track ends within the detector, its momentum can also be determined from range. The momentum resolution is approximately 5% for range and 10% for curvature. It should be noted that the track-finding algorithm also finds tracks created by, for instance, charged pions. As a consequence, the majority of reconstructed events have at least one reconstructed track in them.

Showers are constructed from clusters of hits localized in space and time. The energy of a shower can only be determined by calorimetry, and is given by the summed pulse-height of all the individual hits. For hits that are part of both a track and a shower the estimated pulse-height of the track is subtracted.

Most ν_μ CC events leave a signature in the detector consisting of a long muon track with some hadronic activity at the vertex. For those events, the reconstructed energy is obtained by summing the energy of the most energetic track with the energy of any shower present at the upstream end of the track. The energy scale applied to the showers is based on the assumption that they are hadronic in nature.

As will be shown in the next chapter, the ν_e CC-like events involved in the ν_e appearance search of this thesis have a strong electromagnetic component. A different energy scale is thus derived for them using quasi-elastic ν_e CC events, by calculating their total summed pulse-height in inter-detector calibrated units[9] and

[9] These are the MEUs defined in the next section.

comparing it to their true energy through a linear fit [34]. The procedure is carried out in both detectors separately.

In order to correctly estimate the energy of neutrino interactions it is thus necessary to be able to accurately convert pulse-heights into energies. A description of how this is achieved follows.

3.5 Calibration

There are many factors that affect the measured amount of energy deposited by a particle passing through a strip in the detectors. The goal of the calibration is to remove the variations within each detector in order to relate the energy deposits from different detectors (relative calibration), and to then translate those deposits into energies (absolute calibration). Given that oscillation measurements are done by comparing the energy spectra between the two detectors, the calibration is critical to their success.

3.5.1 Relative Calibration

3.5.1.1 Energy Branch

When calibrating the data we start with raw energy depositions (pulse-heights) measured in ADCs. There are several effects that cause variations within each detector. Before arriving at an inter-detector energy scale each variation must be corrected, as shown in the left side of Fig. 3.16. Due to its end product, we refer to this set of corrections as the "energy branch," and we briefly review each step below. More details are available in [16, 35].

Drift correction

The drift correction accounts for changes in the response of the light detection systems over time. These changes are due mostly to PMT gains drifting with temperature and over time, as well as to scintillator aging.

The variation of the PMT gains over time is directly measured with a light-injection (LI) system that is installed in the two detectors [36]. The system consists of ultra-violet light-emitting diodes (LEDs) placed in pulser boxes, which deliver pulses of light to each module end via fiber up to the cables interspersed with the normal data taking. In order to monitor the gain drifts, each Far (Near) Detector strip is pulsed 300 (1000) times every hour. The average pulse intensity is equivalent to approximately 50 PEs. These LI "drift" data are useful to demonstrate that short-term (<24 h) variations are small and occur mostly due to environmental changes in the detector halls. They are not enough to perform the entire

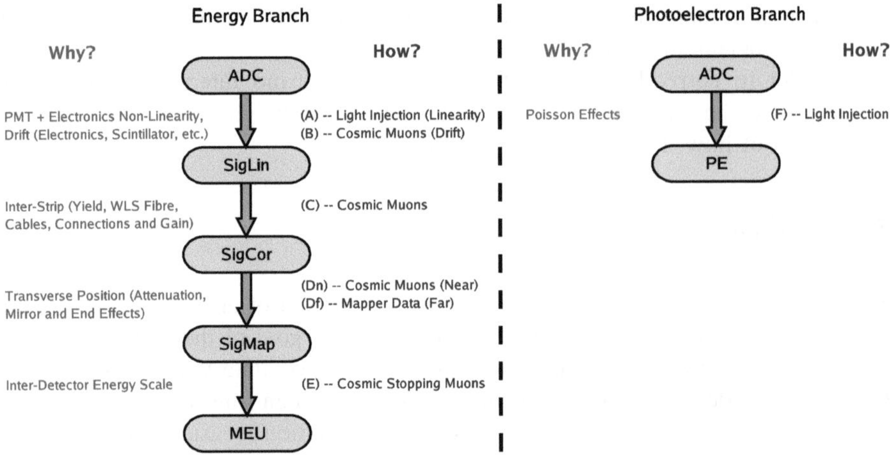

Fig. 3.16 Calibration chains for the MINOS data. The raw energy depositions come in the form of digitized values from the ADCs. Before they can be expressed in terms of an inter-detector energy scale they must be corrected for drift, non-linearity, inter-strip differences and transverse position. SigLin, SigCor and SigMap are the names given in MINOS to the quantities obtained after these corrections have been applied. A conversion to photoelectrons (PEs) is also done by applying the average response per channel to the observed ADC values. Image obtained from [35]

drift calibration however, as the light-injection procedure is insensitive to scintillator aging. Because of this, the drift correction is achieved using the mean signal per plane induced by through-going cosmic ray muons. Studies show that the overall detector response varies by less than 5% over the data-taking period considered in this thesis [35]. The changes in the gains and the scintillator aging effects have gone in opposite directions, and thus have canceled to some degree.

Linearity correction

The response of the PMTs becomes $\sim$5% non-linear for light levels on the order of $\sim$100 PEs and $\sim$50 PEs, for the Near and Far Detectors respectively. Moreover, the nonlinearity of the response of the Far Detector electronics is similar in size to that of the PMTs. The LI system is used to map the non-linearity of the entire readout system, by pulsing each strip 1000 times at different light-levels once a month in both detectors. The LI linearity data are then used to obtain a quadratic parameterization of the PMTs' response as a function of the true illumination, which is applied as a correction.

Strip-to-strip correction

Strips have differences in light output on the order of 30%. The response of a strip depends on a number of factors such as the PMT gain, PMT quantum efficiency, fiber collection efficiency, scintillator light yield, clear fiber length and attenuation

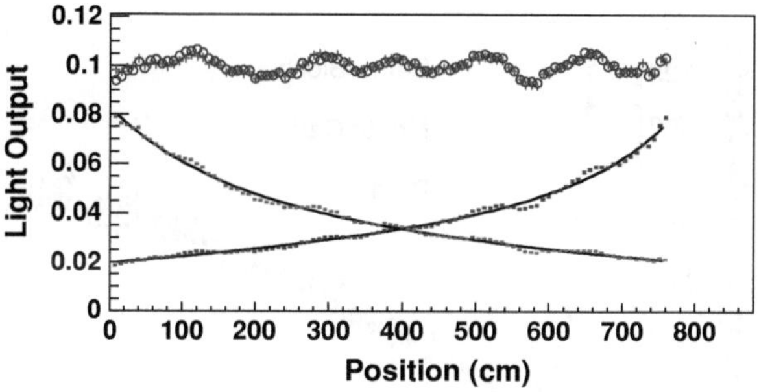

Fig. 3.17 Responses from each end of a typical strip in the Far Detector when illuminated along its length. The least squares fits to the sum of two exponentials are also shown in *black*. The vertical scale is approximately 1/100 the number of photoelectrons at the PMT for a cosmic ray muon passing normally through the module plane. The top series of data points shows the ratios of these data to their respective fits, normalized to 0.1 for display purposes. The small variations in the light output from a smooth exponential behavior are real and result from variations in scintillator, fiber depth and gluing. Image obtained from [16]

in the optical connections. Cosmic ray muons are used to measure the mean light level at each strip end. By accounting for attenuation and the angle of the incident muon, the response is normalized to that of a muon going through the center of the strip at normal incidence. After calibration, the strip-to-strip variations are reduced to 0.5%. This correction is also referred to as the "inter-strip" correction.

Transverse position correction

This correction is used to remove variations in response along the length of the strip (i.e., the transverse position across the plane) which are mostly due to attenuation. A radioactive source was used to map out the response of each scintillator module at many positions along each strip prior to their installation in the detectors. As shown in Fig. 3.17, the attenuation along the WLS fibre can be accurately parameterized by the sum of two exponentials with typical attenuation constants of $\sim$1 and $\sim$7 m. The radioactive source data are used to produce the parameterization used in the Far Detector. In the Near Detector, the transverse position correction parameterizations are derived from the high-statistics cosmic ray muon data. The pulse-height of a strip hit by a track is plotted as a function of the transverse position, and the data is fit to a double exponential. The results are found to be consistent with those obtained using the radioactive source [16].

Inter-detector energy scale

The final step is to relate the intra-detector calibrated energy depositions (referred to as "SigMaps" in Fig. 3.16) of the detectors. This is best accomplished by comparing these energy depositions with the detector's response to stopping muons. Stopping muons must be used because their momenta can be accurately

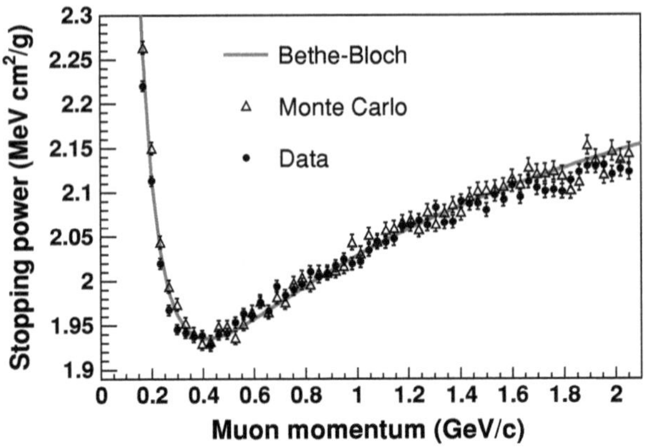

Fig. 3.18 Stopping power of muons as a function of their momentum. The prediction by the Bethe-Bloch equation in polystyrene scintillator agrees very well with the MINOS Far Detector data and the MINOS MC simulation. Both data and MC points have been normalized to the Bethe-Bloch calculation to give the expected stopping power at the minimum ionizing point. Image obtained from [16]

determined from their range. The energy loss of a muon by ionization is very well described by the Bethe-Bloch equation [14], as shown in Fig. 3.18.

All the MINOS detectors[10] have the same scintillator strip width. Therefore, by measuring the detectors' response to muons of a certain track length the inter-detector calibration can be performed. A "track window" technique is used in order to maximize the accuracy, where the response of muons is only measured when their momenta lie between 0.5 and 1.1 GeV. This avoids using the end of the track where the losses by ionization increase very rapidly (as shown in the left side of the curve of Fig. 3.18), which minimizes the dependence on the muon's momentum measurement.

This technique allows for the determination of the inter-detector energy scale with a ~2% uncertainty. The scale is measured in terms of Muon Energy Units (MEUs), which are also known as a Minimum Ionizing Particles (MIPs). One MEU corresponds approximately to the average detector response to a minimum ionizing muon traversing 1 plane of scintillator at normal incidence.

3.5.1.2 Photoelectron Branch

As shown on the right side of Fig. 3.16, there is also a photoelectron branch where the raw energy depositions in ADCs are converted to PEs. This is done by using the LI drift data to measure the average response per photoelectron per channel. From photon statistics we know that if N_{pe} is the mean number of photoelectrons

[10] Including the CalDet, which is described in Sect. 3.5.2.

per pulse per channel,[11] then the mean response λ and the rms width σ are related by

$$N_{pe} = \frac{\lambda^2}{\sigma^2},$$
$$g = \frac{\sigma^2}{\lambda},$$

where g is the gain of the given channel. The above expressions are only correct if the PMT has perfect 1 PE resolution, which is not the case. The more accurate expressions are given by

$$N_{pe} = \frac{\lambda^2}{\sigma^2} \times (1 + w^2),$$
$$g = \frac{\sigma^2}{\lambda} \times (1 + w^2),$$

where w is the fractional 1 PE width, measured on the test stand [32]. The uncertainty in w results in a ~5% uncertainty in the gains at both detectors [37].

3.5.2 Absolute Calibration

Being able to relate the energy depositions between the two detectors is a critical step, but is not the end of the calibration chain. It is also important to express these energy depositions in terms of an absolute energy scale (such as GeV, for example). This way the response of the detectors can be understood for muons, electrons and hadrons of a certain energy.

The absolute calibration is achieved with data from the MINOS Calibration Detector (CalDet) [38]. The CalDet was a small-scale version of the Near and Far Detectors. It had a mass of 12 tons, and it consisted of 60 planes of unmagnetized steel and scintillator, each with dimensions $1 \times 1\ \mathrm{m}^2$. The CalDet used the same 1 cm thick and 4.1 cm wide scintillator strips as the other two detectors. The CalDet steel planes were 0.04 cm less thick, but this is a very small difference that is easily taken into consideration in the calculations.

The CalDet was exposed to protons, pions, electrons and muons in test beams at CERN, and acquired data with both Near and Far Detector electronics. By comparing the calibrated shower energy (in MEUs) to the known energies of the particle beams (in GeV) the absolute calibration of the detector energy response is achieved. The CalDet data were also used to benchmark the hadronic and

[11] The pulse intensity is tuned so that a PMT pixel receives approximately 50 PEs per pulse. Due to variations in the injection fibers, in the readout fibers and in the PMT efficiencies, some PMTs receive a factor of two more or less than the average [35].

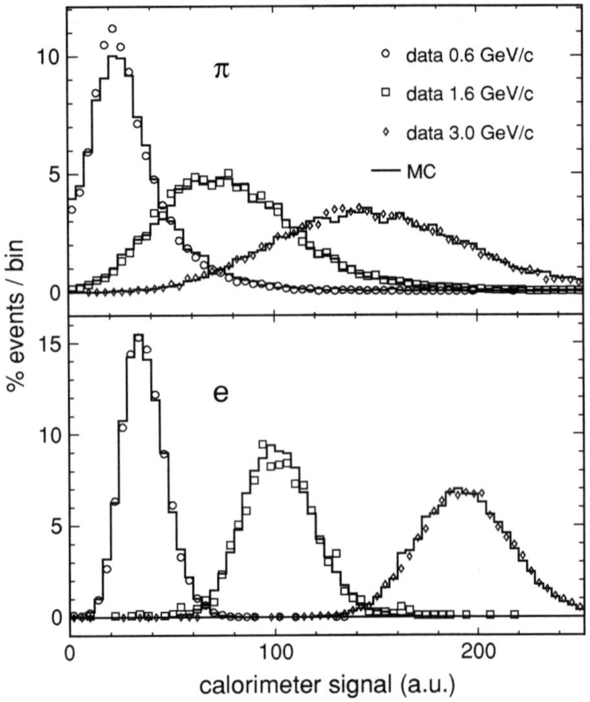

Fig. 3.19 Calorimetric response to pions and electrons at three momenta. The calorimeter-signal scale is in arbitrary units. The data obtained from CalDet is compared to the simulation. Image obtained from [16]

electromagnetic shower simulation and to determine the energy resolution. Figure 3.19 shows the measured detector response to pions and electrons compared to the simulation result. The simulated detector response to electrons agrees with the data to better than 2%. Also, the simulation reproduces the response to pion and proton induced showers to better than 6% at all momenta. The energy resolution was adequately reproduced by the simulation and may be parameterized as $56\%/\sqrt{E} \oplus 2\%$ for hadron showers and $21.4\%/\sqrt{E} \oplus 4\%$ for electromagnetic showers, where the energy E is in GeV.

3.5.3 Implementation in the Simulation

Once the detectors have been calibrated, the information obtained in the calibration process is used in the simulation. Roughly speaking, the situation as described so far is reversed in the MC. Now the energy depositions in GeV are known, and the goal is to accurately translate those into lower-level quantities such as SigMaps and photoelectrons. This is achieved by using the calibration constants derived from the data to "decalibrate" the energy into the lower-level quantities during the MC generation. This way, the MC contains the best knowledge of the detectors' calorimetric response, attenuation, strip-to-strip variations, drift, non-linearity and gains, as obtained from the data.

Technically this is achieved by associating each MC run with a real-world time derived randomly from the period considered in the analysis. The calibration constants from that particular time are then used in the reconstruction of that run. In this fashion, even time-dependent effects such as scintillator aging are accurately taken into account in the simulation.

References

1. http://www-fmi.fnal.gov/
2. D.S. Ayres et al., NOvA proposal to build a 30-kiloton off-axis detector to study neutrino oscillations in the Fermilab NuMI beamline, hep-ex/0503053 (2004)
3. D. Drakoulakos et al., Proposal to perform a high-statistics neutrino scattering experiment using a fine-grained detector in the NuMI beam, hep-ex/0405002 (2004)
4. R.R. Wilson, The tevatron. Phys. Today **30N10**, 23–30 (1977)
5. C.C. Jensen, G.A. Krafczyk, NuMI proton kicker extraction system, in *Proceedings of the 2005 IEEE Particle Accelerator Conference (PAC 05) 21st IEEE Particle Accelerator Conference*. Knoxville, 16–20 May 2005, p. 692
6. P. Adamson et al., A study of muon neutrino disappearance using the fermilab main injector neutrino beam. Phys. Rev. **D77**, 072002 (2008)
7. S. Shukla, J. Marriner, J. Griffin, Slip stacking in the Fermilab Main Injector, prepared for 1996 DPF / DPB summer study on new directions for high-energy physics (Snowmass 96), Snowmass, 25 Jun–12 Jul 1996
8. A.G. Abramov et al., Beam optics and target conceptual designs for the NuMI project. Nucl. Instrum. Methods **A485**, 209–227 (2002)
9. D. Indurthy, S. Kopp, Z. Pavlovich, M. Proga, Profile monitor SEM's for the NuMI beam at FNAL. AIP Conf. Proc. **732**, 332–340 (2004)
10. J. Hylen et al., NuMI technical design handbook, internal NuMI report, 2003
11. Y. Fukuda et al., Evidence for oscillation of atmospheric neutrinos. Phys. Rev. Lett. **81**(8), 1562–1567 (1998)
12. W.W.M. Allison et al., Neutrino oscillation effects in Soudan-2 upward-stopping muons. Phys. Rev. D **72**(5), 052005 (2005)
13. P. Adamson et al., Measurement of neutrino oscillations with the MINOS detectors in the NuMI beam. Phys. Rev. Lett. **101**, 131802 (2008)
14. C. Amsler et al. (Particle Data Group), review of particle physics. Phys. Lett. **B**(667), 1 (2008)
15. A. Pla-Dalmau, Extruded plastic scintillator for the MINOS calorimeters. Frascati Phys. Ser. **21**, 513–522 (2001)
16. D.G. Michael et al., The magnetized steel and scintillator calorimeters of the MINOS experiment. Nucl. Instrum. Methods **A596**, 190–228 (2008)
17. J. Oliver et al., Design and performance of the readout system of the MINOS Far detector. IEEE Trans. Nucl. Sci. **51**, 2193–2195 (2004)
18. P. Adamson et al., First observations of separated atmospheric ν_μ and $\overline{\nu}_\mu$ events in the MINOS detector. Phys. Rev. **D73**, 072002 (2006)
19. T. Cundiff et al., The MINOS near detector front end electronics. IEEE Trans. Nucl. Sci. **53**, 1347–1355 (2006)
20. M. Dorman, Evolution of the ND CC spectrum during runs II and III. MINOS-doc-6187, 2009
21. A. Ferrari, P.R. Sala, A. Fasso, J. Ranft, FLUKA: A multi-particle transport code (Program version 2005), CERN-2005-010

22. R.Brun et al., GEANT detector description and simulation tool, CERN Program Library Long Writeup W5013, 1994
23. Z. Pavlovic, Observation of disappearance of muon neutrinos in the NuMI beam, FERMILAB-THESIS-2008-59
24. H. Gallagher, The NEUGEN neutrino event generator, Nucl. Phys. Proc. Suppl. **112**, 188–194 (2002)
25. H. Gallagher, Event generator tuning in the resonance region for the MINOS experiment. Nucl. Phys. Proc. Suppl. **159**, 229–234 (2006)
26. T. Yang et al., A hadronization model for few-GeV neutrino interactions, arXiv:0904.4043 (2009)
27. T. Sjostrand, S. Mrenna, P. Skands, PYTHIA 6.4 Physics and Manual. J. High Energy Phys. **05**, 026 (2006)
28. Z. Koba, H.B. Nielsen, P. Olesen, Scaling of multiplicity distributions in high-energy hadron collisions. Nucl. Phys. **B40**, 317–334 (1972)
29. R. Merenyi et al., Determination of pion intranuclear rescattering rates in ν_μ-Ne versus ν_μ-D interactions for the atmospheric ν flux. Phys. Rev. D **45**(3), 743–751 (1992)
30. M. Goossens, GEANT: detector description and simulation tool, long writeup W5013, CERN Program Library, CERN, 1993
31. N. Tagg et al., Performance of Hamamatsu 64-anode photomultipliers for use with wavelength-shifting optical fibres. Nucl. Instrum. Methods **A539**, 668–678 (2005)
32. K. Lang et al., Characterization of 1600 Hamamatsu 16-anode photomultipliers for the MINOS Far detector. Nucl. Instrum. Methods **A545**, 852–871 (2005)
33. J.S. Marshall, A study of muon neutrino disappearance with the MINOS detectors and the NuMI neutrino beam, FERMILAB-THESIS-2008-20
34. S. Cavanaugh, G. Pawloski, Cedar-Phy-Daikon energy scale calibration for NueAna, MINOS-doc-3795, 2008
35. MINOS collaboration, calibration position paper for pre-shutdown data, MINOS-doc-3941, 2008
36. P. Adamson et al., The MINOS light injection calibration system. Nucl. Instrum. Methds **A492**, 325–343 (2002)
37. N. Tagg, Calibration plenary, MINOS-doc-5508, 2008
38. P. Adamson et al., The MINOS calibration detector. Nucl. Instrum. Methods Phys. Res. Sect. A Accelerators Spectrometers Detectors Associated Equipment **556**(1), 119–133 (2006)

Chapter 4
Measuring θ_{13} in MINOS

In Chap. 2 we saw that searching for a nonzero θ_{13} is one of the main priorities in the field of neutrino physics. In this chapter, after reviewing the motivation for θ_{13} in the first part of Sect. 4.1, we explain how MINOS can access this mixing angle by searching for ν_e appearance, together with the main challenges involved in that search. Section 4.2 then lays out the strategy for the analysis, by introducing the different methods involved and by giving an overview of their roles. The subsequent chapters, which are referenced throughout Sect. 4.2, provide more detail on each of the steps of the analysis.

4.1 The Search for ν_e Appearance in MINOS

4.1.1 Brief Motivation Review

As explained in Chap. 2, θ_{13} is the only mixing angle in the 3×3 PMNS neutrino matrix [1, 2] that remains unknown. The angle is known to be small, and the most stringent upper limit on it is the one set by the CHOOZ [3] experiment,

$$\sin^2(2\theta_{13}) < 0.15 \quad (90\%\,\text{C.L.}), \tag{4.1}$$

corresponding to the measured value of the atmospheric mass splitting of $|\Delta m^2_{32}| \simeq 2.4 \times 10^{-3}\,\text{eV}^2$.

For the reasons explained in Sect. 2.3.2, making a measurement of θ_{13} is of the utmost importance for neutrino physicists but also for the wider physics community. A very aggressive program of next-generation neutrino oscillation experiments whose primary (and in some cases unique) goal is to measure θ_{13} has been set in motion. This program comprises the two reactor experiments Double CHOOZ and Daya Bay, in addition to the two long baseline oscillation experiments NOνA

J. P. Ochoa Ricoux, *A Search for Muon Neutrino to Electron Neutrino Oscillations in the MINOS Experiment*, Springer Theses, DOI: 10.1007/978-1-4419-7949-0_4,

and T2K. But MINOS has a chance of making this measurement before these new experiments come online, provided that θ_{13} lies closely below the current experimental limit mentioned above.

4.1.2 A Direct Handle on θ_{13}

Electron neutrino appearance constitutes a direct handle onto θ_{13} in MINOS. At MINOS' baseline and in the three-neutrino framework, the probability of a ν_μ oscillating into a ν_e is approximately[1] given by [5]:

$$\begin{aligned} P(\nu_\mu \rightarrow \nu_e) &\approx \sin^2(2\theta_{13})\sin^2\theta_{23}\frac{\sin^2(A-1)\Delta}{(A-1)^2} \\ &+ 2\alpha\sin\theta_{13}\cos\delta\sin 2\theta_{12}\sin 2\theta_{23}\frac{\sin A\Delta}{A}\frac{\sin(A-1)\Delta}{(A-1)}\cos\Delta \\ &- 2\alpha\sin\theta_{13}\sin\delta\sin 2\theta_{12}\sin 2\theta_{23}\frac{\sin A\Delta}{A}\frac{\sin(A-1)\Delta}{(A-1)}\sin\Delta, \end{aligned} \tag{4.2}$$

where

$$\begin{aligned} \Delta &= \frac{\Delta m^2_{31}L}{4E} \simeq \frac{\pm 2.23\,\text{GeV}}{E}, \\ A &= \frac{G_f n_e L}{\sqrt{2}\Delta} \simeq \frac{\pm E}{11\,\text{GeV}}, \quad \text{and} \\ \alpha &= \frac{\Delta m^2_{21}}{\Delta m^2_{31}} \simeq \pm 0.026. \end{aligned}$$

The last three variables change sign for the two mass hierarchies. δ is the CP violating Dirac phase, also referred to as δ_{CP}. Both A and δ change sign when considering neutrinos or antineutrinos.

As explained in Sect. 2.2.1, neutrino oscillations in matter can be either enhanced or suppressed by the Mikheyev–Smirnov–Wolfenstein (MSW) effect [6, 7]. In our case, a resonant enhancement of the oscillations occurs if neutrino masses follow the normal mass hierarchy ($\Delta m^2_{31} > 0$), but a suppression occurs if neutrinos follow the inverted mass hierarchy ($\Delta m^2_{31} < 0$). For antineutrinos this

[1] This expression is good to first order in the mass hierarchy parameter α and to second order in $\sin\theta_{13}$. Also, the probability is calculated based on the assumption that neutrinos travel in matter of constant electron density n_e, which is a very good approximation in our case as neutrinos go only through the earth's crust [4]. Eq. 4.2 is thus accurate for all practical purposes, even though for the analysis we used the full expression for $P(\nu_\mu \rightarrow \nu_e)$, which takes more than one page to write fully but is easy for a computer to evaluate.

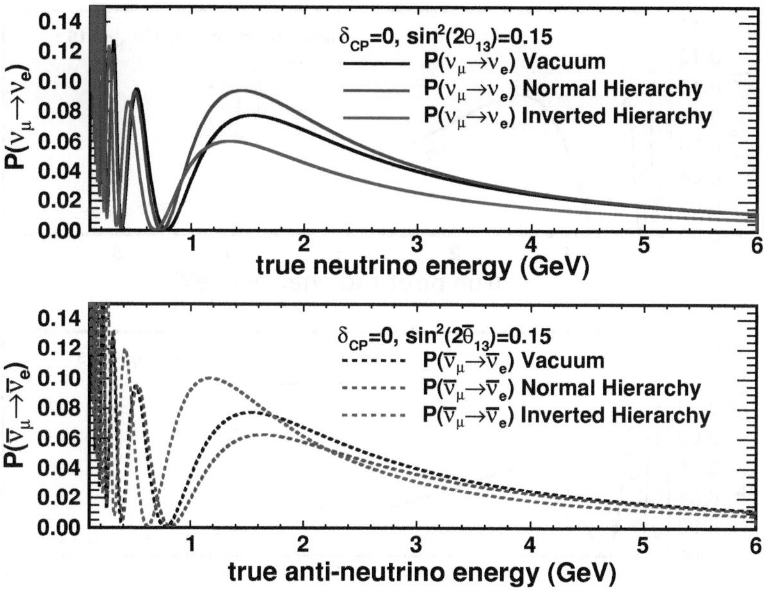

Fig. 4.1 ν_e and $\overline{\nu}_e$ appearance probability in MINOS as a function of true neutrino energy for $\delta = 0$ and $\sin^2(2\theta_{13}) = 0.15$, with standard oscillation parameters $\Delta m^2_{32} = 2.38 \times 10^{-3}\,\mathrm{eV}^2$, $\Delta m^2_{21} = 8.0 \times 10^{-5}\,\mathrm{eV}^2$, $\theta_{23} = \pi/4$, $\theta_{12} = 0.59365$. In MINOS both the selection efficiency and the number of events decrease drastically for $E < 1$ GeV, and thus the first oscillation peak (centered around $E \simeq 1.5\,\mathrm{GeV}$) strongly dominates. For that first peak it can be clearly seen that in the case of the normal mass hierarchy matter effects enhance ν_e appearance for neutrinos while they suppress it for antineutrinos. The situation is reversed in the case of the inverted mass hierarchy

effect is reversed. The enhancement or suppression can be as high as 25% in the energy region of interest ($> 1\,\mathrm{GeV}$). This is illustrated in Fig. 4.1 which shows the ν_e appearance probability for neutrinos and antineutrinos, in the case of both mass hierarchies.

Varying the CP-violating phase δ (also referred to as δ_{CP}) has an impact on $P(\nu_\mu \rightarrow \nu_e)$ of similar magnitude as matter effects have. This is illustrated in Fig. 4.2, which shows how $\nu_\mu \rightarrow \nu_e$ oscillations are suppressed for $\delta \sim \pi/2$ but enhanced for $\delta \sim 3\pi/2$, for both mass hierarchies.

Thus in addition to θ_{13}, the probability for ν_e appearance is a function of two other unknown parameters: the CP violating phase δ and the mass hierarchy of the neutrino masses (i.e., the sign of Δm^2_{31}). This means that any measurement or limit on θ_{13} in MINOS will depend on these two parameters. In particular, the limit or measurement will be more constraining in the case of the normal mass hierarchy, as illustrated in Fig. 4.1. This stands in sharp contrast with reactor experiments where the only unknown parameter driving the disappearance of $\overline{\nu}_e$ is θ_{13}.

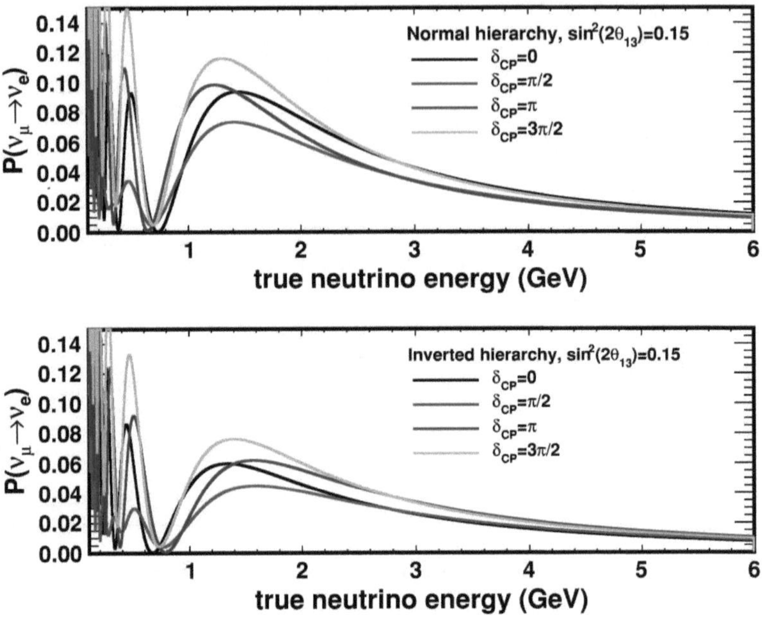

Fig. 4.2 ν_e appearance probability in MINOS as a function of true neutrino energy for $\sin^2(2\theta_{13}) = 0.15$, with standard oscillation parameters $\Delta m^2_{32} = 2.38 \times 10^{-3}\,\text{eV}^2, \Delta m^2_{21} = 8.0 \times 10^{-5}\,\text{eV}^2$, $\theta_{23} = \pi/4, \theta_{12} = 0.59365$. Varying the CP-violating phase δ causes significant changes in the oscillation probability, by as much as approximately $\pm 25\%$ for energies higher than 1 GeV

Nevertheless, as can be seen in Eq. 4.2, no oscillations to ν_e will occur if the angle θ_{13} is zero.[2] Therefore, an observation of ν_e appearance in MINOS would be an unambiguous sign of a nonzero θ_{13}.

MINOS is physically situated in an excellent position to look for ν_e appearance. Eq. 4.2 means, for example, that in the absence of CP violation (i.e., $\delta = 0$), neglecting matter effects and for a value of θ_{13} at the CHOOZ limit, a total of $\sim 58\nu_e$ CC events are expected to appear in the fiducial volume of the Far Detector for a total exposure of 7×10^{20} POTs.[3] This means that for a number of events observed corresponding to the expected signal only, MINOS would be capable of going well below the CHOOZ limit in sensitivity, even for the analysis described in this thesis which corresponds to an exposure of 3.14×10^{20} POTs. Unfortunately,

[2] This statement is not strictly correct when we consider the full expression for $P(\nu_\mu \to \nu_e)$. As a matter of fact, the next term in the expansion is $\alpha^2 \sin^2 2\theta_{12} \sin^2\theta_{23} \frac{\sin^2 A\Delta}{A^2}$ [5], which is not dependent on θ_{13}. However $P(\nu_\mu \to \nu_e)$ in the case of $\theta_{13} = 0$ is ~ 0.002 at 1 GeV and decreases exponentially at higher energies, thus being completely negligible experimentally.

[3] This corresponds to the MINOS recorded dataset as of this writing.

the detectors were not designed for this kind of measurement at this energy, and the search for ν_e appearance is dominated by the backgrounds described in the next section. The issues of how to extract the potential signal from the backgrounds as efficiently as possible, and of how to accurately predict the backgrounds in order to maximize our sensitivity to θ_{13}, are the subject of this thesis in general. In subsequent chapters we will show how, as a result of this work, MINOS could reach a better sensitivity to θ_{13} than CHOOZ by using its full dataset, before the new generation experiments come online.

4.1.3 Backgrounds to the Search

4.1.3.1 Identifying the Neutrino Flavor

As explained in Chap. 3, the MINOS detectors are toroidal magnetized steel-scintillator sampling calorimeters. Each plane has a thickness of 3.54 cm, corresponding to 2.54 cm of steel and 1 cm of plastic scintillator. The scintillator is segmented transversely into 4.1 cm wide strips whose orientation alternates by $\pm 90°$ at every plane.

The neutrino interactions occurring in the detectors are identified based on the topology of the hit strip pattern left by the final state particles. The flavor of the incoming neutrino can only be known for charged-current (CC) neutrino events $\nu_l + \mathrm{Fe} \rightarrow l + X$ through the identification of the charged lepton partner l produced in the interaction. The MINOS experiment was primarily designed to identify ν_μ CC events, where the resulting muon typically leaves a clear long track that extends well beyond the hadronic activity at the vertex. Figure 4.3 shows an example of what a ν_μ CC event looks like in MINOS.

A ν_e CC event however has to be identified based on the electron resulting from the interaction. Electrons are approximately 200 times less massive than muons, and are thus easily deflected by atomic nuclei as they travel through matter. These deflections cause electrons to undergo bremsstrahlung radiation [8], that is, to emit high-energy photons. These high-energy photons subsequently materialize through e^+e^- pair production, as illustrated in Fig. 4.4. This process is repeated many times until the energy of the secondary electrons becomes comparable to their rest mass ($\lesssim$10 MeV), at which point the remaining energy is lost primarily by ionization [9]. The distance over which an electron loses all but $1/e$ of its energy to bremsstrahlung when traveling in a given material is given by the radiation length X_0. The radiation length is also 7/9 of the mean free path for pair production by a high-energy photon, and thus constitutes a good measure of the size of the internal structure of EM showers. Table 4.1 shows the radiation length of electrons in the three different materials that compose the MINOS detectors.

Therefore electrons going through the steel-scintillator layers of MINOS produce a showerlike topology. Electron showers have been studied in CalDet [10],

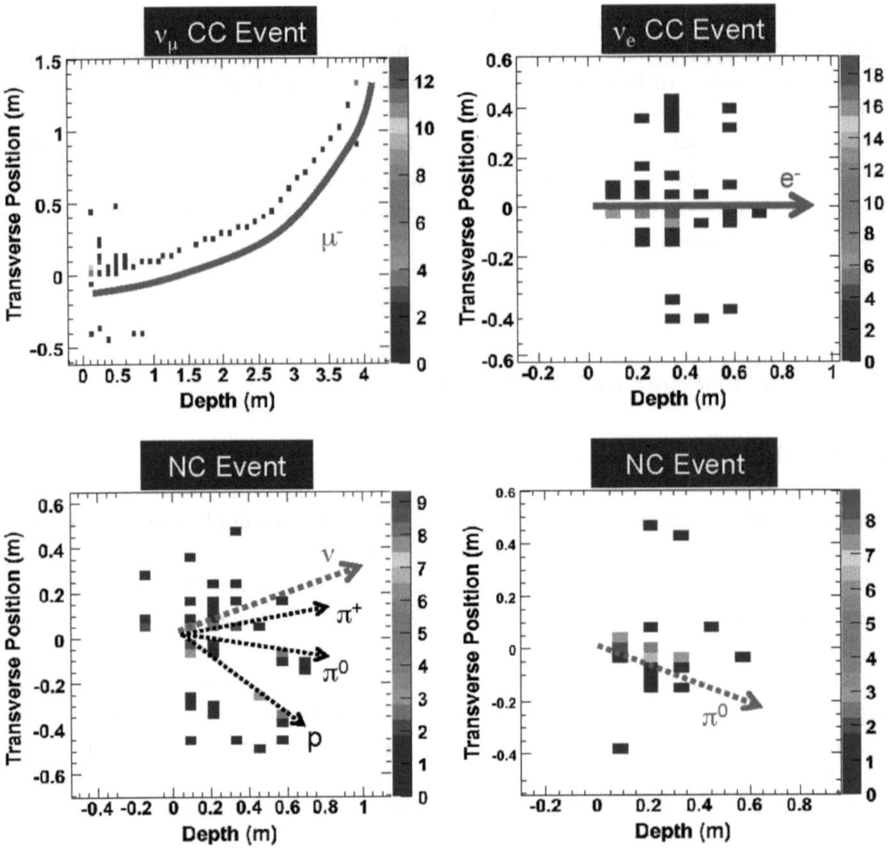

Fig. 4.3 Simulated event topologies in MINOS. The color scale represents the charge deposited per strip in Minimum Ionizing Particle (MIP) units. Most ν_μ CC events leave an unmistakeable signature in the detector consisting of a long muon track with hadronic activity at the vertex. ν_e CC events are usually compact showers with a characteristic electromagnetic profile. NC events tend to be more diffuse than EM showers; a typical example is the bottom-left event where a 1.7 GeV proton is produced alongside a 1.9 GeV π^+ and a 1.2 GeV π^0 (alongside other less energetic particles). Sometimes π^0's that carry an important fraction of the energy transferred to the struck nucleus are produced in NC interactions, giving a significant EM component to the shower, as in the bottom-right event

which has practically the same steel thickness and scintillator width as the MINOS detectors. Figure 4.5 shows the longitudinal and transverse energy deposition profile of electrons in CalDet. Most of the energy of these showers is deposited in a narrow core contained within a few planes and strips. Also, the longitudinal energy deposition has a characteristic distribution that follows a gamma function [11],

$$\frac{dE}{dt} = \frac{dE}{dt} = E_0 b \frac{(bt)^{a-1} e^{-bt}}{\Gamma(a)}, \tag{4.3}$$

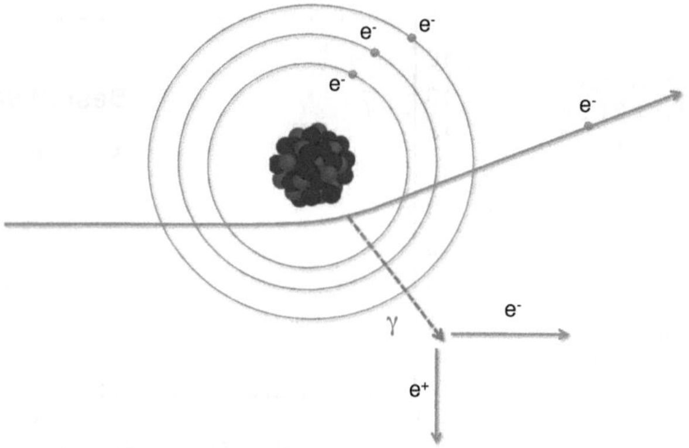

Fig. 4.4 Schematic of the processes responsible for the electromagnetic cascade that occurs when electrons travel in dense matter. The deflection of electrons by the atomic nuclei cause the emission of photons, which in turn materialize through e^+e^- pair production. This process is repeated until the energy of the charged secondaries is low enough in such a way that ionization becomes the dominant mechanism for energy loss

Table 4.1 Radiation lengths of electrons in the three materials that compose the MINOS detectors: steel, plastic scintillator (polysterene) and air

Steel (cm)	Scintillator (cm)	Air (cm)
1.76	47.9	3.04×10^4

The radiation length is both the distance over which an electron loses all but $1/e$ of its energy through bremsstrahlung and also $7/9$ of the mean free path for pair production by a high-energy photon

where E_0 is the total energy deposited, a describes the rise of the shower, b describes the fall of the shower, and t is the depth in radiation lengths. Figure 4.3 shows an example of what a ν_e CC event with these characteristics looks like in MINOS.

4.1.3.2 Hadronic vs. EM Showers

Identifying ν_e CC events in MINOS is a difficult task. As shown in Table 4.2, the distance between two successive scintillator planes is $\sim$1.5 times the effective radiation length[4] of EM showers. The situation is not much better in the transverse

[4] The effective radiation length and the effective Molière radius are calculated considering that a normally incident electron goes through 42.7% of steel, 16.8% of scintillator and 40.5% of air. In the case of the radiation length for example, from the values in Table 4.1 we obtain $X_{\text{eff}} = (0.427/1.76 + 0.168/47.9 + 0.405/3.04 \times 10^4)^{-1} = 4.06\,\text{cm}$.

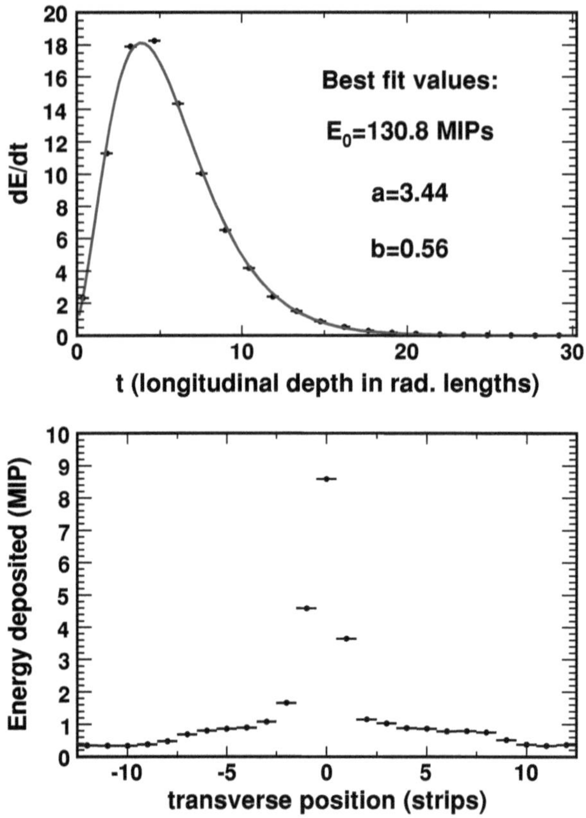

Fig. 4.5 Average longitudinal (*top*) and transverse (*bottom*) energy deposition profiles of 2 GeV electrons in CalDet. The energy is measured in Minimum Ionizing Particle (MIP) units, and *t* is the depth in radiation lengths. The binning corresponds to the physical dimensions of CalDet, that is one plane per bin (in the longitudinal direction) or one strip per bin (in the transverse direction). The longitudinal energy deposition profile fits very well to a gamma distribution

direction, where the strip width corresponds to ~1.1 times the effective Molière radius. Simply put, the MINOS detectors lack the granularity to optimally resolve low-energy (~2 GeV) showers. The detectors were primarily designed to have good enough resolution for ν_μ CC identification, and not ν_e CC identification [12]. The detector's low resolution causes some of the hadronic showers[5] occurring in CC and NC interactions to very closely mimic EM showers.

Normally hadronic showers involve many particles with an average transverse momentum of ~350 MeV each, and as a result tend to be more diffuse than EM showers. The bottom-left section of Fig. 4.3 shows what a typical NC interaction looks like in MINOS. However, π^0's that carry a significant fraction of the energy transferred to the struck nucleus are sometimes produced in hadronic showers. π^0's have a half life in the order of 10^{-17} seconds and decay electromagnetically nearly

[5] Throughout this thesis we use the term "hadronic shower" in its broad sense, i.e., as the shower initiated by the transfer of energy from the neutrino to the struck nucleus in the detector. These showers consist mostly of hadrons although, as discussed in this section, can contain particles that do not interact through the strong interaction.

Table 4.2 Characteristics of EM showers in the MINOS detectors and relevant detector parameters

EM showers in MINOS	Detector parameters
Effective radiation length: 4.06 cm	Plane separation: 5.95 cm
Molière radius: 3.7 cm	Strip width: 4.12 cm

The Molière radius is the radius of a cylinder centered along the shower axis that contains 90% of the shower's energy, and thus constitutes a measure of the transverse size of the showers. The Molière radius and radiation length quoted here are the effective values corresponding to the configuration of the MINOS calorimeters

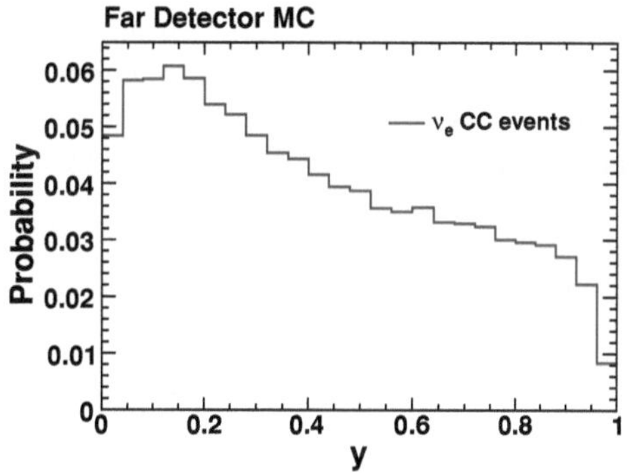

Fig. 4.6 y of simulated ν_e CC events with an energy between 1 and 8 GeV in the Far Detector. Approximately 40% of our signal is such that $y > 0.5$

all of the time, either into two photons (branching ratio of ~98.8%) or into an electron–positron pair with a photon (branching ratio of ~1.2%) [9]. Thus NC showers can have a dominant EM component which can closely mimic an electron shower, particularly when the two photons from π^0 decay are emitted colinearly or when one of the two is not energetic enough to initiate a separately distinguishable shower. The bottom-right event in Fig. 4.3 illustrates this for a case where practically all the visible energy in the shower is carried by the π^0.

Moreover, all CC events have a hadronic shower component to them. The importance of this component is quantified by the y variable, defined as the fraction of the total neutrino energy which was transferred to the struck nucleus to create the hadronic shower,

$$y = \frac{E_\nu - E_l}{E_\nu} = \frac{E_h}{E_\nu}, \tag{4.4}$$

where E_ν is the total neutrino energy, E_l is the energy of the outgoing lepton and E_h is the energy of the hadronic shower. As shown in Fig. 4.6, approximately 40% of the signal events are such that more than half of their energy is in the form of the hadronic shower (i.e., $y > 0.5$).

Thus hadronic showers and ν_e CC events can have both hadronic and EM components to them. As shown in Fig. 4.7, a small subset of NC events can actually have close to 100% EM energy in them, and there is a very significant overlap between NC and ν_e CC events in terms of the fraction of the shower that is

EM. This fact combined with the detector's low resolution seriously complicates the task of signal and background separation.

On top of these difficulties lies the extra complication arising from the fact that the high-energy tail of our spectrum acts as a secondary source of NC events. This is because, during a NC $\nu_l + \mathrm{Fe} \rightarrow \nu_l + X$ interaction, only some of the incoming neutrino energy and momentum is transferred to the target nucleus to produce a shower. In these events only the hadronic shower is visible in the detector, which makes it impossible to determine the energy of the incoming neutrino. This means that even neutrinos with very high energies undergoing an NC interaction are capable of leaving a shower with a visible energy that falls in the region of interest for the ν_e appearance search (from 1 to 8 GeV). In other words, NC events from practically all incoming neutrino energies are a source of background to the analysis. As shown in Fig. 4.8, the high-energy tail of our spectrum contributes approximately half of the total NC events in the background sample.

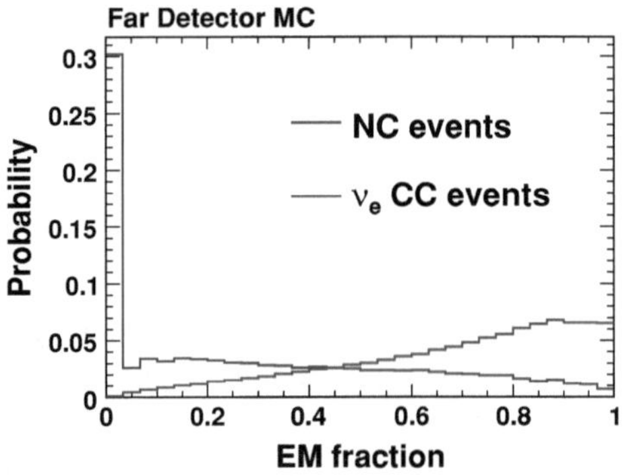

Fig. 4.7 Fraction of the total visible shower energy that is electromagnetic for simulated NC and ν_e CC events that leave small showers with energies between 1 and 8 GeV, scaled to unity. Only approximately 30% of NC events do not produce any π^0's

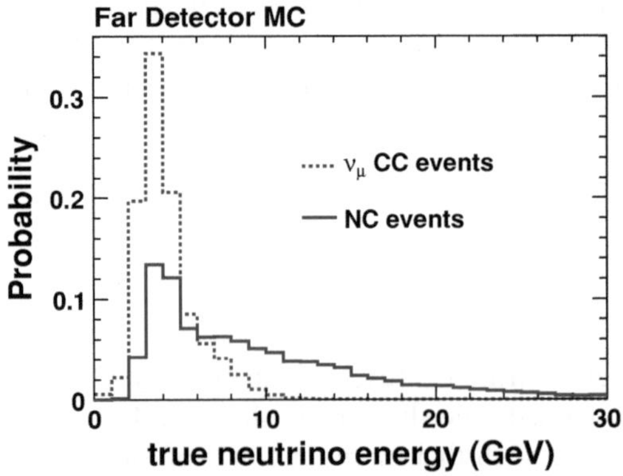

Fig. 4.8 True energy of the simulated ν_μ CC and NC events that comprise the background to the search for ν_e appearance in MINOS, normalized to unity. Events are required to be small showers with a reconstructed energy between 1 and 8 GeV. For CC events we can measure the incoming neutrino energy but not for NC events, and consequently approximately half of the NC events in our background sample originate from the high energy tail ($E_{true} \gtrsim 7$ GeV)

4.1.3.3 Summary of Backgrounds

Because of the reasons just explained, NC events are the most important source of background to the search for ν_e CC appearance in MINOS, but not the only one. High-y ν_μ CC events where the track is contained within the hadronic shower come next.

Thirdly, the beam contains an intrinsic ν_e component that accounts for 1.2% [13] of the beam and which originates primarily with the decay of secondary muons in the decay pipe. The muon parents of these ν_e's are produced together with the muon neutrinos (that make up most of the beam) in the decay of positively charged pions from the target:

$$\pi^+ \rightarrow \mu^+ + \nu_\mu$$
$$\hookrightarrow e^+ + \nu_e + \overline{\nu}_\mu.$$

Lastly, ν_τ CC events $\nu_\tau + \text{Fe} \rightarrow \tau^- + X$ originating from the oscillations of ν_μ neutrinos in the beam can also mimic ν_e CC events. This is particularly so when the resulting τ lepton decays into an electron or into a low multiplicity hadronic system with π^0 production. As shown in Table 4.3, this will be the case approximately half of the time. The other half of the time the τ^- will decay into a muon (thus making the event look almost identical to a ν_μ CC) or into a hadronic system such as the one shown in Fig. 4.9. The relevance of this background in the search for ν_e appearance is further reduced by the fact that only tau neutrinos with an energy of $\gtrsim$3.4 GeV can undergo a CC interaction in the detector. As a matter of fact, the ν_τ CC cross-section does not become important until $\sim$5 GeV or higher, at which point only $\sim$20% or less of the ν_μ's oscillate. Because of this ν_τ CC events are only a small ($<$5%) source of beam-related background to our search.

4.2 The Overall Strategy for the Analysis

4.2.1 The Keys to Maximizing the Reach

As discussed in the previous section, the main backgrounds to the ν_e appearance search in MINOS are NC events, ν_μ CC events with short tracks, intrinsic beam ν_e CC events and ν_τ CC events. When put together these backgrounds are significantly

Table 4.3 Decay modes of the τ^- lepton that are relevant to our discussion [9]

Decay mode	Branching ratio (%)
$e^-\overline{\nu}_e\nu_\tau$	17.8
$h^- \geq 1\pi^0\nu_\tau$	36.5

h^- stands for π^- or K^-. With the exception of the $\mu^-\overline{\nu}_\mu\nu_\tau$ decay mode, all other decay modes not mentioned here involve hadronic systems with, typically, higher multiplicity and/or negligible EM content

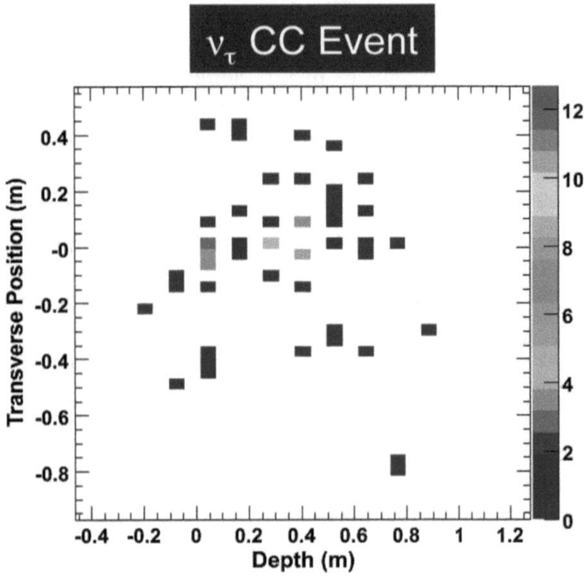

Fig. 4.9 Topology of a ν_τ CC interaction in MINOS where the resulting τ^- lepton decayed into a high multiplicity hadronic system. The color scale represents the charge deposited per strip in Minimum Ionizing Particle (MIP) units

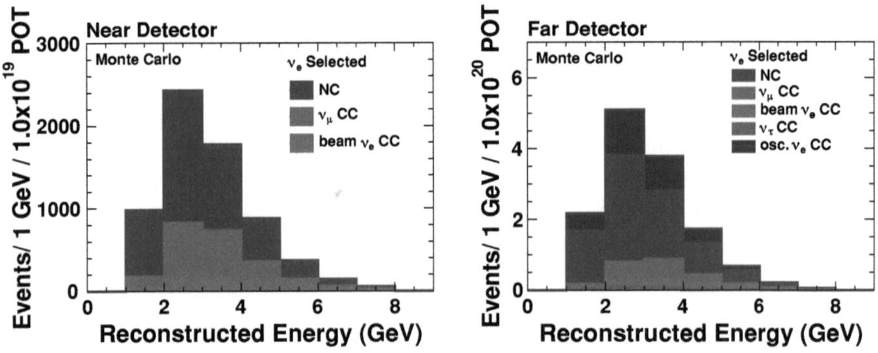

Fig. 4.10 Reconstructed energy spectra for selected events in the Near Detector (*left*) and in the Far Detector (*right*), for one particular ν_e selection algorithm. The spectra were obtained ignoring matter effects and CP violation effects, and assuming $\sin^2(2\theta_{13}) = 0.15$ (at the CHOOZ limit), with standard oscillation parameters $\Delta m^2_{32} = 2.38 \times 10^{-3}\,\mathrm{eV}^2$ and $\theta_{23} = \pi/4$. In this particular example it can be seen how the relative amount of ν_μ CC background diminishes almost by $\sim$40% when going from the Near Detector to the Far Detector, due to oscillations

larger than the signal expected at the CHOOZ limit, as shown on the right-hand side of Fig. 4.10.

The reach of the analysis thus depends on the ability to identify an excess over the predicted backgrounds. This means that our reach in θ_{13} rests on two pillars:

- Our ability to predict the backgrounds in the Far Detector as accurately as possible.
- Our ability to separate the signal from the backgrounds through our ν_e selection.

Of these two items, the second one is more critical when it comes to maximizing our sensitivity to θ_{13}. This is because even in the ideal case where the backgrounds are perfectly predicted in the Far Detector, there remains a statistical error associated with them. Background fluctuations can potentially obscure the signal, and the only way to reduce this effect is by suppressing the background relative to the signal as much as possible. Because of this, a substantial effort was devoted to develop an optimized new approach for selecting ν_e CC events that performs better than previously applied, more conventional approaches.

4.2.2 *Identifying* ν_e *CC Events*

All the information about a given MINOS event is contained in the time, readout charge and position of the strips hit during the event. Apart from determining if an event occurs concurrently with a beam spill, the timing information is not very useful as all beam events are forward-going. Therefore event identification has to be performed based solely on the topology of the energy deposition pattern left by the neutrino interaction. Examples of typical event topologies in MINOS were shown in Fig. 4.3.

Our situation in MINOS motivates an approach that is different from the traditional one of building reconstructed variables and feeding them to a multivariate discrimination technique. Only an average of 22 strips are hit for every ν_e CC candidate event in the energy region of interest, which is not a large number. It is precisely because the amount of information in each event is not very large that it is hard to separate the signal from the background. Therefore, in order to maximize the use of all the available information in each event, the selection technique must be able to use the strip information alone to perform event identification. We achieve this through the Library Event Matching (LEM) selection, which consists of three basic steps:

1. Compare the strip hit pattern of the input event to a large set of simulated ν_e CC and neutral current (NC) events.
2. Select the N best matches.
3. Compute the value of a discriminant from the information of the N best matches (Fig. 4.11).

In LEM the task of ν_e CC identification is effectively reduced to a problem of strip hit pattern recognition. The method works because the signal events are compact, which means that the different strip hit patterns from the signal events can be sampled by a finite amount of simulated data. Also, by avoiding the loss of information that occurs when going from raw to reconstructed data, the method is less reconstruction dependent and optimal in principle. As expected from these considerations, LEM achieves a higher sensitivity to θ_{13} than any of the other selection techniques attempted in MINOS.

LEM is described in detail in Chap. 5.

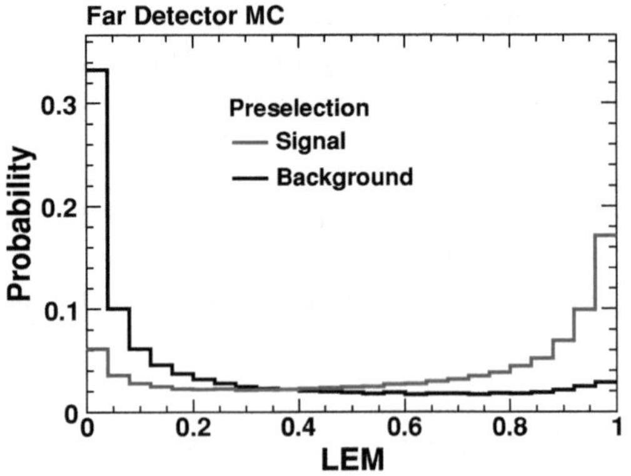

Fig. 4.11 LEM pid distribution in the Far Detector, normalized to unity

4.2.3 Predicting the Backgrounds

4.2.3.1 General Strategy

Even with the separation achieved by the LEM selection method, the backgrounds still dominate over the signal expected at the CHOOZ limit. In order to be able to distinguish a signal excess over those backgrounds, they need to be predicted in the Far Detector as accurately as possible.

As shown in the beginning of Chap. 6, the Monte Carlo (MC) does a relatively poor job in simulating the signal-like events in our background. In general, the simulation greatly overestimates the amount of signal-like background in our data. MINOS however is a two detector experiment, and we do not need to predict the Far Detector background using the MC: we can just measure it in the Near Detector.

There is however a very significant difference in event rates between the Far and Near detectors. This difference is due mostly to

- the difference in solid angles covered by the two detectors,
- neutrino oscillations altering the beam's composition,
- detector specific differences.

The first item, which is responsible for the largest difference in event rates between the two detectors (on the order of 10^4), is illustrated graphically in Fig. 4.12.

The three items mentioned above are well simulated in the MC. In addition, as shown in Chap. 6, the simulation's inability to accurately simulate hadronization processes with high EM content does not affect the difference in event rates between the Far and Near detectors. Therefore, the strategy of our analysis is to directly measure the background rates in the Near Detector and to only rely on the simulation to predict the corresponding Far Detector rates. This way the MC is

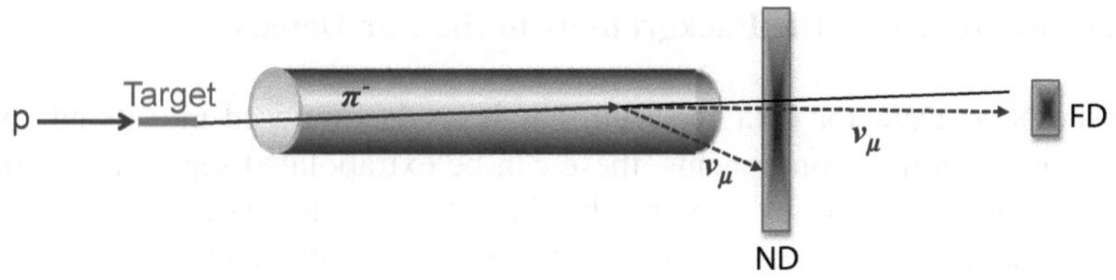

Fig. 4.12 Schematic of the NuMI beamline and the two detectors. Even if the Far Detector is approximately five times more massive than the Near Detector here it is displayed smaller in order to illustrate how, due to the widening of the beam as it travels, the two detectors cover different solid angle regions of it. Also, the Near Detector sees an extended line source (the decay pipe) while the Far Detector sees it as a point source for all practical purposes. Furthermore, the beam's composition changes as it travels due to oscillations, and the detectors have some intrinsic differences (mostly in their readout). All of these factors create a difference in event rates between the two detectors in the order of 10^4

only used to provide the necessary corrections in energy smearing and acceptance between the two detectors, which are well understood.

4.2.3.2 Near Detector Decomposition

The process of predicting a Far Detector rate from a Near Detector rate is referred to as the extrapolation. In order for our strategy to work, each type of background needs to be extrapolated separately. This is because, as illustrated in Fig. 4.12, the difference in event rate between the two detectors depends heavily on the type of event considered. For example, ν_μ CC events undergo oscillations, while NC and beam ν_e events do not. An illustration of this can be seen by comparing the relative background compositions of the left and right plots in Fig. 4.10. In that example it is particularly easy to see how the relative importance of the ν_μ CC background component is strongly diminished in the Far Detector with respect to the Near Detector, due to oscillations. Furthermore, events such as intrinsic beam ν_e's whose parents are created along the decay pipe extrapolate differently than events whose parents are created in the target.

In the Near Detector data however we can only measure the total sum of the backgrounds when applying our ν_e selection method. In order to decompose the measured Near Detector energy spectra into its NC, ν_μ CC and intrinsic beam ν_e components, we have developed two independent data-driven methods:

- the Muon Removed Charged Current (MRCC) method,
- the Horn-on/off (HOO) method.

Both of these techniques are explained in detail in Chap. 6. Also shown in that chapter is how both methods require the intrinsic beam ν_e component as an input. This component is obtained from the simulation tuned to the ν_μ CC data, but it is also cross-checked by counting the antineutrinos from μ^+, which are created alongside the beam ν_e's in the decay pipe.

4.2.3.3 Extrapolating the Backgrounds to the Far Detector

Once the Near Detector data spectrum has been decomposed into its NC, ν_μ CC and intrinsic beam ν_e components, these can be extrapolated separately to the Far Detector. This is done by having the MC provide the necessary corrections between the two detectors for each background component i at each bin of reconstructed energy E_{reco}:

$$\mathrm{FD}_i^{\mathrm{pred}}(\mathrm{E}_{\mathrm{reco}}) = \mathrm{ND}_i^{\mathrm{data}}(\mathrm{E}_{\mathrm{reco}}) \frac{\mathrm{FD}_i^{\mathrm{MC}}(\mathrm{E}_{\mathrm{reco}})}{\mathrm{ND}_i^{\mathrm{MC}}(\mathrm{E}_{\mathrm{reco}})}. \quad (4.5)$$

The necessary corrections provided by the MC are due to beamline geometry and oscillations, but also due to small differences in the detectors. The signal and background Far Detector predictions can be seen in Chap. 7. It should be noted that the following standard values of the oscillation parameters are used to derive the Far Detector predictions: $\Delta m^2_{32} = 2.43 \times 10^{-3}\,\mathrm{eV}^2, \theta_{23} = \pi/4, \Delta m^2_{21} = 8 \times 10^{-5}\,\mathrm{eV}^2, \theta_{12} = 0.59365$. When simulating a signal, a value of $\sin^2(2\theta_{13}) = 0.15$ (corresponding to the CHOOZ limit) is used, and CP violation and matter effects are ignored. These values are also used throughout this thesis, unless otherwise indicated.

Even though the two detectors are identical in terms of the environment where the neutrino interactions occur (strip size and steel width), there are differences in their readout, in neutrino intensity and in their energy calibration. A corollary of Eq. 4.5 is that Far–Near differences that are not properly accounted for in the simulation are a source of systematic error in our background prediction. All of these Far–Near differences, as well as other sources of systematic error, are addressed in Chap. 8. The strategy generally followed is to generate a special simulation sample where a particular Far–Near difference is changed within its bounds and the effect on the total background prediction recorded.

4.2.4 A Blind Analysis

It is important to note that the development and tuning of the selection algorithms, the Near Detector spectral decomposition, the extrapolation to the Far Detector and the systematic studies were all finalized without knowing whether or not a ν_e appearance-like signal existed in the MINOS data. The Far Detector data was analyzed in order to ensure its quality, but only by looking at low-level quantities.

Once all the Far Detector predictions were completed and before the Far Detector data was searched for ν_e appearance, three Far Detector sidebands known to contain no signal were used to ensure that the machinery of the analysis was working correctly. The background predictions for the sidebands were done using the same methods and with the same code as for the main analysis. It was not until the sidebands were looked at in the Far Detector data that a search for a ν_e signal was performed. The discussion of the results for the Far Detector sidebands and for the ν_e appearance search can be found in Chap. 9.

References

1. B. Pontecorvo, Inverse beta processes and nonconservation of lepton charge, JETP **34**, 172–173 (1958)
2. Z. Maki, M. Nakagawa, S. Sakata, Remarks on the unified model of elementary particles, Prog. Theor. Phys. **28**, 870–880 (1962)
3. M. Apollonio et al. (the CHOOZ Colla oration), Search for neutrino oscillations on a long base-line at the CHOOZ nuclear power station, Eur. Phys. J. **C**(27), 331–374 (2003)
4. I. Mocioiu, R. Shrock, Matter effects on neutrino oscillations in long baseline experiments, Phys. Rev. D **62**(5), 053017 (2000)
5. E.K. Akhmedov et al., Series expansions for three-flavor neutrino oscillation probabilities in matter, J. High Energy Phys. **2004**(04), 078 (2004)
6. L. Wolfenstein, Neutrino oscillations in matter, Phys. Rev. D **17**(9), 2369–2374 (1978)
7. S.P. Mikheyev, A.Y. Smirnov, Resonance enhancement of oscillations in matter and solar neutrino spectroscopy, Sov. J. Nucl. Phys. **42**, 913–917 (1985)
8. T. Ferbel, *Experimental Techniques in High-Energy Nuclear and Particle Physics* (World Scientifc, Singapore, 1991)
9. C. Amsler et al. (Particle Data Group), Review of particle physics, Phys. Lett. **B**(667), 1 (2008)
10. P. Adamson et al., The MINOS calibration detector, Nucl. Instrum. Methods Phys. Res. A **556**(1), 119–133 (2006)
11. E. Longo, I. Sestili, Monte Carlo calculation of photon-initiated electromagnetic showers in lead glass, Nucl. Instrum. Methods **128**(2), 283–307 (1975)
12. The MINOS Collaboration, A Long-Baseline Neutrino Oscillation Experiment at Fermilab, Fermilab Proposal P875 (1995)
13. D.G. Michael et al., The Magnetized steel and scintillator calorimeters of the MINOS experiment, Nucl. Instrum. Methods **A596**, 190–228 (2008)

Chapter 5
A Novel Approach for Selecting ν_e CC Events

Having established the strengths and weaknesses of the ν_e appearance search in MINOS, we proceed in this chapter to describe in detail the selection technique that is used in this thesis. The philosophy of our approach together with a description of the basic preselection cuts employed in the analysis are first laid down. The technical characteristics of the new selection method are then reviewed, followed by its optimization and performance.

5.1 The Philosophy of Our Approach

As reviewed in Chap. 4, the search for a nonzero θ_{13} is one of the highest priorities in the field of neutrino physics, and MINOS has a chance to make this measurement if the θ_{13} mixing angle lies in the vicinity of the current experimental limit set by CHOOZ [1]. The challenge resides in the granularity of the MINOS detectors, which is far from optimal when it comes to low-energy ($\sim 2\,\text{GeV}$) shower resolution. Consequently, as seen in Sect. 4.2.1, our reach in θ_{13} depends primarily on the capability to distinguish the signal from the background. The background consists primarily of hadronic showers originating from NC and ν_μ CC events.

All the useful information about a given MINOS event is contained in the readout charge and position of all the strips hit during the event. From this information, reconstructed variables can be built that quantify different aspects of the shape of the showers. For instance, a "Molière radius" variable can be constructed, which is the radius around the shower axis that contains 90% of the shower's visible energy. With the exception of the technique described in this chapter, all of the ν_e identification techniques that exist to date in MINOS combine different reconstructed variables. Such is the case of the Artificial Neural Network (ANN) selection, described in [2].

J. P. Ochoa Ricoux, *A Search for Muon Neutrino to Electron Neutrino Oscillations in the MINOS Experiment*, Springer Theses, DOI: 10.1007/978-1-4419-7949-0_5,

We follow a different and complementary approach. Only an average of 22 strips are hit for every ν_e CC candidate event in the energy region of interest. This number is small enough that it motivates the development of a technique that uses all the available basic information, namely the strip information alone, to perform event identification. The goal is to reduce the task of ν_e CC identification to a problem of strip hit pattern recognition.

This idea is implemented in the Library Event Matching (LEM) selection, where each event is compared to a very large number of simulated signal and background events and a discriminant constructed from the information of the best matches. The method works because the signal events are compact, which means that the different strip hit patterns from the signal events can be sampled by a finite amount of simulated data. At the end of this chapter we show that, by making better use of all the available information in each event, LEM gives the best sensitivity to θ_{13} among all the other selection methods ever attempted in MINOS.

5.2 A Preselection for ν_e CC Events

5.2.1 Selecting Events in the Fiducial Volume

In order to identify events as well as possible, we need to ensure their completeness and quality. For instance, the track of a ν_μ CC interaction occurring near the end of the detector may leave without being reconstructed, thus increasing the event's likeness to a ν_e CC event. Actually, any event occurring near the physical boundaries may be incompletely reconstructed and thus misidentified. Furthermore, ν_μ CC events originating in the surrounding rock are very abundant and can only be removed by requiring that the event vertex does not occur too close to the front of the detector. This is why it is necessary to remove events that occur too close to the edges, or in regions where the calibration cannot be trusted (which also happens to be close to the edges).

For our analysis, only events that have a vertex as determined by the standard MINOS event reconstruction with a longitudinal position z and a distance r away from the beam's center satisfying the following conditions are considered:

- Far Detector: $0.5\,\text{m} < z < 14.3\,\text{m}$ or $16.3\,\text{m} < z < 28\,\text{m}$, $0.5\,\text{m} < r < 3.7\,\text{m}$.
- Near Detector: $1\,\text{m} < z < 5\,\text{m}$, $r < 0.8\,\text{m}$.

These cuts are illustrated graphically in Fig. 5.1.

5.2.2 Removing the Obvious Background

In order to get the most out of any advanced separation technique, it is advantageous to first remove the events that are obvious backgrounds to the search. The

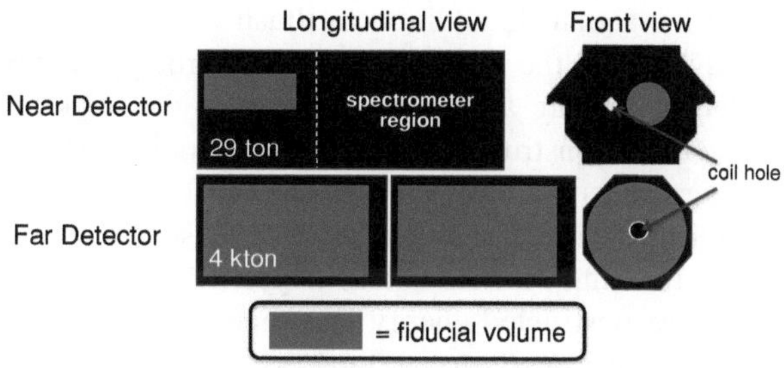

Fig. 5.1 A schematic view of the fiducial volume considered in both detectors. The Near Detector's fiducial volume amounts to 29 ton, and the Far Detector's to 4 kton

advantages are twofold. First, processing time is saved by not having to calculate a PID for the events that fail the preselection. Second, it allows the more sophisticated selection algorithm (LEM in this case) to concentrate its power on those background events that look closer to signal. This way the training need only involve the preselected events.

Most events in the MINOS detectors are ν_μ CC interactions, of which the great majority have clear tracks in them. We cannot, however, remove all events with tracks in them as the tracker algorithm frequently finds tracks within showers, including those of ν_e CC interactions. Also, in some cases a charged particle (such as a $\pi^\pm$) can be created in the hadronic shower of the ν_e CC interaction, giving a clear track. Nevertheless, the tracks found in ν_e CC events are shorter than for other events and, as can be seen in Fig. 5.2, rarely cross more than 25 planes. If we consider the number of track-like planes (i.e., the number of planes crossed by a track that have no shower hits in them) we find that they very infrequently exceed 16 in the case of ν_e CC events. Therefore, by making these two cuts, a very significant portion of the background can be removed while retaining essentially

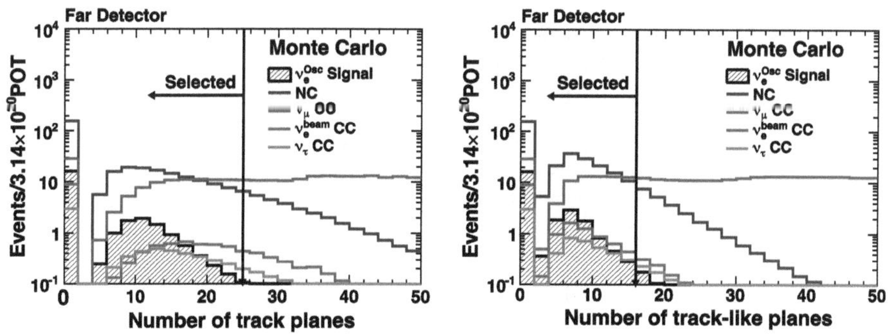

Fig. 5.2 Distributions of the number of track planes and track-like planes for different event types in the fiducial volume of the Far Detector. A track-like plane is a plane with no hits other than the ones associated with the track. By cutting at track planes <25 and track-like planes <16 a very significant portion of the NC and ν_μ CC background is removed at the expense of practically no signal. The signal corresponds to a value of θ_{13} at the CHOOZ limit, neglecting matter effects and CP violation

all of the signal. In addition to illustrating these two cuts, Fig. 5.2 shows how ν_μ CC events are the ones that are most strongly removed by these cuts.

Other cuts are applied in addition to the track-related cuts just mentioned. At least one reconstructed shower is required to be in each event in order to avoid having events with just a short track. Also, to remove badly reconstructed events, we reject events that do not have at least five contiguous planes with a 0.5 Minimum Ionizing Particle (MIP) deposition in them. Furthermore, we reject events with a reconstructed energy less than 1 GeV, where NC events vastly dominate. Events with a reconstructed energy higher than 8 GeV are also rejected. This is because practically no signal events reach that energy, but also because it is the only way to remove the higher energy intrinsic beam ν_e events which, at those energies, originate mostly from kaon decays. All of these cuts are illustrated in Fig. 5.3.

A number of cuts are applied to the data in order to ensure its integrity. Spills that fail the data quality cuts are removed from the sample and do not count toward the total POT exposure. In both detectors, the high voltage must be operating

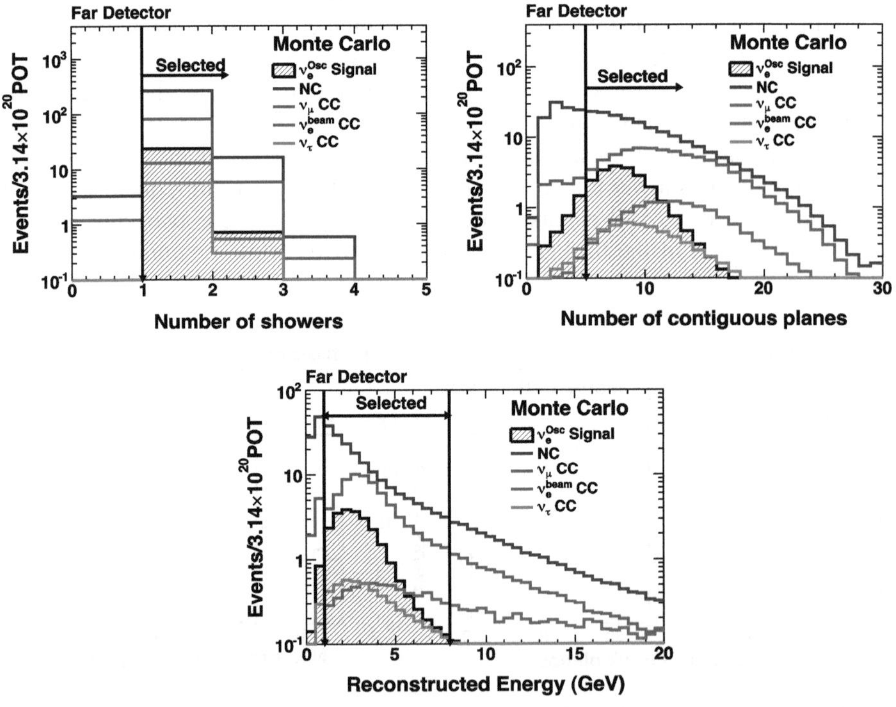

Fig. 5.3 Number of reconstructed showers per event, contiguous planes with at least 0.5 MIP deposition in each, and reconstructed energy spectra for different event types in the fiducial volume of the Far Detector. These plots were done after the track length cuts of Fig. 5.2 have been applied. The signal corresponds to a value of θ_{13} at the CHOOZ limit, neglecting matter effects and CP violation

normally, the electronics must have a small number of dead channels, and the magnetic field coils have to be energized with their nominal currents. Also, periods where the spill data could be contaminated with light-injection data are removed. A series of cuts are also applied in order to ensure the quality of the beam. These cuts ensure that parameters such as the beam configuration, the horn-current, the beam size, the position of the beam center and the proton intensity are within their normal bounds.

In the Far Detector we apply additional cuts to remove cosmic ray events that occur concurrently with a beam spill. If an event has a track with an angle of 0.6 rad or less with respect to the vertical, and extends more than 2 m in the vertical direction, it is rejected as a cosmic. Alternatively, if the event consists only of a shower at a very steep angle then it is also removed. These cosmic rejection cuts, combined with the data quality cuts, remove only 3% of the spills, and allow us to keep the expected cosmic background in the Far Detector as low as $\sim$0.2 events.

5.2.3 Performance

In the end, only 21% of the signal is removed by the preselection cuts, the majority of which is due to the contiguous planes cut. Many of the signal events lost with that cut would not have been recovered anyway, due to their sparse nature. By applying the preselection cuts however the signal-to-background ratio is improved dramatically, going from 1:55 to 1:12. As seen in Fig. 5.4, the background composition before and after preselection also changes very drastically, as the preselection is particularly effective in removing ν_μ CC events.

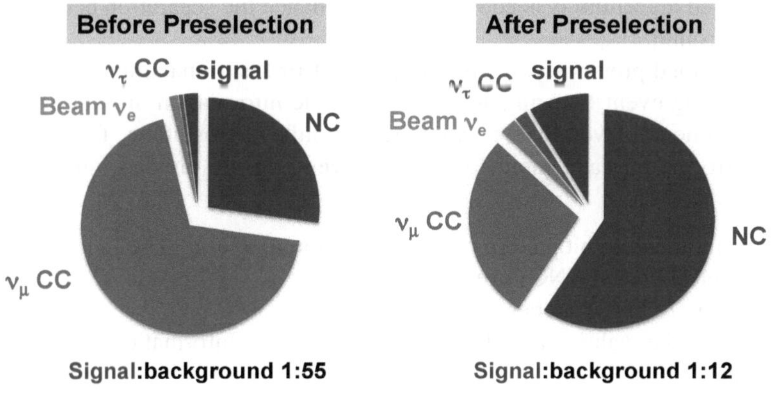

Fig. 5.4 Event type composition in the Far Detector before and after the preselection cuts have been applied. The signal corresponds to a value of θ_{13} at the CHOOZ limit, neglecting matter effects and CP violation

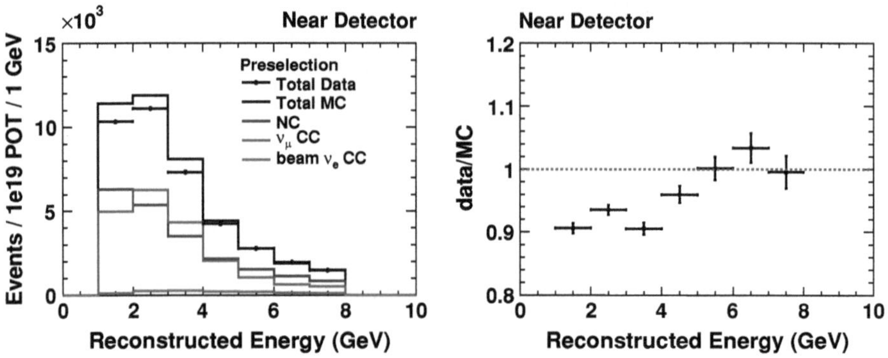

Fig. 5.5 Energy spectrum of preselected events in the Near Detector (*left*) and corresponding data to MC ratio (*right*). The data and the MC agree to within 10% at the preselection level

Figure 5.5 shows the energy spectrum of the preselected events in the Near Detector, for both data and MC. The agreement between data and MC is relatively good at the preselection level. A detailed breakdown of preselected events can be found at the end of this chapter in Table 5.2.

5.3 The Workings of the LEM Algorithm

5.3.1 Basic Principle

Once the obvious background has been removed by the preselection, the remaining events mostly consist of well-reconstructed small showers. At this point a more sophisticated algorithm is needed to further enhance the separation between signal and background.

As explained previously, the philosophy of LEM is to maximize our sensitivity by performing event ID using all of the available information in each event at the most fundamental level, that is, using the strip information alone. The process that LEM performs on each input (candidate) event can be divided into three basic steps:

1. Compare the strip hit pattern of the input event to a large set of simulated ν_e CC and neutral current (NC) events.
2. Select the N best matches.
3. Compute the value of a discriminant from the information of the N best matches.

This process is shown schematically in Fig. 5.6. The set of events to which each input event is compared is called "the library" and it consists of 30 million events, as explained below. The discriminant that is constructed from the N best matches

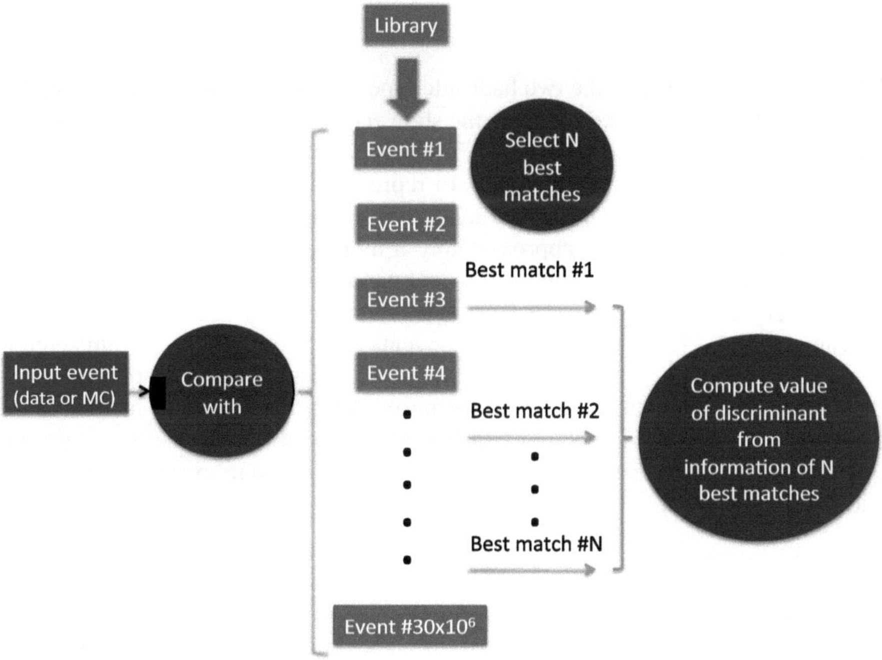

Fig. 5.6 Schematic view of the LEM algorithm. Each input event (represented by the box at the left) is compared to a library consisting of 30M signal and background events (represented by the vertically stacked boxes). At the end of the comparison the N best matches are recorded, and a discriminant is constructed from their information. For instance, the simplest discriminant that can be built is *the fraction of the N best matches that are* ν_e *CC events*

information can be as simple as *the fraction of the N best matches that are* ν_e *CC events*, although a more sophisticated one can be built, as shown in Sect. 5.4.3.

The LEM approach possesses several advantages. The first one is that the method is, in principle, optimal. In contrast with the situation where reconstructed variables are obtained from the raw event, LEM uses all the information kept in each event[1] when doing the matching.

The second advantage of the LEM is that the method is largely reconstruction free, as only the strip information provided by the event builder is used. This makes the selection less senstive to changes in the reconstruction code.

The method is however technically and computationally challenging. It involves performing a very large number of comparisons which leads to massive data flows during the analysis. The technical issues that must be addressed in order to ensure the success of the method are dealt within Sects. 5.3.2–5.3.4. The technical performance is analyzed in Sect. B.2.

[1] Systematic error constraints made it necessary for strips with a total charge less than 3.0 photoelectrons (PEs) to be ignored by the algorithm, as explained in Sect. 5.3.3.

5.3.2 The Library

It has been shown [3] that the two hadronic types of background (NC events and ν_μ CC events with short tracks buried in the shower) look very similar to each other in the MINOS detectors. Therefore the LEM library need only contain ν_e CC events to represent the signal and NC events to represent the backgrounds. 10M ν_e CC and 20M NC events were produced and reconstructed in the Far Detector at the Caltech farm in a period of approximately 6 months.[2] The latest versions (at the time of the analysis) of the Monte Carlo and the reconstruction were used to create these library events.

The same 3.0 PE cut used in the input events (see Sect. 5.3.3) is administered to the library events. A preselection is applied to the library events in order to ensure the quality of the events, and to remove those events that are not of interest to the analysis. The library preselection consists of a fiducial volume cut to ensure the quality of the events, alongside some physics cuts to ensure events have the size and energy we are interested in:

$$5 < N_{\text{planes}} < 30, 7 < N_{\text{strips}} < 80, N^V_{\text{strips}} > 2, N^U_{strips} > 2 \text{ and } 50 < Q_{tot} < 3,000 \text{ PE}.$$

In order to optimize the use of computing resources, input events in both the Near and Far Detectors are compared to the same Far Detector library events. This is achieved by requiring all events to go through a compacting stage where the detector-specific differences are removed. One of the first steps in this compacting process is to correct for light attenuation effects, as described in the next section.

5.3.3 Compacting Events

5.3.3.1 Light Attenuation

As will be seen in Sect. 5.3.4, it is necessary to account for photon statistics when comparing events. The most natural way to accomplish this is to consider the energy depositions measured in photoelectrons and to use Poisson probabilities to quantify their width. As described in Sect. 3.5.1 however, the photoelectron calibration branch in MINOS consists simply of applying the gains to the raw energy depositions in ADCs. Consequently, the energy depositions in photoelectrons lack the corrections due to light attenuation.[3]

---

[2] The relative proportion of signal-to-background events in the library is justified in Sect. 5.4.2.

[3] In the "energy branch" described in Sect. 3.5.1 and whose end result is the reconstructed energy used in this analysis, the attenuation corrections are accounted for by the "strip-to-strip" and the "transverse position" corrections.

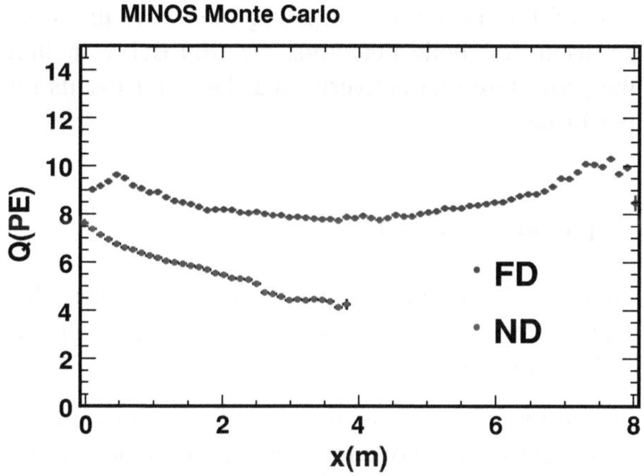

Fig. 5.7 The average total charge per strip Q deposited by muons as a function of the distance from the west end x, as read from the PMTs. For the Far Detector, Q is the sum of the charge from both strip ends. For the Near Detector, light is read out from the west end only. The longest Far Detector strips are 8 m long, whereas the longest Near Detector strips are 4 m long. In the Far Detector the charge is approximately 15% lower if the interaction occurs in the center of the strip than if it occurs near one of the edges. Due to the single ended readout, the Near Detector charge is highest near the west end and diminishes by $\sim$50% if the interaction occurs near the east end. Overall, the charge (or, as we shall refer to in this document, the "light level") is higher in the Far Detector than in the Near Detector

Light attenuation in the fiber causes for the charge distribution of an event as measured in the PMTs to be strongly dependent on the event's location in the detector. This is clearly illustrated in Fig. 5.7, which shows the charge of muons as a function of position along the strip. In the Far Detector light is read out from the two strip ends, and the total charge can vary by approximately $\pm$8% with respect to the average. In the Near Detector light is reflected in the east end and only read out from the west. This causes larger variations in the observed charge, in the order of $\pm$25% with respect to the strip's center.

Therefore two raw events cannot be compared with each other on an equal footing without first correcting for light attenuation. These corrections are applied by the LEM algorithm during the compacting stage, described in the following section. The specifics are somewhat intricate, and are thus part of the "Some technical aspects of LEM" appendix. Section B.1 in particular deals with all the attenuation corrections that are involved.

An additional difficulty arises from the fact that the photoelectron light level in the Near Detector is lower than in the Far Detector, as evidenced by Fig. 5.7. In order for some of the systematics to cancel (for instance, the systematic associated with the mismodeling of hadronic showers in the simulation), it is advantageous for LEM to select the same events in the two detectors. As discussed in Sect. B.1.3, there are several possible approaches to this challenge. The preferred choice is to correct the Near Detector charges for attenuation while also scaling the light level to that of Far Detector events. In this way, the same Far Detector libraries can be used for both detectors. Section B.1.3 includes a complete description of the procedure that converts Near Detector events into Far Detector events and its performance.

5.3.3.2 The Compacting Procedure

As previously mentioned, the compacting stage incorporates all the attenuation corrections and the light level scaling of Near Detector events. The compacting consists of the following steps:

1. Strips with a total raw charge less than 3.0 PE are thrown away. The primary motivation of this cut is to remove most of the crosstalk and the badly modeled low pulse-height hits (a detailed explanation can be found at Sect. 8.1.3). Consequently, the cut has to be applied before any attenuation corrections are done.
2. The charge weighted mean position (centroid) of the event is calculated in the U and V views. This is done iteratively, by calculating a centroid with all strips first and then recomputing the centroid after rejecting outliers.
3. The total charge for each strip is corrected for attenuation using the location of the charge weighted centroid as a reference (according to Eqs. (B.6) and (B.9)).
4. The event is moved to a standard arbitrary location in the Far Detector (plane = 100, strip = 100).
5. Strips in the same plane are grouped according to their distance from the centroid, as described in Fig. 5.8.
6. All the strips and their modified charges are stored in an array, alongside some very basic information such as energy and event type (in the case of the MC). The rest of the event's information is thrown away.

The primary purpose of the compacting procedure is to get events to the point where they are ready to be compared with others. By disposing of the information of each event that is not useful to the algorithm, significantly diminishing the amount of memory each event occupies, the compacting stage also serves the very important purpose of improving the algorithm's technical performance (time and memory wise). A discussion on the technical constraints involved in the implementation of the LEM algorithm can be found in Section B.2.

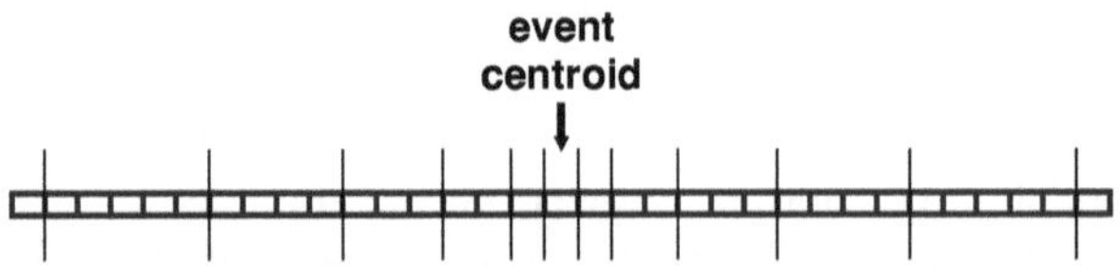

Fig. 5.8 A schematic view of the strips surrounding the event's charge weighted centroid after a transverse cut on a plane. Strips in between the thin lines are grouped when compacting the event

The goal of the strip grouping (step 5) is to ensure that the matching concentrates on the core of the event, which is where the most significant information is for ν_e CC events. At the current library size, the performance of the LEM selection in terms of sensitivity to θ_{13} degrades by approximately 7% without the strip grouping [4]. This suggests that as the algorithm strives harder to get a good match for the outlier hits it sacrifices some of the performance in the core of the events. It should be noted that alternative grouping patterns have not been explored due to time constraints. Future improvements in sensitivity might thus be achieved by optimizing the grouping without having to increase the library size.

Figure 5.9 shows an example of a ν_e CC event being compacted. The application of the attenuation corrections and the event's displacement to a standard location allow for events to be compared on an equal basis. In addition, compact events are approximately 60 times smaller (in terms of disk space) than their raw counterparts. Thus the compacting also serves the very important purpose of allowing for a much faster execution time of the algorithm by greatly speeding up the library I/O.

5.3.4 Event Comparison

5.3.4.1 The Comparison Metric

Light detection fluctuations are accurately described by Poisson statistics. If N_e photoelectrons are expected on average for a given energy deposition in a strip, the corresponding standard deviation is $\sqrt{N_e}$. These fluctuations must be taken into account when comparing two events, as even if they originate from exactly the same energy deposition at the strip level they still look different at the PMTs. One must therefore ask how likely it is for two events to have originated from the same hit pattern at the energy deposition level, in order to obtain a quantitative measure of their resemblance.

Let us consider a particular strip i, hit during two independent events A and B. Let n_A^i and n_B^i be the numbers of photoelectrons read out in the PMT corresponding to strip i for events A and B respectively. The likelihood that n_A^i and n_B^i photoelectrons were generated from the same expected amount of energy deposited in the strip is given by

$$S^i(n_A^i, n_B^i) = \int_0^\infty P(n_A^i; \lambda)P(n_B^i; \lambda)d\lambda \qquad (5.1)$$

where $P(n; \lambda)$ stands for the Poisson probability of obtaining n photoelectrons with a mean of λ.

In general, S^i becomes smaller as the light level increases (even if $n_A = n_B$). Because of this, simply summing the S^i terms of all the hits involved in the

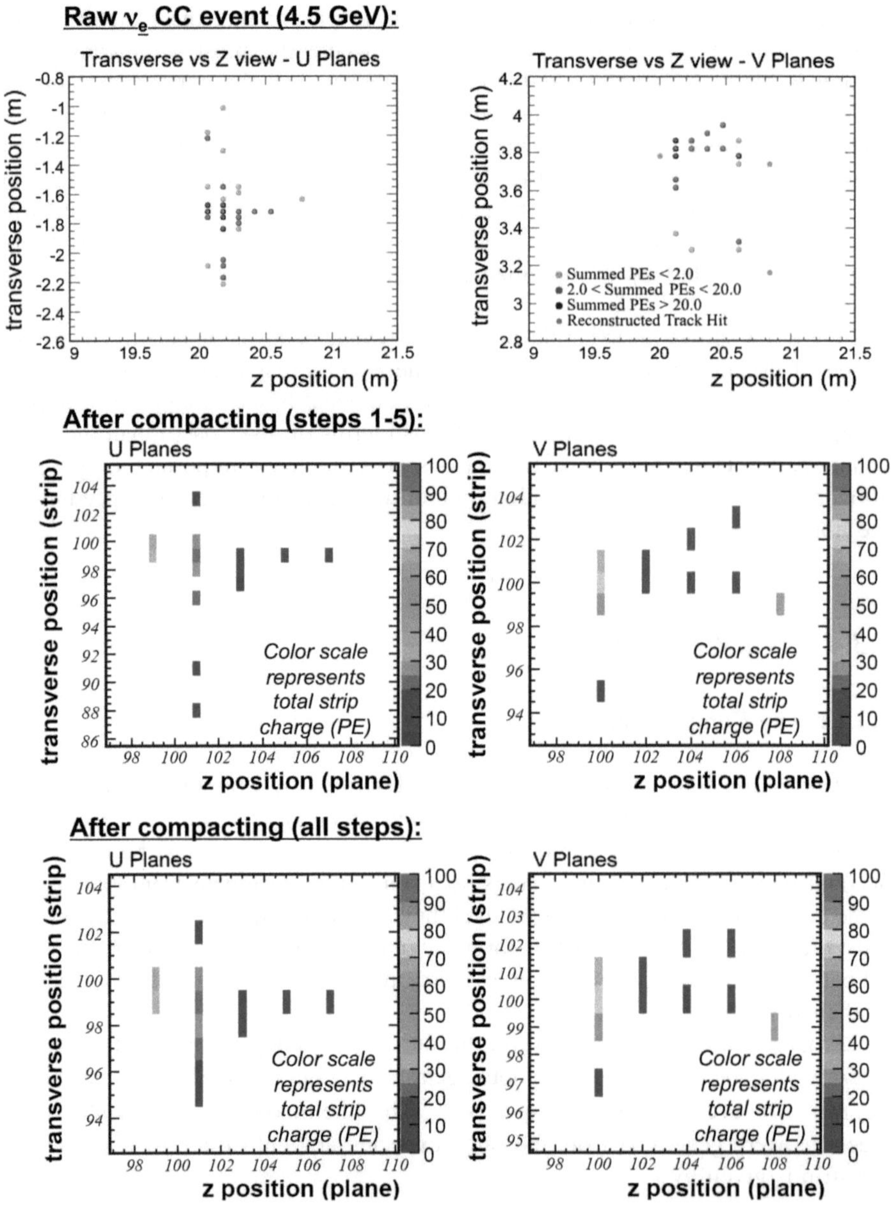

Fig. 5.9 A 4.5 GeV ν_e CC event in the Far Detector being compacted by the LEM algorithm

comparison between an input event and a library event would give more importance to the successful matching of lower light level hits. An equal treatment of all hits in the comparison can be achieved if the S^i terms are instead multiplied together or, as is done in our case, if the $\ln(S^i)$ terms are summed. Therefore the likelihood that two events A and B come from the same hit pattern of energy deposition is given by

$$\mathcal{L} = \sum_{i=1}^{N^{\text{strips}}} \ln\left(S^i\right) = \sum_{i=1}^{N^{\text{strips}}} \ln\left(\int_0^{\infty} P(n_A^i;\lambda)P(n_B^i;\lambda)d\lambda\right), \tag{5.2}$$

where the above sum is carried out over the set of all N^{strips} strips that result from the union of events A and B (i.e., any strip with n_A or n_B not equal to zero is included in the sum).

The log likelihood in Eq. 5.2 presents the main advantage that it is able to correctly account for Poisson fluctuations when comparing events. The main disadvantage resides in that it does not account for the fact that small spatial fluctuations are unimportant [5]. But this is compensated by the grouping of strips occurring during the compacting stage which averages the information of several strips into one and the fact that a very big library of events is used (see Sect. 5.3.2).

5.3.4.2 Implementation in the Algorithm

In the LEM algorithm the different values of $\ln(S(n_A, n_B))$ are calculated as a function of n_A and n_B and stored in a $\sim$30 megabyte (MB) table. The integral is evaluated numerically with a step size of 0.1 PE and a range from 0 to 500 PE. For each input event, the difference between the $\mathcal{L}$ calculated with respect to a library event and the one calculated with respect to itself,

$$-\Delta(\mathcal{L}) = -(\mathcal{L}^{\text{lib}} - \mathcal{L}^{\text{self}}), \tag{5.3}$$

is used to determine the quality of the matching. The left side of Fig. 5.10 shows the $-\Delta(\mathcal{L})$ distribution for the best matches of ν_e CC events. Also, Fig. 5.11 illustrates the results attained by the matching procedure with the example of an unusual looking ν_e CC event.

It is important for the algorithm to be robust against the imperfections associated with the process of moving all events to the same standard location in the detector. For instance, due to spatial or charge fluctuations, two physically identical events may end up with slightly different charge weighted centroids and thus shifted from each other. Because of this, input events are compared with each library event a total of three times. One comparison is done having the input event exactly as it comes out of the compacting stage. Two more comparisons are done with the entire compact input event shifted in the longitudinal direction by ± 1 plane.[4] The comparison giving the smallest $-\Delta(\mathcal{L})$ is the one used by the

[4] Due to the strip grouping mentioned in Sect. 5.3.3 it is very hard to do something analogous in the transverse direction.

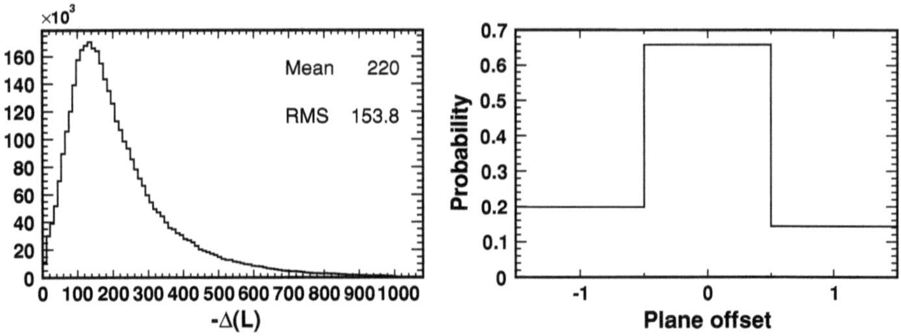

Fig. 5.10 (*left*) Distribution of the $-\Delta(\mathcal{L})$ between ν_e CC input events and their corresponding best matches. (*right*) Probability that a ν_e CC input event finds its best match at a given plane offset, which is measured with respect to the position of the original compact input event

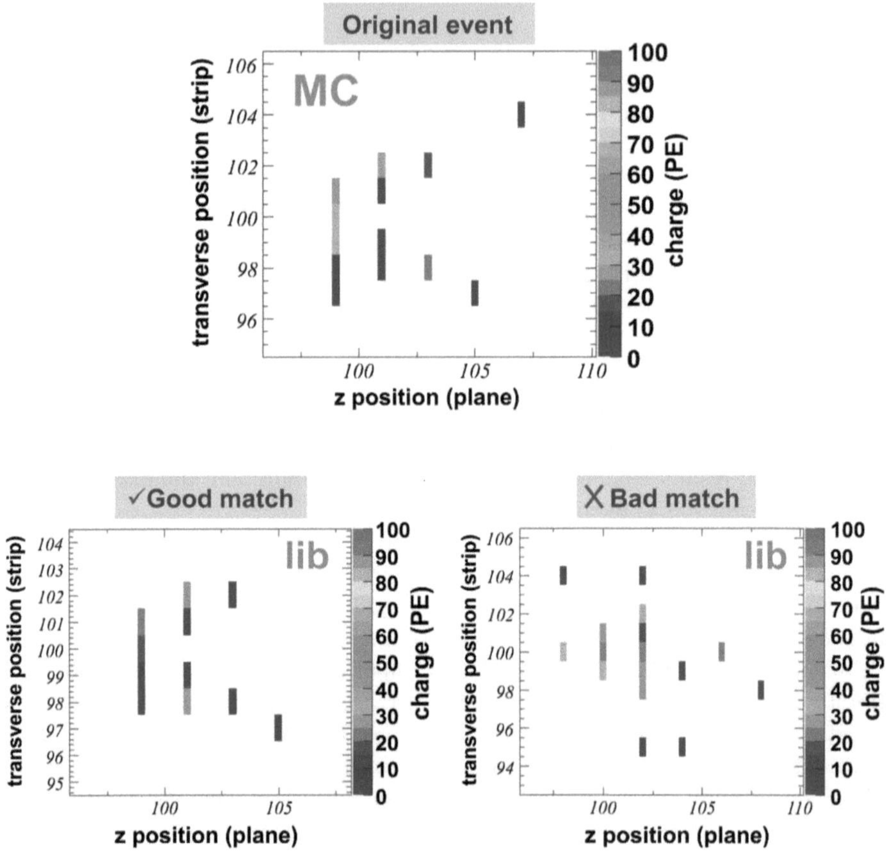

Fig. 5.11 A good and a bad match to a simulated ν_e CC with an unusual topology, according to our matching procedure. The good match is the best match found in the 30M library, and is also a ν_e CC event. The bad match is just one of many events in the library giving a very high $-\Delta(\mathcal{L})$ when compared to this input event

algorithm to rank the library event among the list of best matches. As can be seen on the right plot of Fig. 5.10, approximately two-thirds of the time a typical ν_e CC input event finds its best match without being shifted longitudinally. The other third of the time the ν_e CC input event finds its best match after being shifted by ± 1 plane.

5.3.4.3 Selecting the Events Worth Matching

In order to save processing time no comparisons with the library events are attempted for input events that are not relevant to the MINOS ν_e CC analysis. Input events are required to meet a relaxed version of the preselection described in Sect. 5.2:

- $0.5\ \mathrm{GeV} < \mathrm{E_{reco}} < 12\ \mathrm{GeV}$,
- $N_{\mathrm{planes}}^{\mathrm{track}} < 30$,
- $N_{\mathrm{planes}}^{\mathrm{track-like}} < 20$,
- (Near Detector only) the event vertex must be in the extended fiducial volume, defined as the volume satisfying $r < 1.0\,\mathrm{m}$ and $0.5\,\mathrm{m} < vtx_z < 6.5\,\mathrm{m}$, where $r = \sqrt{vtx_x^2 + vtx_y^2}$ and vtx_i represents the i coordinate of the event vertex,

where $N_{\mathrm{planes}}^{\mathrm{track}}$ is the number of planes crossed by the main reconstructed track in the event, and $N_{\mathrm{planes}}^{\mathrm{track-like}}$ is the number of planes in the event that are free from the presence of non-track associated hits. In addition, more time is saved by avoiding a comparison with those library events that are not similar enough to the input event. Two events A and B are compared only if they satisfy the following criteria:

- $|N_{\mathrm{planes}}^A - N_{\mathrm{planes}}^B| \leq x$, where $x = 4$ if $N_{\mathrm{planes}}^{avg} = \frac{N_{\mathrm{planes}}^A + N_{\mathrm{planes}}^B + 1}{2} > 20, x = 3$ if $N_{\mathrm{planes}}^{avg} > 15$ and $x = 2$ otherwise,
- $2|\frac{N_{\mathrm{strips}}^A - N_{\mathrm{strips}}^B}{N_{\mathrm{strips}}^A + N_{\mathrm{strips}}^B}| \leq 0.2$,
- $2|\frac{Q_{total}^A - Q_{total}^B}{Q_{total}^A + Q_{total}^B}| \leq 0.2$,

where N_{planes} and N_{strips} denote the number of planes and the number of strips for a given event, and Q_{total} stands for the total charge of the event as calculated by the LEM algorithm after the attenuation corrections have been applied.

Due to these conditions, events with long muon tracks or high energy will, in general, never get any matches assigned by the LEM algorithm. But of those signal ν_e CC events that pass the official ν_e preselection and constitute the goal of the analysis, 99.8% are successfully compared to the library by the LEM algorithm, as shown in Table 5.1.

Table 5.1 Percentage of Far Detector events that are compared to the library in LEM

	ν_e CC	NC	ν_μ CC	Beam ν_e	ν_τ CC
In fiducial volume	95.3%	69.5%	9.2%	49.0%	65.2%
Passing ν_e preselection	99.8%	97.3%	95.1%	99.1%	96.0%

5.4 Making the Most of the LEM Selection

5.4.1 The Information from the Best Matches

After all the comparisons with the library events have been done, a discriminant can be constructed from the information of the N best matches (i.e., taking the best, second-best, ..., up to the Nth-best match). The amount of information available is actually very rich as practically everything can be known for each best match.

A simple discriminant could be constructed that relied uniquely on the event type of the best matches. One such discriminant that was tried was *the fraction of the best Nmatches that are ν_e CC*. But it is possible to achieve better results by using more information. For example, the recoil system in a ν_e CC interaction may carry a large fraction of the total energy available, leaving behind a signature—a hadronic shower—nearly indistinguishable from a NC event. Thus, it is advantageous to consider y, the fraction of available energy carried by the recoil system, when constructing a discriminant.

It is also beneficial to use the fractional charge matched Q_{frac}. This quantity is calculated for every comparison and is defined by

$$Q_{frac} = \frac{Q_{\text{matched}}}{Q_{\text{matched}} + Q_{\text{unmatched}}}, \qquad (5.4)$$

where Q_{matched} is the sum of the charges of all strips that are shared between the two events, and $Q_{\text{unmatched}}$ is the sum of the charges of all strips that are not shared between the two events.

Q_{frac} is thus a measure of how well an input event matches a given library event. Its separation power originates from the fact that hadronic showers are more diffuse than EM showers. It is thus more common for ν_e CC events to have best matches where most or all of the strips are matched. The variable could perhaps be improved in the future by somehow taking into account how close the charges of the matched strips are. But as implemented here, LEM already has a good performance and is particularly effective at getting rid of ν_μ CC events (see Fig. 5.15).

For the reasons given above all the discriminants that the LEM uses are based on these three properties of the matched events:

- The event type (ν_e CC or NC),
- The hadronic y(of the ν_e CC best matches only),
- The fractional charge matched Q_{frac}.

Plots of these variables are shown in subsequent sections. There are many other properties that could be used and future versions of the LEM might include them, but these three already give very good results, as shown in Sect. 5.4.3.

5.4.2 Optimizing the Selection Basics

5.4.2.1 A Simple LEM Selection

There remain two free parameters in the LEM selection that can be optimized:

- The number of best matches N used to calculate the discriminant.
- The ν_e CC to NC event proportion in the library.

To study the performance of LEM as a function of these parameters it is useful to have a simple selection that can be quickly modified and optimized. The properties described in the previous section can be combined into two variables:

- $fracCC(y<0.5)$: the fraction of the best N matches that are ν_e CC with $y<0.5$.
- $Q_{frac}^{mean}(y<0.5)$: the mean fractional charge matched Q_{frac}^{mean} of those ν_e CC best matches with $y<0.5$ (among the best N best matches).

The y cut at 0.5 yields optimal results. Note that it is possible to divide the N best matches into smaller sets for which these variables are calculated. For instance, it is possible to calculate $fracCC(y<0.5)$ and/or $Q_{frac}^{mean}(y<0.5)$ for the best $N/2$ matches and then for the next $N/2$. But doing so gives no appreciable increase in performance.

Cuts are applied to these variables in order to select the ν_e CC candidates. A quick way of evaluating the performance of this simple selection neglecting systematics is to use a figure of merit (fom) which quantifies how significant the number of selected signal events $N_{\rm sig}$ is compared to the statistical fluctuations of the background $N_{\rm bg}$:

$$\text{fom} = \frac{N_{\rm sig}}{\sqrt{N_{\rm bg}}}. \tag{5.5}$$

5.4.2.2 Number of Best Matches

The performance of this prototype LEM selection seems largely insensitive to the choice of the number of best matches N. As shown in Fig. 5.12, varying N from 25 to 200 results in changes in fom of less than 1.5%.

An optimum is, however, expected as the quality of the matching decreases with increasing N, due to the finiteness of the library. Such an optimum seems to occur at approximately $N = 50$ which is the value that is used consistently unless mentioned otherwise. The optimum is expected to be a function of the library size.

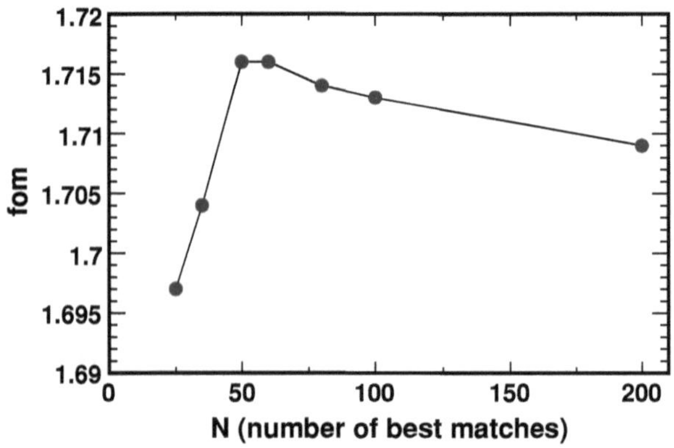

Fig. 5.12 Figure of merit (fom) as a function of the number of best matches N used to calculate $fracCC(y<0.5)$ and Q^{mean}_{frac}. The cut values for the two variables are optimized for each value of N based on the fom. A nonstandard preselection is used and standard oscillation parameters are assumed

5.4.2.3 Library Composition

One can also use this simple selection to test the performance of LEM in terms of the library size and composition. The numbers of ν_e CC and NC events in the library are $N^{std}_{\nu_e} = 10\,\text{M}$ and $N^{std}_{NC} = 20\,\text{M}$ respectively, and the best 200 matches are stored for each input event. Since not all the best matches are used in the selection, it is possible to simulate smaller library samples $N^{sim}_{\nu_e(NC)}$ by randomly ignoring best matches according to the desired proportion $N^{sim}_{\nu_e(NC)}/N^{std}_{\nu_e(NC)}$. For instance, if we want to simulate the effect of having only 5M ν_e CC events in the library instead of the standard 10M, it suffices to randomly ignore half of the ν_e CC best matches for each input event when calculating the two variables used in the selection.

The performance of LEM as a function of the ν_e fraction and the total library size is shown in the top plot of Fig. 5.13. The ν_e fraction is defined as $N^{sim}_{\nu_e}/(N^{sim}_{\nu_e} + N^{sim}_{NC})$. The selection is performed using $N = 20$ best matches, which implies that only values satisfying $N^{sim}_{\nu_e} > 1\text{M}$ and $N^{sim}_{NC} > 2\text{M}$ can be simulated in this fashion. These constraints together with the obvious requirement that $N^{sim}_{\nu_e} < N^{std}_{\nu_e}$ and $N^{sim}_{NC} < N^{std}_{NC}$ are the reason why there are many bins in the plot for which a fom is not calculated.

The bottom left plot of Fig. 5.13 shows the fom of the selection as a function of the total library size for a constant ν_e fraction $= 0.3$, while the bottom right plot shows the fom as a function of the ν_e fraction for a constant library size of 10M. As expected, the performance of the selection improves with the total library size.[5] It seems also that the optimal value for the ν_e fraction is ~ 0.3. This means that the value of ν_e fraction $= 1/3$ that is being used yields results that are very close to optimal. The original choice of $N^{std}_{\nu_e} = 10\text{M}$ and $N^{std}_{NC} = 20\text{M}$ is very reasonable.

[5] While the fom of Fig. 5.13 improves by only $\sim 2.5\%$ when the total library is increased from $\sim 4\text{M}$ to $\sim 26\text{M}$, the super-fom of Eq. (5.12) improves by more than 6%. Consequently, it is worth having all 30M events in the library.

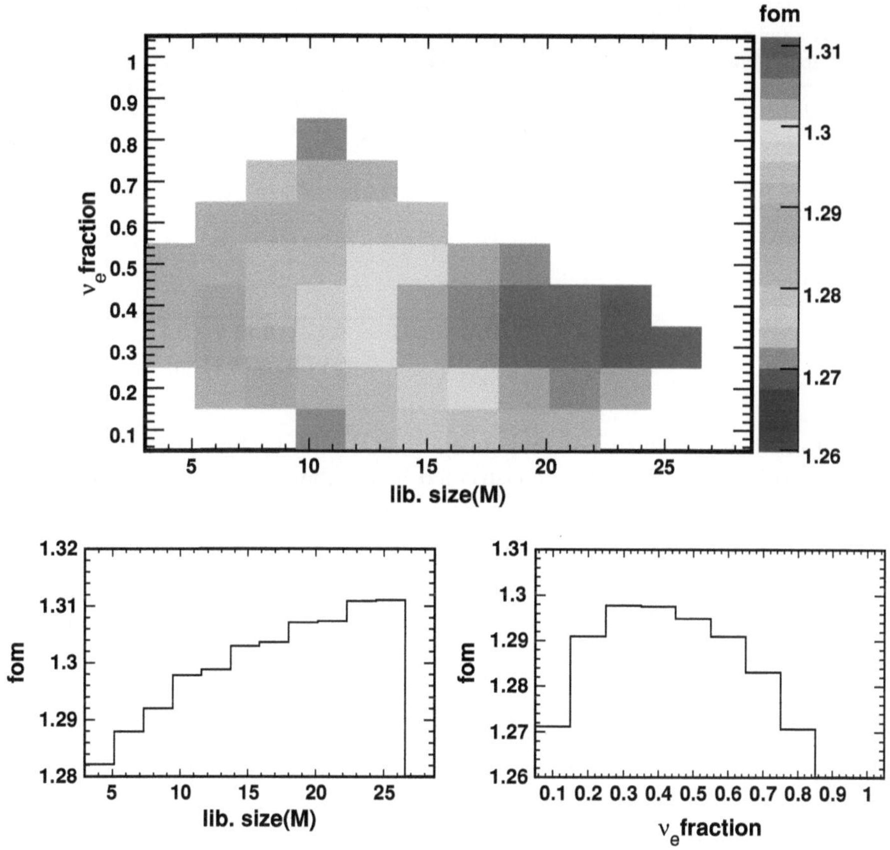

Fig. 5.13 (*top*) Figure of merit (fom) as a function of the ν_e fraction and the total library size. The ν_e fraction is defined as the ratio between the number of ν_e CC events in the library over the total library size. The cut values for the two variables used in the selection are optimized at every bin based on the fom. A nonstandard preselection is used as well as standard oscillation parameters. *(bottom left)* A slice of the two-dimensional histogram on the top for a constant ν_e fraction of 0.3. *(bottom right)* A slice of the two-dimensional histogram on the top for a constant total library size of 10M

It is also interesting to plot the fom of the selection as a function of the inverse of the total library size, for a constant ν_e fraction $= 0.3$. As seen in Fig. 5.14 significant improvements in sensitivity can still be gained by going to very large library sizes.

5.4.3 The LEM PID

The best possible signal and background separation is usually achieved by combining different variables through a multivariate approach rather than by simply

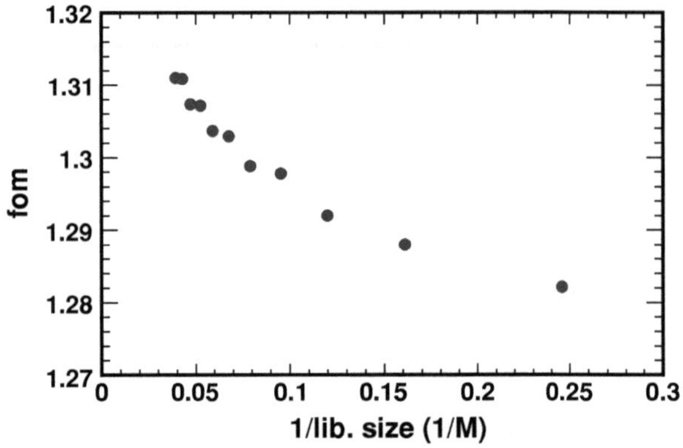

Fig. 5.14 Figure of merit (fom) as a function of the inverse total library size

cutting on them. Extensive studies reveal that a more sophisticated discriminant referred to as LEM pid can be constructed by combining the three variables that correspond to the three properties of Sect. 5.4.1:

- f_{50}: The fraction of the best 50 matches that are ν_e CC with $y<0.9$.
- y_{50}: The mean y of the best 50 matches that are ν_e CC with $y<0.9$.
- q_{50}: The mean fractional charge matched Q_{frac}^{mean} of the best 50 matches that are ν_e CC with $y<0.9$.

The y cut is necessary as ν_e CC events with $y > 0.9$ look practically the same as NC events. The particular value of 0.9 was chosen after several attempts, being the value at which the best separation is obtained. The three variables are shown in Fig. 5.15 for the Far Detector. The Near Detector plots are shown in Fig. 5.16 and display an obvious data-MC disagreement in the signal-like regions (i.e., high f_{50}, low y_{50} and high q_{50}). The fact that y_{50} displays the strongest disagreement suggests the problem with the simulation is a physics one and not a detector modeling one.

The LEM pid is obtained by combining these three variables in a simple likelihood. Let x_i be the value for variable i and $P^i_{\mathrm{sig(bg)}}(x_i)$ the value of the signal (background) probability density function for variable i at x_i. Then the LEM pid is given by

$$\text{LEM pid} = \frac{\prod_{i=f_{50},y_{50},q_{50}} P^i_{\mathrm{sig}}(x_i)}{\prod_{i=f_{50},y_{50},q_{50}} P^i_{\mathrm{sig}}(x_i) + \prod_{i=f_{50},y_{50},q_{50}} P^i_{\mathrm{bg}}(x_i)}. \tag{5.6}$$

With the above likelihood, an improvement in sensitivity to θ_{13} of approximately 4% is achieved with respect to the simple selection described in Sect. 5.4.2. If, in addition, the likelihood is binned in energy with a bin width of 0.5 GeV, an additional improvement in sensitivity of 4% is gained. Figure 5.18 shows the probability density functions for f_{50}, y_{50} and q_{50} as a function of reconstructed energy that are used to construct the LEM pid. As expected, the signal and

Fig. 5.15 The three variables that are combined into the LEM pid for all events passing the preselection, in the Far Detector. The total background and the signal curves are scaled to unity. The individual background components are scaled so that their proportion to the total background remains correct

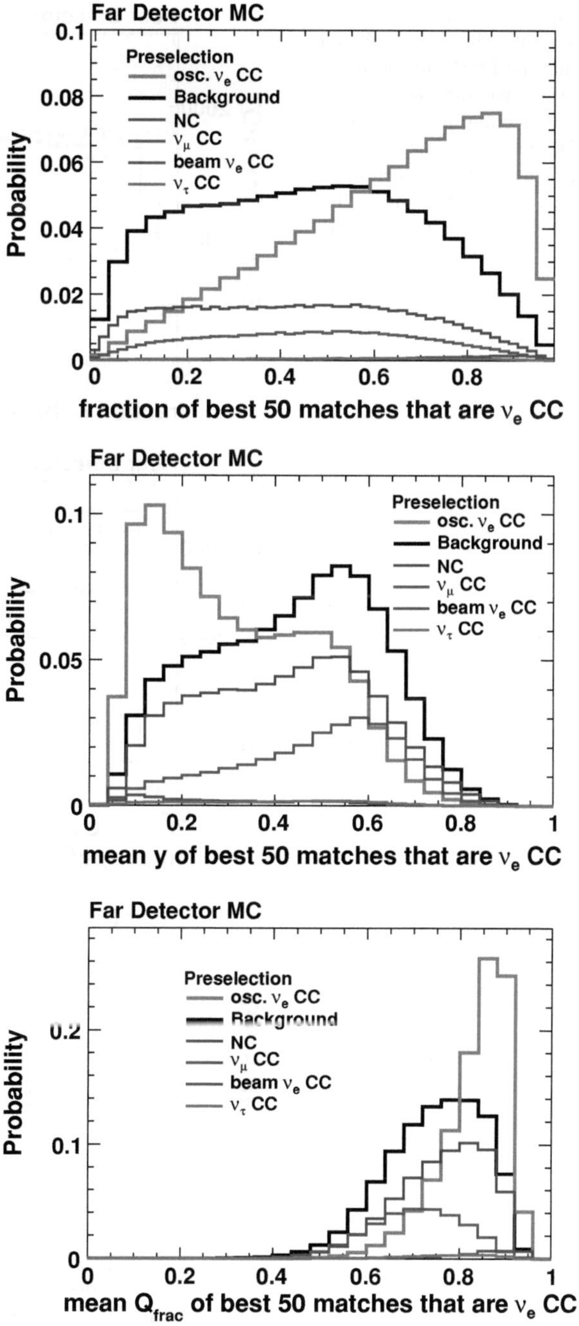

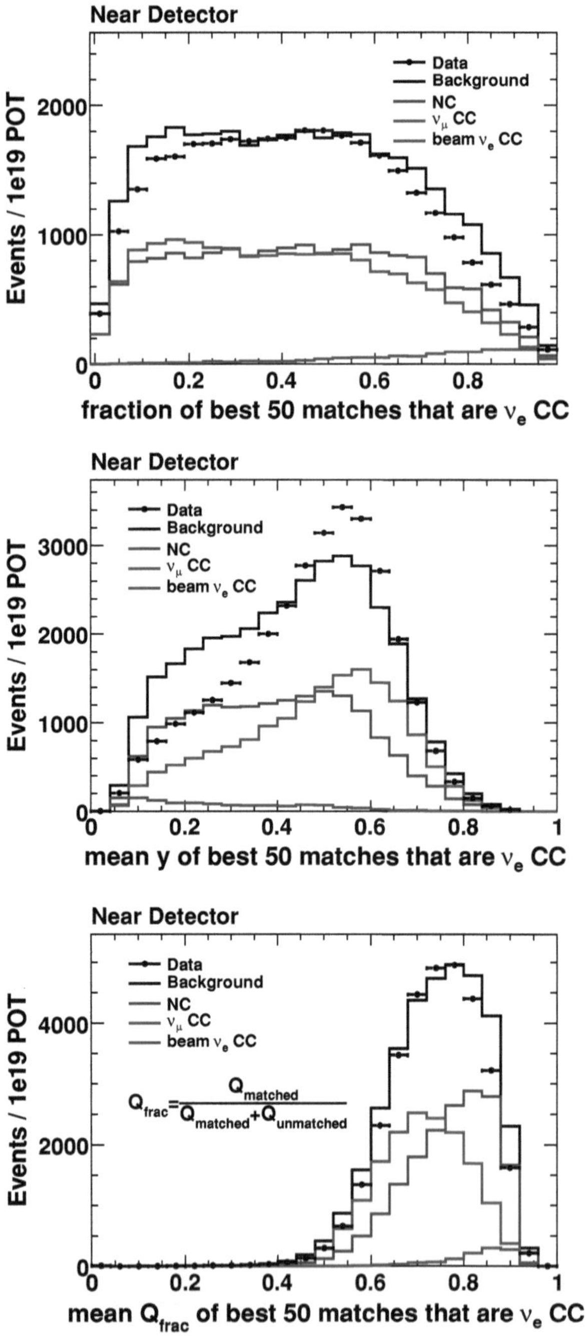

Fig. 5.16 The three variables that are combined into the LEM pid for all events passing the preselection, in the Near Detector. There is a large disagreement between the data and the MC mostly for the signal-like background

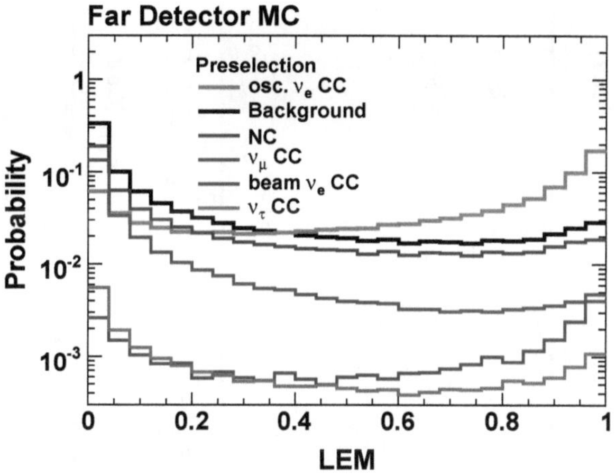

Fig. 5.17 The LEM pid in the Far Detector, on semilogarithmic scale. The total background and the signal curves are scaled to unity. Each of the individual background components is scaled so that its proportion to the total background remains correct. A similar plot in linear scale can be found in Fig. 4.11

background separation becomes easier as energy increases. This is particularly clear for the f_{50} variable. The resulting pid is shown for the Far Detector in Fig. 5.17 and for the Near Detector in Fig. 5.19. In the Near Detector there is a deficit of events in the data with respect to the simulation that worsens with increasing LEM pid. As mentioned in Sect. 4.2.3, this is mostly due to the mis-modeling of signal-like hadronic showers in the simulation. These observations are consistent with what the other selection methods observe.

5.4.3.1 Obtaining a Quick Data-Based Far Detector Prediction

The only step remaining in getting LEM ready for the analysis is to determine the optimal cut value. This requires an estimation of the expected Far Detector signal and background. While signal events are well modeled in MINOS,[6] Fig. 5.19 shows unambiguously that the simulation cannot be trusted to accurately predict the Far Detector background, particularly at high cut values of the LEM pid where the discrepancy between the data and the MC approaches 50%. Following the strategy laid down in Sect. 4.2.3, we must use the Near Detector data directly to predict the Far Detector background in order to perform the cut value optimization. The Near Detector data can be extrapolated to the Far Detector in the following way:

$$\mathrm{FD}_i^{\mathrm{pred}}(\mathrm{E}_{\mathrm{reco}}) = \mathrm{ND}_i^{\mathrm{data}}(\mathrm{E}_{\mathrm{reco}}) \frac{\mathrm{FD}_{\mathrm{i}}^{\mathrm{MC}}(\mathrm{E}_{\mathrm{reco}})}{\mathrm{ND}_i^{\mathrm{MC}}(\mathrm{E}_{\mathrm{reco}})}, \tag{5.7}$$

[6] This is shown in detail in Sect. 7.2.2.

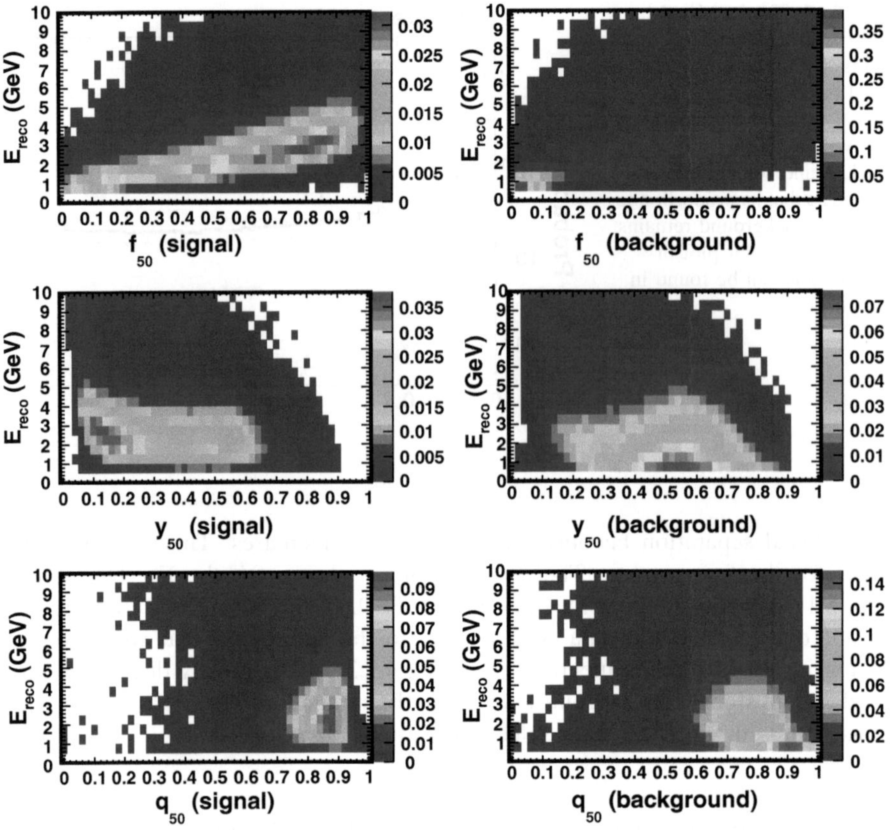

Fig. 5.18 Two-dimensional histograms of reconstructed energy as a function of the three variables used to construct the LEM pid. The Z-axis scale represents the number of Far Detector events per bin per 10^{19} POT. The first column is for signal events and the second column for background events, which includes all four types of background. The ν_e preselection is applied. All events are weighted by their oscillation probability according to standard oscillation parameters except for $\sin^2(2\theta_{13}) = 0.5$. As expected, the signal and the background separate more clearly as the energy increases

where i denotes a particular background component (NC, for instance), and FD and ND are the number of events selected (or predicted) in the Far Detector and Near Detector respectively at a particular reconstructed energy E_{reco}. Chapter 7 includes a discussion on the intricacies of the extrapolation. Chapter 8 contains a complete assessment of the overall systematic error on the Far Detector background prediction.

As explained in Sect. 4.2.3, each background component extrapolates differently to the Far Detector (i.e., the ratio $\frac{FD_i^{MC}(E_{reco})}{ND_i^{MC}(E_{reco})}$ differs for every i), making it necessary to know ND_i^{data} for each i. Chapter 6 describes the two formal data-based methods

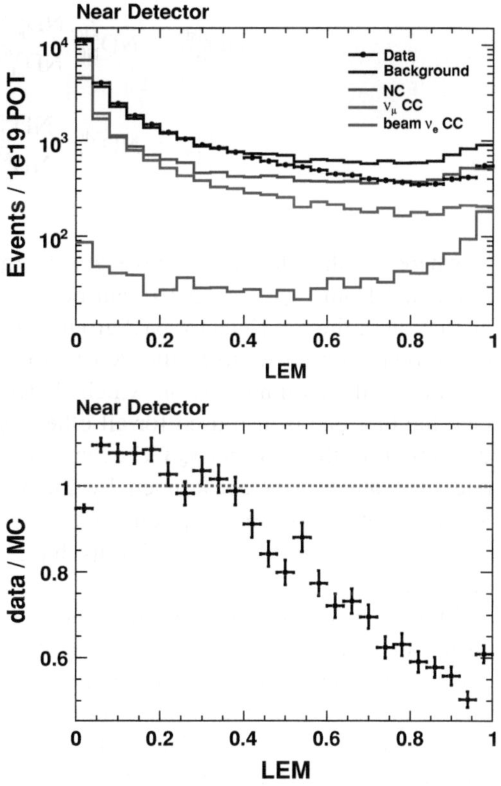

Fig. 5.19 The LEM pid in the Near Detector (*top*), and the corresponding ratio of data to MC (*bottom*). Only events that pass the ν_e preselection are shown. A very significant discrepancy between the data and the simulation exists in the signal region

that we use to decompose the Near Detector spectrum. These methods however are somewhat elaborate and computationally expensive, and consequently are only applied to one particular cut value of the pid. It is thus necessary to have a way to decompose and extrapolate the Near Detector data in a quick and automatic way, not just for the optimization at hand but also to conduct various other studies. We achieve this by making two assumptions:

- The data/MC discrepancy originates entirely from the NC and ν_μ CC components (i.e., the beam ν_e and ν_τ components are well simulated in the MC).
- The energy-dependent correction that must be applied to the simulated NC and ν_μ CC components to achieve perfect data/MC agreement is the same for both components.

Based on these two assumptions the number of selected events in the Near Detector data for each background component can be approximated by

$$\text{ND}^{\text{data}}_{\text{beam}\nu_e} \simeq \text{ND}^{\text{MC}}_{\text{beam}\nu_e}, \tag{5.8}$$

$$\mathrm{ND}_{\mathrm{NC}}^{\mathrm{data}} \simeq \mathrm{ND}_{\mathrm{NC}}^{\mathrm{MC}} \frac{\mathrm{ND}_{\mathrm{all}}^{\mathrm{data}} - \mathrm{ND}_{\mathrm{beam}\nu_e}^{\mathrm{MC}}}{\mathrm{ND}_{\mathrm{all}}^{\mathrm{MC}} - \mathrm{ND}_{\mathrm{beam}\nu_e}^{\mathrm{MC}}}, \tag{5.9}$$

$$\mathrm{ND}_{\nu_\mu\mathrm{CC}}^{\mathrm{data}} \simeq \mathrm{ND}_{\nu_\mu\mathrm{CC}}^{\mathrm{MC}} \frac{\mathrm{ND}_{\mathrm{all}}^{\mathrm{data}} - \mathrm{ND}_{\mathrm{beam}\nu_e}^{\mathrm{MC}}}{\mathrm{ND}_{\mathrm{all}}^{\mathrm{MC}} - \mathrm{ND}_{\mathrm{beam}\nu_e}^{\mathrm{MC}}}, \tag{5.10}$$

where all the quantities above are energy dependent. The values of $\mathrm{ND}_i^{\mathrm{data}}$ calculated from Eqs. (5.8–5.10) can be used in Eq. 5.7 to obtain the data-based prediction of the Far Detector background. This method to predict the Far Detector background is referred to as the PORP[7] method.

Both of the assumptions on which PORP rests are very solid. As a matter of fact, the first one is also used for all other Near Detector decomposition methods. It stems from the fact that ν_e CC events are in general well modeled and that the rate of beam ν_e's is constrained by the ν_μ CC data, as explained in Chap. 6. Similarly, the second assumption is based on the fact that ν_μ CC and NC showers are very alike in MINOS and thus have a very similar degree of data-MC discrepancy.

Moreover, studies [6] show that the impact that accurately decomposing the Near Detector spectrum has on the actual Far Detector prediction is relatively small. For instance, even if the extreme case where the intrinsic beam ν_e component in the Near Detector background is known to 30% and the NC and ν_μ CC components to worse than a factor of two as a result of the decomposition, the final error on the total Far Detector background prediction is only $\sim$6%, which is approximately half the overall systematic error. The reason for this is that the decomposition is subject to the constraint that the sum of the components must equal what is observed in the data:

$$\mathrm{ND}_{\mathrm{beam}\nu_e}^{\mathrm{data}} + \mathrm{ND}_{\mathrm{NC}}^{\mathrm{data}} + \mathrm{ND}_{\nu_\mu\mathrm{CC}}^{\mathrm{data}} \doteq \mathrm{ND}_{\mathrm{all}}^{\mathrm{data}}. \tag{5.11}$$

Therefore if one of the components is small, the other two must be increased to compensate, yielding the same total Far Detector prediction to within 6%.

Because of these reasons, the PORP method gives very accurate results. As shown in Sect. 7.1.2, the PORP prediction typically falls right between the predictions of the two formal decomposition methods, differing from either of them by 3% at the most. For all practical purposes then, PORP is a perfectly valid Near Detector decomposition method and can be used to perform studies that require a mapping of the data-based prediction for all cut values, such as the cut value optimization.

[7] The acronym is formed with the initials of the method's creators, who are the author (Pedro Ochoa) and Ryan Patterson.

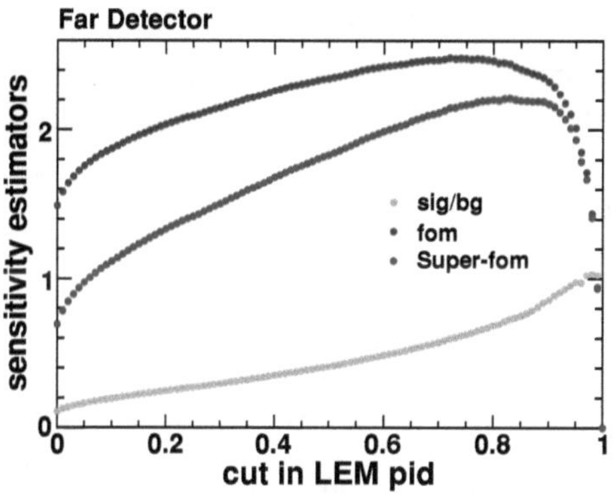

Fig. 5.20 Super-fom, fom and signal-to-background ratio as a function of the cut in LEM pid. All the quantities are data-based with PORP and are calculated using standard oscillation parameters, with $\sin^2(2\theta_{13}) = 0.15$ (at the CHOOZ limit), and assuming an exposure of 3.14×10^{20} POTs. The cut at LEM pid = 0.80 yields the highest super-fom of 2.2

5.4.3.2 Cut Value Optimization

Combining the simulated signal with the Far Detector prediction obtained using the PORP scheme just described, we can find the cut value for which the performance is the highest.

In this case, the performance of the selection needs to be evaluated using a more sophisticated figure of merit that involves the impact of systematic errors. This can be achieved through what we refer to as a super-fom, which is a measure of how significant the number of selected signal events N_{sig} is compared to the total error in the background. By accounting for both the statistical and the systematic error in the background prediction, the super-fom is a good measure of the sensitivity to θ_{13}:

$$\text{super-fom} = \frac{N_{\text{signal}}}{\sqrt{\sigma_{\text{stat}}^2 + \sigma_{\text{syst}}^2}} = \frac{N_{\text{signal}}}{\sqrt{N_{\text{bg}} + (0.14 N_{\text{bg}})^2}}, \tag{5.12}$$

where a systematic error on the background of 14% is assumed, based on the results of Chap. 8. The super-fom as a function of cut in LEM pid is shown on Fig. 5.20. Also shown is the fom defined in Sect. 5.4.2 as $N_{\text{sig}}/\sqrt{N_{N_{\text{bg}}}}$, as well as the signal-to-background ratio. As expected, the signal-to-background ratio rises monotonically with increasing cut value. Also, the super-fom and the fom converge as N_{bg} goes to zero, although it is clear that the simple fom is not good enough to base the optimization on, as it peaks at a lower cut value than the super-fom does.

The cut value for the LEM pid that yields the best super-fom is 0.80. Other more sophisticated figures of merit that do not rely on Gaussian statistics have been used but they yield the same optimal cut value [7]. For reasons that will be evident later, we also consider throughout this thesis a cut at LEM pid > 0.65.

5.5 Characteristics of LEM

5.5.1 Selected Events

The Near Detector energy spectra of LEM selected events for a cut at 0.65 and at 0.80 are shown in Figs. 5.21 and 5.22 respectively. The data and the MC disagree to approximately 40% in both cases. The disagreement is worst in the 2–4 GeV energy region, although it is still larger than 20% in the higher-energy region.

The number of data-based Far Detector predicted events at the fiducial, preselection and LEM levels can be found in Table 5.2. As seen there, with a cut at LEM pid at 0.80 we can achieve a background rejection of 98.9% while keeping a

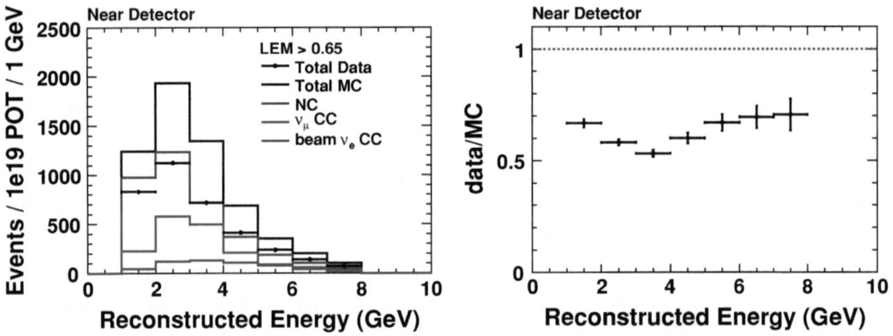

Fig. 5.21 Energy spectrum of Near Detector selected events for a cut at LEM pid > 0.65 (*left*) and the corresponding data to MC ratio (*right*). The integrated data to MC ratio is 60.2%

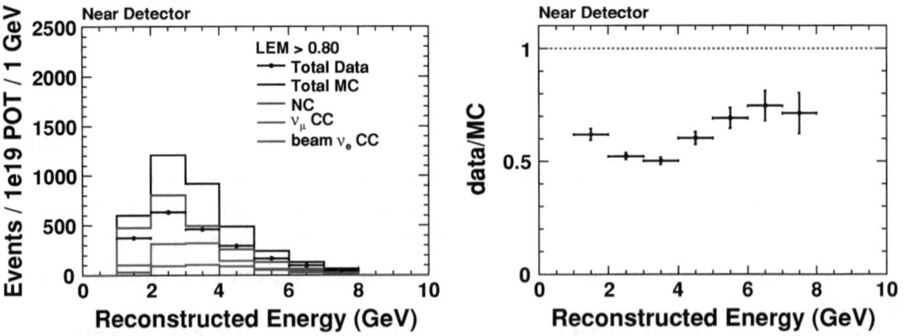

Fig. 5.22 Energy spectrum of Near Detector selected events for a cut at LEM pid > 0.80 (*right*) and the corresponding data to MC ratio (*left*). The integrated data to MC ratio is 56.7%

Table 5.2 Number of Far Detector predicted events with the corresponding selection efficiencies for an exposure of 3.14×10^{20} POT

Selection		osc. ν_e CC	Total bg.	NC	ν_μ CC	ν_τ CC	Beam ν_e
Fiducial vol.		25.7	1,222.5	340.5	857.8	8.1	16.2
Preselection	Events	20.4	209.5	136.3	63.3	4.4	5.6
	Efficiency	79.2%	17.1%	40.0%	7.4%	54.0%	34.2%
LEM > 0.65	Events	11.4	21.6	14.7	3.2	1.0	2.7
	Efficiency	44.4%	1.8%	4.3%	3.7%	12.7%	16.9%
LEM > 0.80	Events	8.9	12.9	8.3	1.8	0.7	2.2
	Efficiency	34.5%	1.1%	2.4%	2.1%	8.6%	13.6%

The efficiencies are calculated with respect to the number of selected events when only the fiducial volume cuts are applied. The values for the two LEM cuts are data-based, obtained with PORP. The values for the fiducial and preselected samples however are obtained directly from the raw MC, given that the data-MC disagreement is small for those samples (see Fig. 5.5). Standard oscillation parameters are used. The signal corresponds to $\sin^2(2\theta_{13}) = 0.15$ and $\delta_{CP} = 0$, while ignoring matter effects

Table 5.3 Number of π^0's in the final state for signal and background events, according to the MC

Selection	Event type	No π^0(%)	Single-π^0(%)	Multi-π^0(%)
Fiducial	Signal	68.6	25.2	6.1
	Background	36.7	39.5	23.8
Preselection	Signal	68.8	25.2	6.0
	Background	35.9	40.4	23.7
LEM > 0.65	Signal	76.1	20.2	3.7
	Background	18.1	46.6	35.3
LEM > 0.80	Signal	79.6	17.6	2.8
	Background	19.6	46.4	34.0

signal selection efficiency of 34.5% with respect to the fiducial volume cuts. Also, the number of selected background events quickly falls with increasing LEM pid cut. With the cut at 0.80, the signal-to-background ratio approaches 70%, which was previously unheard of for a ν_e CC selection in MINOS.

It is also interesting to look at the π^0 content shown in Table 5.3. Going from fiducial volume cuts to preselection does not change the composition in terms of π^0 content (and thus EM fraction), most likely because the preselection only extracts showers of certain size without making any explicit distinction in terms of shape and topology. Applying the PID however clearly affects the event composition. As expected, more than 80% of the background events selected with LEM contain at least one π^0 in the final state. For the signal the situation reverses and LEM preferentially selects events with no π^0's in them. This is most likely due to the fact that an event with an electron not emitted colinearly with the two photons from π^0 decay looks diffuse and thus more similar to a high-multiplicity hadronic shower.

Table 5.4 Deep inelastic scattering (DIS), resonant (RES), quasi-elastic (QE) and coherent π^0 production (COH) event breakdown for signal and background events, according to the MC

Selection	Event type	DIS (%)	QE (%)	RES (%)	COH (%)
Fiducial	Signal	42.1	23.7	33.5	0.7
	Background	85.2	3.1	11.0	0.7
Preselection	Signal	41.9	23.8	33.6	0.7
	Background	87.7	2.4	9.3	0.7
LEM > 0.65	Signal	25.0	35.9	38.2	0.9
	Background	84.6	4.1	7.9	3.4
LEM > 0.80	Signal	20.7	40.2	38.1	1.0
	Background	82.3	5.2	7.7	4.8

Table 5.5 Number of PORP Far Detector predicted events corresponding to an exposure of 3.14×10^{20} POT, as well as the corresponding systematic error on the background, signal-to-background ratio and super-fom

Selection	osc. ν_e CC	Total bg.	σ_{bg}	sig/bg	Super-fom
LEM > 0.65	11.42	21.64	12%	0.53	2.14
LEM > 0.80	8.87	12.95	14%	0.68	2.20
ANN > 0.70	10.29	27.26	7%	0.38	1.85

The super-foms are calculated in accordance to the total systematic errors which are derived in Chap. 7. The optimal cut value for ANN is 0.7. Standard oscillation parameters are used. The signal corresponds to $\sin^2(2\theta_{13}) = 0.15$ and $\delta_{CP} = 0$, without matter effects

The interaction type of selected events broken down into deep-inelastic scattering (DIS), quasi-elastic (QE), resonant (RES) and coherent-π^0 production (COH) can be found in Table 5.4. As for the π^0 content, the interaction type composition does not change when going from fiducial level to preselection, but does when applying the PID. The LEM selected background consists mostly of DIS events, whereas for the signal QE and resonant RES events are preferentially selected in approximately equal proportions.

5.5.2 Preliminary Relative Sensitivity to θ_{13}

In order to get a preliminary idea of the performance of LEM relative to other selections, Table 5.5 contains the signal and event breakdown for LEM and the ANN selections, together with some performance indicators. The ANN selection is a neural network described in [2], which consists of 11 different variables. It is important to note that ANN is the best performing ν_e CC selection in MINOS after LEM, which is why it is instructive to compare the two.

As seen in Table 5.5 LEM clearly has better sensitivity to θ_{13} than ANN does. In the case of the cut at 0.80, LEM selects 86% of the signal but less than half of the background of ANN. Even with the lower cut at 0.65, LEM selects 11% more signal with 80% of the background. The relative sensitivity to θ_{13} can be evaluated with the super-fom, which in this case indicates that the performance of the LEM selection with the cut at 0.80 is approximately 18% better than that of ANN. The fact that LEM selects significantly less background is a considerable advantage, as the impact of the total error in the background diminishes accordingly.

A more precise estimation of the sensitivity and the exclusion contours that can be obtained with LEM is included at the end of Chap. 8, once all the systematic error estimations and the predictions are finalized. Nevertheless, LEM is clearly the most sensitive ν_e CC selection in MINOS to date.

References

1. M. Apollonio et al. (the CHOOZ Collaboration), Search for neutrino oscillations on a long base-line at the CHOOZ nuclear power station. Eur. Phys. J. C27, 331–374 (2003)
2. T. Yang, A study of muon neutrino to electron neutrino oscillations in the MINOS experiment, FERMILAB-THESIS-2009-04.
3. A. Holin, Nue Muon Removal Background Estimation, MINOS-doc-4018, 2008.
4. J. P. Ochoa-Ricoux, The status of the new MCNN, MINOS-doc-4888, 2008.
5. R. B. Patterson, MCNN : Exploring micro and macro properties of the method, MINOSdoc-4208, 2008.
6. R. B. Patterson and J. P. Ochoa-Ricoux, The impact of background composition on nue sensitivity, MINOS-doc-3760, 2007.
7. J. P. Ochoa-Ricoux, Selection Optimization for the nue appearance analysis, MINOSdoc-4207, 2008.

Chapter 6
Measuring the Backgrounds in the Near Detector

This chapter describes the data-driven methods used to determine the backgrounds to sufficient accuracy, which as mentioned earlier is one of the pillars on which the analysis rests. We begin with some data-MC comparisons that motivate the use of the Near Detector to directly measure the backgrounds. Given that each background component must be extrapolated separately to the Far Detector, two data-based methods are introduced in Sects. 6.3 and 6.4 that decompose the observed Near Detector background energy spectrum. Both of these methods require the intrinsic beam ν_e component as an input, which is why a discussion on this component's origin and modeling uncertainty is included in Sect. 6.2.

6.1 The Need for the Near Detector

6.1.1 Uncertainties in the Simulation

While describing LEM in the previous chapter it became evident that there is a very significant discrepancy between the data and the simulation concerning the number of signal like events in the LEM selected background. This discrepancy is reproduced in Fig. 6.1 with an emphasis on the high PID region, where the discrepancy is the largest and exceeds 40%.

Despite its size, the disagreement between the data and the MC is not surprising given the large uncertainties present in the simulation. Chapter 7 contains a complete description and an assessment of all the systematic errors involved in the analysis. Table 6.1 summarizes that information in the case of the Near Detector, and shows how the total systematic error on the Near Detector MC prediction is actually higher than 50%. The biggest contributor is, by far, the overall uncertainty in the hadronic model.

The total systematic error in the Near Detector simulation is presented in Fig. 6.2 as a function of energy. There it can be seen that the data-MC discrepancy,

J. P. Ochoa Ricoux, *A Search for Muon Neutrino to Electron Neutrino Oscillations in the MINOS Experiment*, Springer Theses, DOI: 10.1007/978-1-4419-7949-0_6,

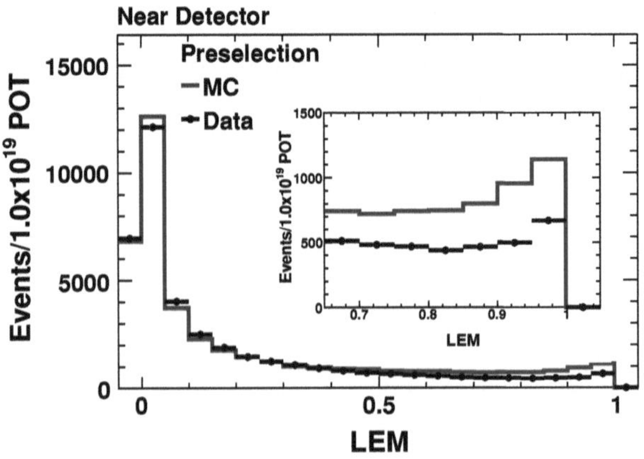

Fig. 6.1 LEM distribution of preselected events in the Near Detector. The plot contains the same data as Fig. 5.19 but in a different format. The simulation clearly overestimates the amount of signal-like background in the data by more than ~40% in the high PID region. The events at LEM pid <0 are the preselected events that do not get any matches in the LEM library

Table 6.1 Breakdown of systematic errors in the Near Detector simulation for LEM selected events with a cut at 0.65. A description of each item can be found in Sect. 8.1. The total error in the MC prediction is larger than 50%. The largest sources of error are physics related, particularly in the hadronic model

Type	Description	Min	Max
Physics	Beam blux	−9.5%	9.5%
	Cross-sections	−3.5%	4.0%
	Hadronic model	−47.3%	47.3%
	Hadron mult.	−2.9%	1.1%
	Intranuke	−15.9%	15.9%
Detector	Had/EM energy	−1.7%	2.3%
	Abs. energy	−9.3%	8.3%
	Calibration	−4.3%	6.0%
	Low pulse-height	−5.0%	5.0%
	Crosstalk	−2.3%	2.3%
	Total	−52.3%	51.3%

despite its large size when evaluated in terms of a percentage, actually corresponds to less than a 1σ discrepancy and is thus not unexpected. This is still the case for the tighter LEM cut (at 0.80), when even more signal-like events are selected. Also, the systematic error diminishes with increasing energy. This is mostly due to the fact that the physics of hadronic showers are better known at higher energies.

Figure 6.3 shows the systematic error as a function of LEM pid due to the uncertainties in the beam flux as well as in the hadronic and cross-section

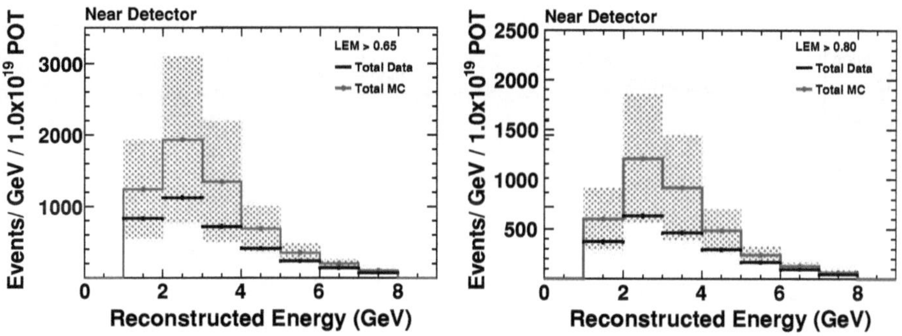

Fig. 6.2 Energy spectra of LEM selected events for a cut at 0.65 (*left*) and at 0.80 (*right*). The band represents the total systematic error on the simulation at the 1σ level for each energy bin. For both cuts the data is within the bounds of the systematic error. Note that the vertical scale is different for the two plots

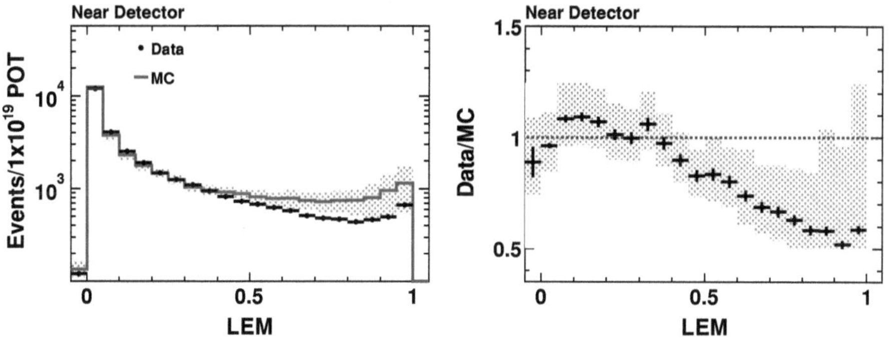

Fig. 6.3 LEM pid of events passing the preselection in the Near Detector (*left*), and the corresponding data to MC ratio (*right*). The orange band represents the systematic error due to uncertainties in the beam flux, the hadronic model and the cross-sections. The data-MC discrepancy is practically covered by these three sources of systematic error to less than 1.5σ. Also, the systematic error gets proportionately larger with increasing PID

models.[1] As expected, the data-MC discrepancy along the whole range of pid values is almost covered by these three sources of systematic error, except at the high-LEM region. Also, the systematic error grows with increasing pid value, which shows that the hadronic showers that are the least well known are precisely the ones that constitute the background to the ν_e appearance search. This matches the behavior that we see in the data, where the discrepancy gets larger with increasing pid cut. Figure 6.4 shows the same systematic error envelope for

[1] Due to the way the systematics are assessed, it is very hard to include all the others in these plots. Nevertheless, Figs. 6.3 and 6.4 are representative of what happens to the overall systematic error, as the hadronic model systematic is the dominant one by far.

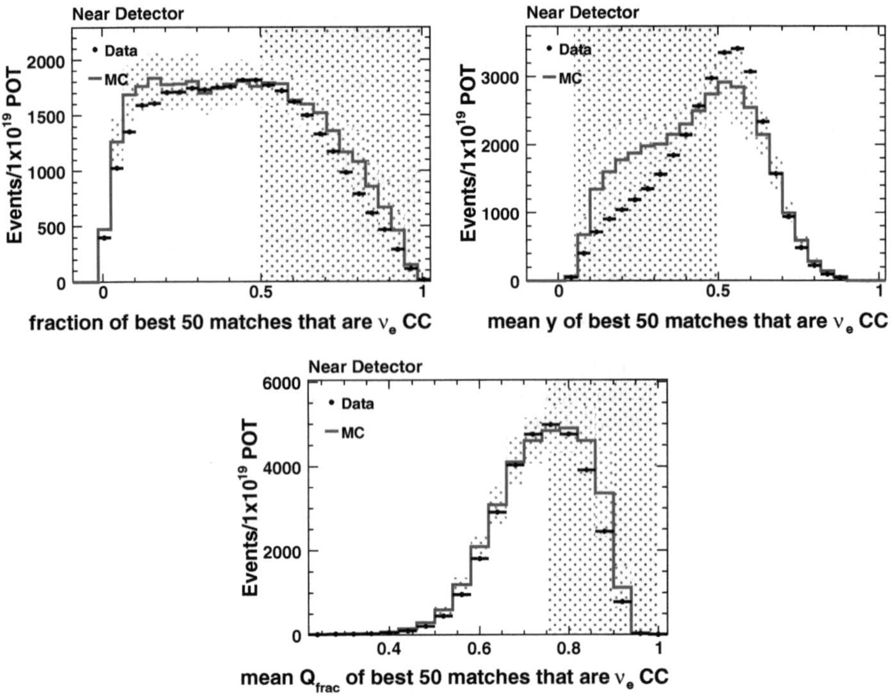

Fig. 6.4 The three variables that compose the LEM pid for events that pass the preselection in the Near Detector: f_{50} (*top left*), y_{50} (*top right*) and q_{50} (*bottom*). The region shaded in blue represents the signal-like region. The orange band represents the systematic error due to uncertainties in the beam flux, the hadronic model and the cross-sections

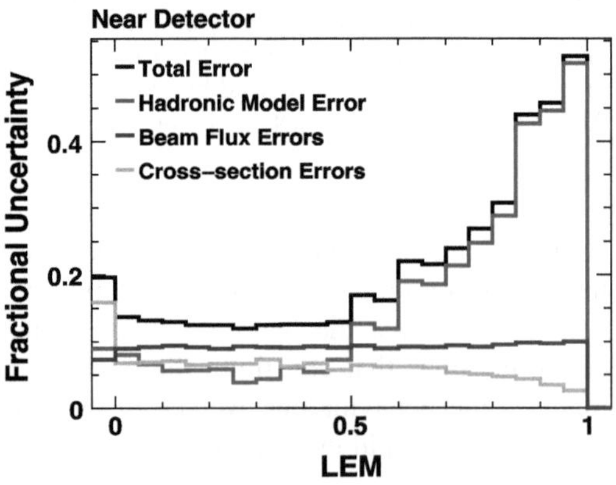

Fig. 6.5 Fractional systematic uncertainty of preselected events as a function of LEM pid. Only the contributions due to uncertainties in the beam flux, the hadronic model and the cross-sections are included. The error in the hadronic model rises dramatically in the high-pid region

the three variables that make up the LEM pid. Again, the error band increases in the signal-like regions (high f_{50}, low y_{50} and high q_{50}), and covers most (and in some cases all) of the data-MC discrepancy.

The increase in the overall systematic error as a function of LEM pid is due to a corresponding increase in the hadronic model error. The dramatic rise of the hadronic model error as a function of pid is shown in Fig. 6.5, indicating that the hadronization of the signal-like background events in our analysis is simply not well understood. This is not surprising given the circumstances surrounding the AGKY hadronization model described in Sect. 3.4.2. The AGKY model performs a phenomenological extrapolation between the KNO-based model and the PYTHIA/JETSET model in the invariant mass regime which is of interest to our analysis, but which is also where there is a lack of external data to constrain such an extrapolation. A detailed account of the uncertainties involved in the AGKY model and their impact on the analysis can be found in Sect. 8.1.2.

6.1.2 The Near Detector's Role in the Analysis

The fact that the MC is unable to accurately reproduce the backgrounds to the ν_e appearance search is not a problem for the analysis, as these backgrounds can be directly measured in the Near Detector. The measured Near Detector background rates are then converted into Far Detector rates in the following way:

$$\mathrm{FD}_i^{\mathrm{pred}}(\mathrm{E_{reco}}) = \mathrm{ND}_i^{\mathrm{data}}(\mathrm{E_{reco}}) \frac{\mathrm{FD_i}^{\mathrm{MC}}(\mathrm{E_{reco}})}{\mathrm{ND}_i^{\mathrm{MC}}(\mathrm{E_{reco}})}. \tag{6.1}$$

The preceding expression is used for every background component[2] i at each bin of reconstructed energy $\mathrm{E_{reco}}$. $\mathrm{ND}_i^{\mathrm{data}}$ is the number of events observed at the Near Detector and can be obtained, for instance, using the PORP decomposition method of Sect. 5.4.3. The quantity $\frac{\mathrm{FD_i}^{\mathrm{MC}}}{\mathrm{ND_i}^{\mathrm{MC}}}$, which converts a Near Detector rate into a Far Detector rate, is referred to as the Far-Near ratio, and needs to be obtained from the simulation. Chapter 7 deals with all of this in detail.

Even though the simulation cannot accurately predict the number of background events in the ν_e appearance search, it can still be used to obtain the Far-Near ratios. The dominant sources of systematic error in the simulation are uncertainties in the physics processes rather than in the detector modeling, as illustrated in Table 6.1. This is a critical point, as it means that many of these uncertainties affect the selection very similarly in the two detectors and thus cancel to first order when taking the Far-Near ratios.

Chapter 8 deals with the systematic errors in detail. Table 6.2 gives a preview by showing what happens to the number of selected events in both detectors as the different aspects of the simulation are varied within their uncertainties. As shown by the Δ column, which represents the difference between the Far Detector and Near Detector columns, the uncertainties listed in the table cancel to within a few

[2] With the exception of ν_τ CC events, which originate through ν_μ CC oscillations and are thus unobserved at the ND.

Table 6.2 Breakdown of systematic errors that cancel to first order between the two detectors, for LEM selected (with a cut at 0.65) NC and beam ν_e events. The quantity Δ is the difference between the Far and Near systematics. $M_A(QE)$, $M_A(RES)$ and KNO are the cross-section parameters that are used in the NEUGEN simulation. T1–T6 represent the different uncertainties in the hadronic model. Despite the large changes in the number of selected events in each detector, the Far-Near ratio does not change by more than a few percent when all of these different sources of systematic errors are considered. All of of these sources of systematic error along with the ones not mentioned here are addressed in detail in Chap. 7.

		NC			Beam ν_e		
Systematic	Shift	ND(%)	FD(%)	Δ(%)	ND(%)	FD(%)	Δ(%)
Beam flux	$\pm 1\sigma$	±9.8	±9.0	∓0.8	±9.0	±13.8	±4.8
$M_A(QE)$	−15%	−0.1	−0.1	+0.0	−4.4	−4.0	+0.4
	+15%	+0.1	+0.2	+0.1	+4.5	+4.2	−0.3
$M_A(RES)$	−15%	−2.0	−2.1	−0.1	−6.3	−6.3	+0.0
	+15%	+2.9	+2.9	+0.0	+6.5	+6.6	+0.1
KNO	±50%	±2.3	±2.5	±0.2	±6.4	±6.9	±0.5
Hadronic model	T1	−37.4	−36.6	+0.8	−1.6	−2.0	−0.4
	T2+	+10.8	+11.0	+0.2	+0.3	+0.2	−0.1
	T2−	−10.4	−10.6	−0.2	−0.3	−0.2	+0.1
	T3	−19.6	−21.2	−1.6	−0.7	+0.4	+1.1
	T4	−15.5	−16.3	−0.8	−0.7	−0.8	−0.1
	T5	−14.1	−15.0	−0.9	−0.7	−0.8	−0.1
	T6	−14.5	−16.1	−1.6	−1.2	−0.9	+0.3
Intranuke	1σ	−17.3	−16.8	+0.5	+3.3	−2.2	−5.5
Hadronic/EM energy	$+1\sigma$	+1.9	+1.3	−0.6	−1.7	−1.5	+0.2
	-1σ	−2.6	−1.9	+0.7	−0.8	+0.1	+0.9
Abs. energy	$+1\sigma$	+7.5	+7.9	+0.4	+8.4	+7.3	−1.1
	-1σ	−8.4	−7.9	+0.5	−11.4	−6.7	+4.7

percent.[3] This is particularly true for the NC events, which are the dominant component in the background. The table includes two systematics related to detector modeling that also cancel, which are the hadronic/EM energy and the absolute energy scale systematics. All of the systematics in the table are described in detail in Chap. 7.

Therefore, the strategy of our analysis is to measure the backgrounds in the Near Detector and extrapolate them to the Far Detector using the MC. Even if the simulation does a relatively poor job in reproducing the topology of the background events in the analysis, it can still be used to provide the Far-Near ratios required to convert the measured Near Detector rates into Far Detector rates to a very good accuracy.

For the strategy to work however it is necessary to decompose the measured Near Detector data spectrum into its NC, ν_μ CC and beam ν_e components. The

[3] Please note that Δ is not exactly the same as the error on the extrapolation given in Chap. 7, given that we are neglecting the energy dependence of these shifts. Nevertheless, Δ is a good estimator of the extrapolation error.

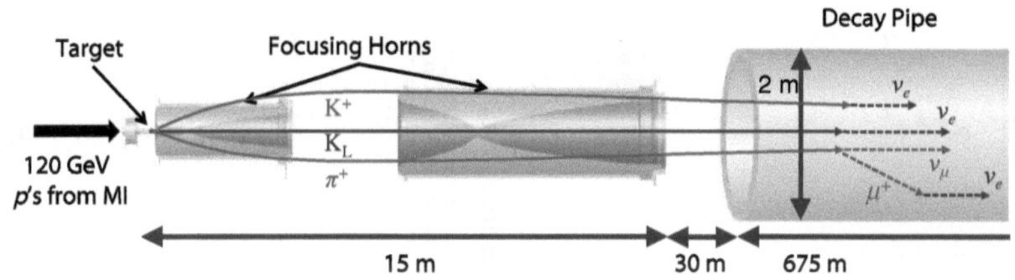

Fig. 6.6 Schematic of the NuMI beamline, illustrating the main three processes that yield ν_e CC events. The dominant production mechanism is through μ^+ decay, although the decays of K_L and K^+ mesons also generate ν_e's. Note that not all the particles involved in each decay are represented in the picture

reason is that each of these components extrapolates differently to the Far Detector,[4] as explained in Sect. 4.2.3. The PORP decomposition method, which has already been introduced in Sect. 5.4.3, is used whenever a quick prediction is required (such as in the optimization of the LEM pid). To obtain the final Far Detector background predictions in the analysis, two independent data-driven decomposition methods, HOO and MRCC, have been employed. These methods are respectively described in Sects. 6.3 and 6.4. Since both of these methods require the beam ν_e component as an input, the next section describes how this component is obtained in the Near Detector.

6.2 The Intrinsic Beam ν_e Component

6.2.1 *Origin*

Approximately 1.3% of the NuMI beam-induced events in MINOS are ν_e CC events [1]. The intrinsic beam ν_e's that produce these events (also referred to as beam ν_e's) originate primarily with the decay of secondary muons in the decay pipe.

$$\begin{aligned} \pi^+ &\rightarrow \mu^+ + \nu_\mu \\ &\hookrightarrow e^+ + \nu_e + \overline{\nu}_\mu. \end{aligned} \tag{6.2}$$

The muon parents are produced alongside the muon neutrinos that make up most of the beam with the decay of positively charged pions from the target.

Beam ν_e's also originate in the K_{e3} decays of kaons:

$$K^+ \rightarrow \pi^0 + e^+ + \nu_e, \tag{6.3}$$

$$K_L \rightarrow \pi^- + e^+ + \nu_e. \tag{6.4}$$

Figure 6.6 illustrates the three main modes in which electron neutrinos are produced in the beam. Note that, in an analogous way, beam $\overline{\nu}_e$'s are also produced

[4] In other words, the Far-Near ratio is different for each background component.

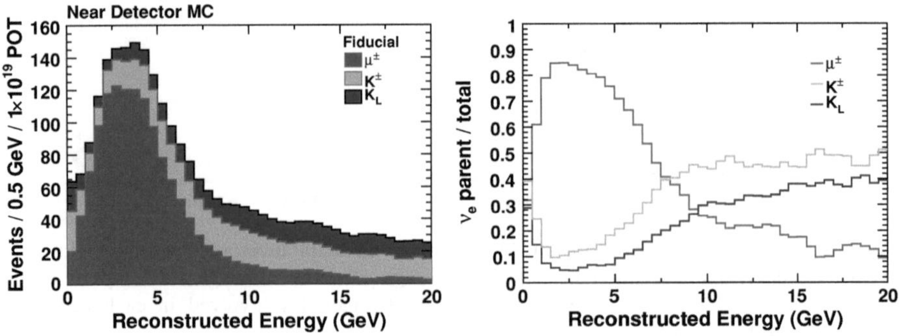

Fig. 6.7 Simulated energy spectra of beam ν_e's (and ν_e's) that interact inside the fiducial volume, broken down by parent type (*left*) and the corresponding propositions of each parent as a function of energy (*right*). Below 8 GeV, approximately 71% of beam ν_e's orginate from muon decay.

in the beam by the decay of μ^-, K^- and K_L particles. Due however to the fact that most of the negative particles produced in the target are defocused by the horns, only approximately 5% of all electron neutrinos in the beam below 8 GeV are really $\overline{\nu}_e$'s. In this thesis we refer as beam ν_e's to both ν_e's and $\overline{\nu}_e$'s in the beam. Note also that there are other channels in which electron neutrinos are produced in the beamline (such as $\pi^+ \rightarrow e^+ + \nu_e$). These channels are included in the simulation but give negligible contributions to the overall ν_e flux.

Because of the much more abundant production of pions vs. kaons in the target at low energies, the primary channel for ν_e production is the one represented by Eq. 6.2. This can be seen in Fig. 6.7, which shows the energy spectra of beam ν_e's from their three different parents. In the energy region of interest to the analysis (<8 GeV), approximately 71% of beam ν_e's originate from muon decay, 19% from $K^\pm$ decay and only 10% from K_L decay. These proportions change drastically at higher energies.

6.2.2 Beam ν_e's in the Simulation

In the Near Detector decomposition methods that are subsequently described in this chapter, the beam ν_e component in the background is taken directly from the simulation. There are several reasons for that.

First, electron neutrinos are well modeled in the MINOS MC. This is evidenced by the data-MC comparisons of electron events done at Caldet (see Sect. 7.2.2), as well as by the size of the systematic errors for beam ν_e's shown in Table 6.2. It is particularly remarkable that while the hadronic model uncertainties have a very large impact on NC (and ν_μ CC) events, beam ν_e's are mostly unaffected by them. This suggests that, as expected, the EM component largely dominates in the ν_e events selected by LEM.

Second, approximately 90% of all beam ν_e's originate ultimately from the decay of π^+ (through secondary muon decay) or K^+ mesons, which are also the parents of practically all ν_μ CC events in the energy region of interest. Therefore, $\pi^\pm$ and $K^\pm$ production at the target is well constrained by the fit described in Sect. 3.4.2 which is done to ν_μ CC (and $\overline{\nu}_\mu$ CC) data in various beam configurations.

Because of these reasons, beam ν_e's are predicted by the simulation to within 20% in the Near Detector, as evidenced by the systematic error estimations of Sect. 8.1.4. As shown earlier, this is significantly better than what the simulation can do to predict the hadronic background (NC and ν_μ CC events).

Finally, beam ν_e's represent only ~10% of the total background selected by LEM. Therefore the error on this component does not have a strong impact in the Near Detector decomposition.

6.2.3 Cross-Check with Antineutrinos

While the systematic error on beam ν_e's is relatively small, it is important to have an independent data-based cross-check to ensure that the actual beam ν_e rate is consistent with the tuned MC prediction that is used in the rest of the ν_e appearance analysis. This can be achieved with muon antineutrinos. Given that most beam ν_e's originate from muon decay alongside an antineutrino, as described by Eq. 6.2, measuring the rate of antineutrinos from muon decay constitutes a measurement of most beam ν_e's.

Unfortunately the antineutrinos from muon decay are a small component of the whole antineutrino spectrum, as shown in Fig. 6.8. Nevertheless, given that it is precisely this component that is the most significantly affected when the distance

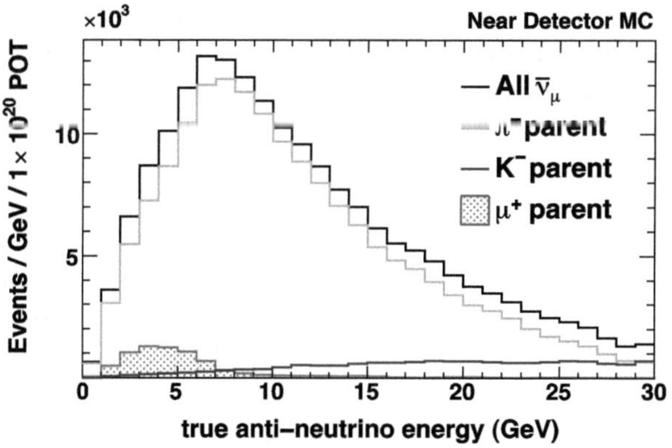

Fig. 6.8 True energy spectrum of antineutrinos in the Near Detector fiducial volume, broken down by antineutrino parent. A very small amount of antineutrinos originates from K_L decay and is not shown here. Most antineutrinos originate from π^- decay and, at high energies, from K^- decay. A small amount originates from secondary μ^+ decay in the beam pipe, as described by Eq. 6.2

between the target and the horns is changed, we can measure it by comparing the antineutrino spectra in two different beam configurations.

A measurement of the beam ν_e rate from muon decay has been carried out by comparing the antineutrino spectra in the LE and pHE beam configurations. All of the details of the analysis are available in Appendix C. The measured rate is $1.57 \pm 0.37\text{(stat)} \pm 0.41\text{(syst)}$ times the expectation from the tuned simulation. If the two sources of uncertainty are added in quadrature, this measurement confirms the tuned MC expectation at one standard deviation.

The measurement is limited by the amount of data available in the pHE configuration, which is only 1.41×10^{20} POT. If more pHE or pME data became available in the future, the measurement could be carried out with significantly more precision. This is particularly so in the case of pME data, for which the impact of the systematic uncertainties would be greatly reduced, as explained in Appendix C. Nevertheless, the current measurement still provides an independent cross-check of the beam ν_e flux estimation and is relevant because it consists of a method that could be exploited to great advantage by future long baseline neutrino experiments that have a near detector capable of muon charge-sign determination and an adjustable beam focus.

6.3 The Horn-On/Horn-Off Decomposition Method

6.3.1 Basic Concept

As explained in Sect. 6.1.2, each background component must be extrapolated to the Far Detector separately. This means that the measured Near Detector data energy spectrum must be separated into its NC, ν_μ CC and beam ν_e components.

The horn-on/horn-off (HOO) method utilizes the data taken in the horn-off configuration to carry out the Near Detector decomposition. When the horns in the NuMI beam are turned off, the mesons produced in the target are no longer focused. When that happens, only the neutrinos originating in the decays of the

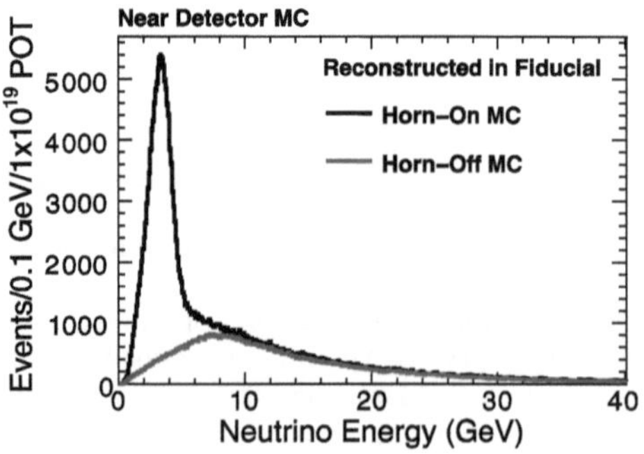

Fig. 6.9 Reconstructed energy spectra for neutrinos in the Near Detector fiducial volume, in the horn-on and horn-off configurations. When the horns are turned off, the main focusing peak of the spectrum disappears and only the high-energy tail persists

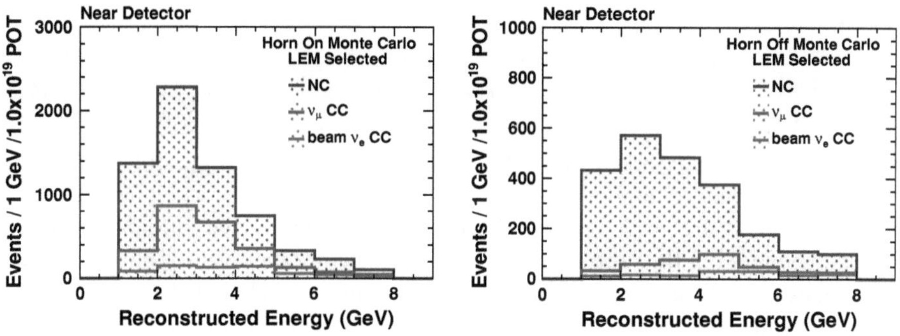

Fig. 6.10 Reconstructed energy spectra of LEM selected events (with a LEM pid cut at 0.65) in the horn-on and horn-off configurations, as simulated in the Near Detector MC. Applying the LEM selection to the data taken in the horn-off configuration results in a NC-enriched sample

mesons that go straight through the center of the horns remain, giving the high-energy tail of the spectrum. This is shown in Fig. 6.9.

When the ν_e selection is applied to the horn-off data, the events that originate from the main focusing peak are lost. This includes some NC events, but also most of the ν_μ CC and beam ν_e events, thus yielding a sample dominated by the NC events that leak from the high-energy tail. Figure 6.10 shows the energy spectra of selected events in the horn-on and horn-off configurations as given by the simulation. The proportion of NC to ν_μ CC events is very different in the two configurations, with the horn-off data having a much larger share of NC events in general.

The HOO method utilizes this large difference in background composition between the horn-on and horn-off samples, in addition to the fact that the ratios of events between the two configurations is well known, to decompose the Near Detector spectrum.

6.3.2 Description of the Method

The number of events selected in the horn-on configuration N^{on} is the sum of the NC, ν_μ CC and beam ν_e components:

$$N^{\text{on}} = N^{\text{on}}_{\text{NC}} + N^{\text{on}}_{\text{CC}} + N^{\text{on}}_{\text{e}}. \tag{6.5}$$

Similarly, we can express the number of events selected in the horn-off configuration N^{off} as the sum of its components:

$$\begin{aligned} N^{\text{off}} &= N^{\text{off}}_{\text{NC}} + N^{\text{off}}_{\text{CC}} + N^{\text{off}}_{\text{e}} \\ &= r_{\text{NC}} N^{\text{on}}_{\text{NC}} + r_{\text{CC}} N^{\text{on}}_{\text{CC}} + r_{\text{e}} N^{\text{on}}_{\text{e}}, \end{aligned} \tag{6.6}$$

where

$$r_i = \frac{N_i^{\text{off}}}{N_i^{\text{on}}} \tag{6.7}$$

is the ratio of events between the horn-off and horn-on configurations for a particular background component i.

The r_i ratios are well modeled, as the uncertainties in the simulation largely cancel when taking the ratio of events. This is certainly the case for physics related systematics (such as uncertainties in the hadronic model, intranuke, and cross-sections), but also for some detector specific systematics (such as cross-talk modeling), given that the two samples are obtained in the same detector. Energy-dependent uncertainties are an example of systematics that do not cancel between the two detectors, given that the energy spectra of selected events are very different in the two configurations. This effect can be mitigated however by performing the separation in bins of 0.5 GeV.

Furthermore, we saw that the beam ν_e component is reasonably well modeled in the MC. If the r_i ratios and the number of selected beam ν_e events $N_e^{\text{on/off}}$ are obtained from the simulation, Eqs. 6.5 and 6.6 constitute a system of two equations with two unknowns: $N_{\text{NC}}^{\text{on}}$ and $N_{\text{CC}}^{\text{on}}$. Solving this system at every bin of reconstructed energy is equivalent to decomposing the Near Detector spectrum.

6.3.3 Uncertainties in the Decomposition

The error in the HOO decomposition stems from three sources:

- the statistical error in the samples used,
- the error in the number of selected beam ν_e events,
- the error in the r_i ratios.

The amount of MC and the amount of horn-on data employed[5] are far greater than the amount of horn-off data that is available for the analysis, which is 5.52×10^{18} POT. Because of this, the statistical error in the decomposition is dominated by the amount of data taken in the horn-off configuration.

The error in the predicted number of beam ν_e events, when integrated over all energies, is approximately 20% in the Near Detector (see Sect. 8.1.4). For the Near Detector decomposition however, rather than performing a detailed estimate of the beam ν_e systematic error as a function of energy, a flat systematic error of 30% is assumed at each energy bin. This is certainly a conservative approach, but one that

---

5 The horn-on data used corresponds to 4.5×10^{20} POT of exposure, the horn-on MC to 1.8×10^{20} POT and the horn-off MC to 3.3×10^{19} POT.

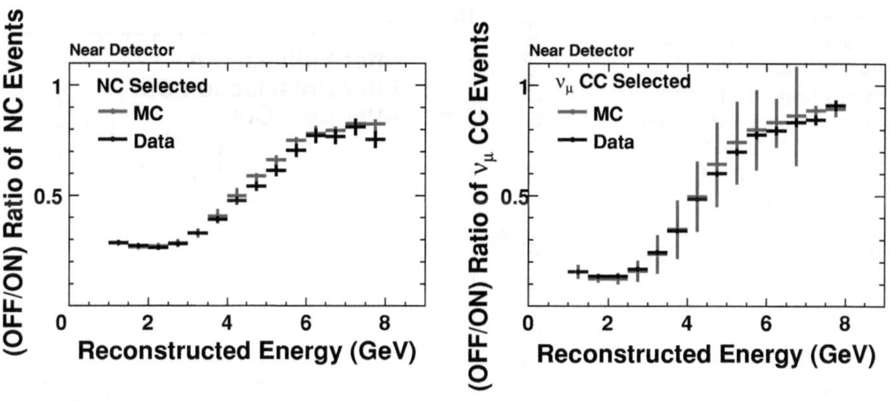

Fig. 6.11 $r_{\rm NC}$ (*left*) and $r_{\rm CC}$ (*right*) after fiducial volume cuts. The LEM cut value used here is 0.65. Here a ν_μ CC event is an event with a clearly defined track, and a NC event is an event with just a shower. The ratios are well modeled in the simulation. The error shown for the MC is the sum of the statistical and systematic errors

bears almost no impact on the resulting decomposition uncertainty given the small size of the beam ν_e component with respect to the total background.

Ideally the systematic uncertainty in the ratios r_i would be assessed through a direct comparison between data and MC. This is evidently not possible after applying the ν_e selection, since the problem we are precisely trying to solve is that we cannot identify the interaction type of selected events in the data. Nevertheless, a direct data-MC comparison can be done for ν_μ CC-like or NC-like events, where an ν_μ CC-like event is an event with a well-distinguished track, and NC-like events are all others. As shown in Fig. 6.11 there is very good agreement between the data and the MC, particularly in the region between 2 and 4 GeV, which is where most of the background events are.

Figure 6.12 shows how the ratios do not change much after the application of the different cuts, with the exception of $r_{\rm CC}$. The big change in $r_{\rm CC}$ that occurs when applying the preselection cuts is well understood. The preselection is particularly effective at removing obvious ν_μ CC events with well-defined tracks, as explained in Sect. 5.2. In the absence of the focusing peak, the horn-off sample is enhanced with higher energy (and consequently longer) muons. Therefore, when the preselection is applied, proportionately more ν_μ CC events are removed in the horn-off sample compared to the horn-on sample, causing $r_{\rm CC}$ to diminish.

Since the ratios are properly reproduced in the MC at the fiducial level (Fig. 6.11) and they do not change very significantly when the successive cuts are applied (Fig. 6.12), we can be confident that they are adequately modeled at the selection level. Changes in the ratios after the application of the successive cuts, as well as small discrepancies between the data and the MC, are incorporated as systematic errors. In particular, when the difference in the ratios between

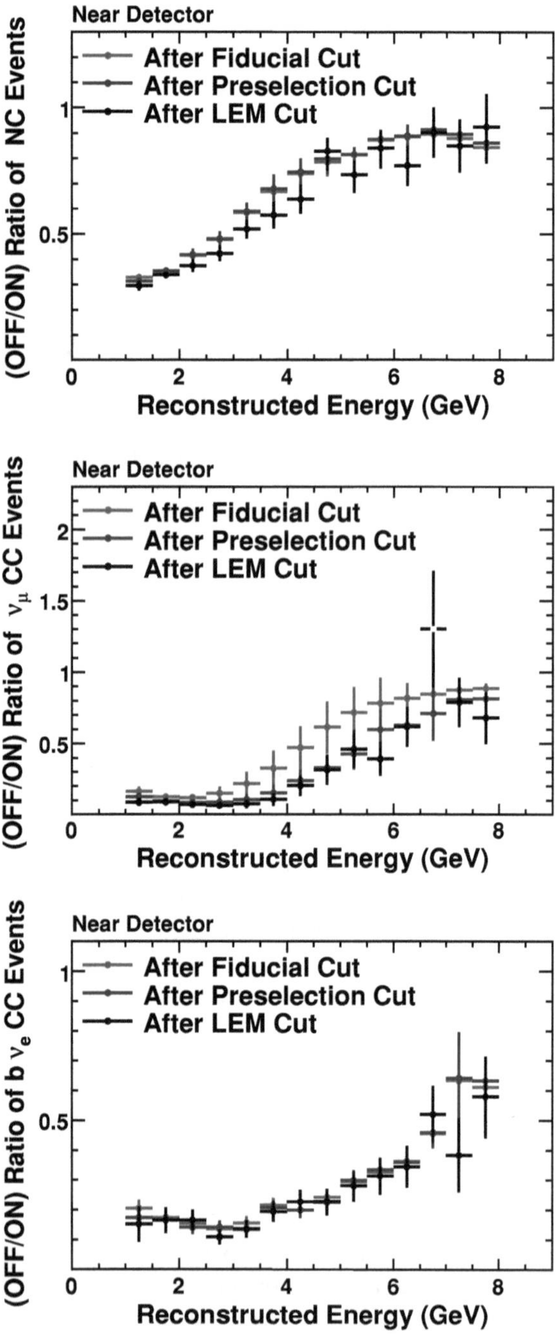

Fig. 6.12 r_{NC} (*top*), r_{CC} (*middle*) and r_{e} (*bottom*), for true NC, ν_μ CC and beam ν_e events respectively. The ratios are obtained from the MC after the application of different cuts. The LEM cut value used to make these plots is 0.65. The ratios shown here at the fiducial level are very similar but not exactly the same as the ones from Fig. 6.10, given that the ν_μ CC-like and NC-like samples do not correspond exactly to the true NC and true ν_μ CC samples used here. With the exception of r_{CC}, the ratios do not change very significantly after the application of the different cuts

successive cuts does not appear to be statistical only, an additional systematic error is added until $\chi^2/\text{dof} \leq 1$. The same procedure is done when the difference between the data and the MC cannot be explained by statistics alone, the only difference being that the error implemented in this case is asymmetric given that the sign of the data-MC difference is known. All the technical details on the estimation of the decomposition's systematic errors can be found in [2].

This approach to estimate the systematic errors is certainly conservative. This is particularly so in the case of r_{CC}, for which the change observed after the application of the preselection cuts is well understood and expected in the data. However, our studies show that a better treatment of the error on r_{CC} makes a negligible difference in the uncertainty of the final Far Detector prediction [2, 3].

Table 6.3 Near Detector decomposition as obtained from the HOO decomposition method, for an exposure of 1×10^{19} POT. The errors quoted for the MC values are all statistical. The total quoted for HOO row is the total number of selected events in the Near Detector data, by construction, and the error quoted in that case is just statistical. For the NC and ν_μ CC components the error is the one resulting from the HOO decomposition. The beam ν_e component is obtained from the simulation and the error quoted in the HOO row corresponds to 30% of the expectation

		Total	NC	ν_μ CC	Beam ν_e
LEM > 0.65	MC	6432 ± 20	4017 ± 16	1825 ± 10	590 ± 6
	HOO	3528 ± 28	2073^{+260}_{-258}	865^{+351}_{-216}	590 ± 177
	1-HOO/MC	45.1%	48.4%	52.6%	–
LEM > 0.80	MC	4073 ± 16	2520 ± 13	1089 ± 8	464 ± 5
	HOO	2068 ± 21	1021^{+213}_{-205}	582^{+284}_{-176}	464 ± 139
	1-HOO/MC	49.2%	59.5%	46.6%	–

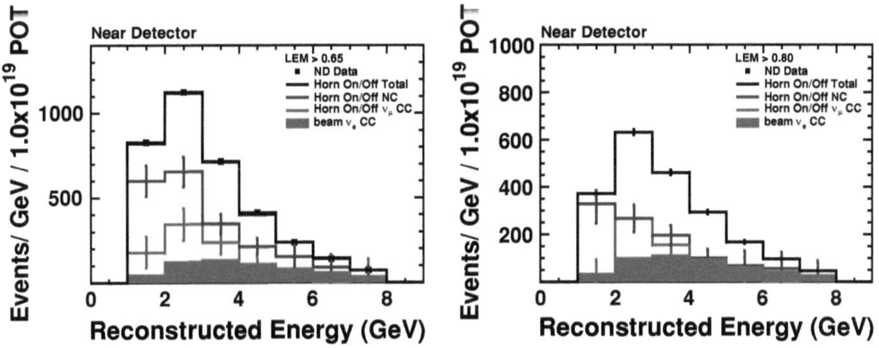

Fig. 6.13 Energy spectra of LEM selected events decomposed by HOO with a cut at 0.65 (*left*) and at 0.80 (*right*)

6.3.4 Results

The results of the HOO decomposition in the Near Detector can be seen in Table 6.3, and they are also illustrated in Fig. 6.13. HOO predicts the NC and ν_μ CC components respectively to 12% and 40% in the case of the cut at 0.65, and to 21% and 49% for the cut at 0.80. The uncertainty on the ν_μ CC component is significantly larger because of the greater systematic error on the r_{CC} ratio, as explained in the previous section. Also, the error in the separation grows with increasing cut value. This is because as the number of selected events diminishes, the impact of the statistical error in the horn-off data increases. The correction that HOO does on the MC is in the order of 50% for both cut values, although slightly larger for the cut at 0.80. In both cases however, the correction applied on the NC and ν_μ CC components is similar, which was expected and is one of the assumptions in the PORP decomposition method explained in Sect. 5.4.3. It is also interesting to note that the ν_μ CC component as determined by HOO is comparable in size to the beam ν_e component, especially for the cut at 0.80.

6.4 The Muon-Removal Decomposition Method

6.4.1 Basic Principle

Another method used to decompose the observed Near Detector energy spectrum is the "Muon-Removal in Charged Current events" (MRCC) method [4]. In the MRCC technique, ν_μ CC events with well-defined tracks are identified and the track removed. The hadronic shower that remains in the event is then processed through the reconstruction. This procedure is illustrated in Fig. 6.14. MRCC events constitute an almost pure sample of hadronic showers which can be studied separately from the standard sample and which can be used to obtain a data-based correction to the MC.

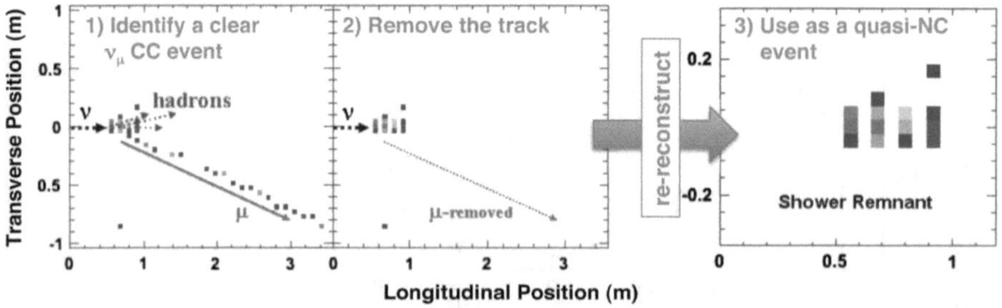

Fig. 6.14 Graphical illustration of the MRCC technique. MRCC events constitute a pure sample of quasi-NC events, and can thus be used to provide a data-based correction to the NC background component from the simulation or as an independent sideband

6.4.2 Obtaining a Sample of Muon-Removed Events

6.4.2.1 The Muon-Removal Algorithm

The MRCC algorithm operates on the same reconstructed events that are used for the rest of the analysis. The goal of the algorithm is to remove the muon's contribution to the event's energy deposition pattern without interfering with the hadronic shower contribution. Since the algorithm needs to be applied on both data and MC, it cannot rely on the truth information to remove the muon, and it must rather rely on the information of the reconstructed track(s) in each event. Very frequently, however, both the shower and the track contribute to the same hits, particularly in the part of the track that is contained within the hadronic shower. Also, the reconstruction sometimes mistakenly assigns pure shower hits to the track.

Because of this, the MRCC algorithm does not directly remove every hit belonging to the primary reconstructed track[6] of each event. Track hits are either removed, kept or scaled according to the following criteria:

- If the energy deposition is consistent with the amount deposited by muons (between 0.3 and 1.2 MEU),[7] the hit is removed.
- If the energy deposition is larger 1.2 MEU and is less than 80% of the total energy deposited in that plane, the hit is considered to be part of the shower and is kept.
- If the energy deposition is larger than what is typically deposited by muons ($>$ 1.2 MEU) and is more than 80% of the total energy deposited in that plane, the energy deposition of an average muon (one MEU) is effectively deducted from it.

In all cases the energy deposition per hit is calculated by taking into account the slope of the track, given that steeper tracks deposit more energy per plane. Once the muon hits have been removed, the hadronic shower remnant is processed through the reconstruction. This is an important step, as it ensures that any reconstruction biases present for NC-like events will be reproduced in the MRCC events. An example of a ν_μ CC event and its corresponding MRCC event in the Far Detector data can be seen in Fig. 6.15. This particular MRCC event has a very EM-like topology and is selected by LEM as a ν_e candidate.

6.4.2.2 Ensuring the Quality of Muon-Removed Events

Whenever a MRCC event is obtained, the information of the original event is kept. The purpose is twofold:

---

[6] The primary track in an event is the longest one. If an event has no track then the MRCC algorithm skips it.

[7] MEU stands for Muon Equivalent Unit. In this thesis we also refer to this quantity as a Minimum Ionizing Particle (MIP).

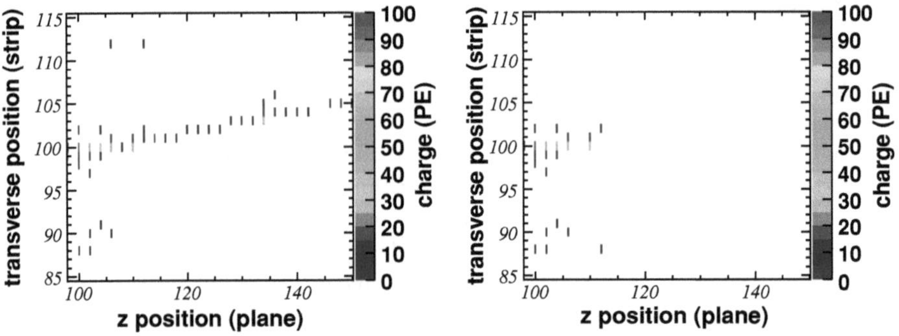

Fig. 6.15 An event in the Far Detector data before (*left*) and after (*right*) muon-removal. Due to the EM nature of the hadronic shower remnant, this MRCC event is selected by LEM as a ν_e candidate. The clusters of hits below and above the main hadronic shower are due to cross-talk, and are entirely ignored by the LEM algorithm due to the 3.0 PE cut that is applied before events are compared to the library. Sometimes hits that occur in the same snarl (defined in Sect. 3.4.3) and that are not included in the original event end up being included in the muon-removed event. An example of this is the rightmost hit in the lower cross-talk cluster. Also, hits that were present in the original shower remnant can sometimes be lost in the MRCC event, such as the two cross-talk hits in the top

- It allows for the application of quality cuts to ensure that a pure sample of authentic MRCC events is used in our studies. An authentic MRCC is an event where a muon track was removed from a true ν_μ CC event as opposed to, for instance, a π^+ track removed from a NC event.
- It allows to check if the hadronic shower remnant is not split or contaminated during reconstruction. The example in Fig. 6.15 illustrates how hits can be added or lost during reconstruction.

The MRCC quality cuts consist of a fiducial volume cut on the original event's vertex, a check that the original track was successfully fit by the tracker, and the application of the standard cuts used in the main ν_μ CC disappearance analysis [5]. In this way, the parents of the MRCC events that are used for our studies are the same events used in the MINOS ν_μ CC disappearance analysis.

The integrity of the MRCC events after the reconstruction is ensured by looking at their purity and completeness. The completeness of an MRCC event is the pulse-height weighted fraction of the hits in the original shower remnant which are also in the final MRCC event, and thus gives a measure of how much charge is lost during the reconstruction. The purity of a MRCC event is the pulse-height weighted fraction of the hits in the MRCC event that are also in the original shower remnant, and thus gives a measure of how much charge was added during the reconstruction. Figure 6.16 shows the purity and the completeness of MRCC events in the Near Detector, at the fiducial and preselection levels. After the application of the quality cuts, 98% of the MRCC events are authentic (i.e., they have a ν_μ CC parent). When the preselection is also applied, the average MRCC event has a completeness of 99% and a purity of 96%.

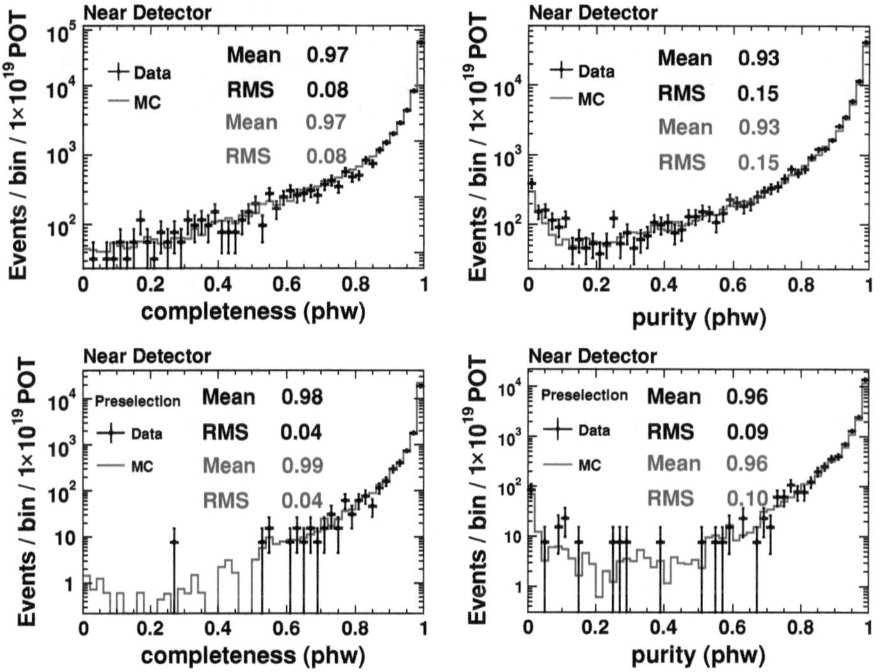

Fig. 6.16 Completeness and purity of MRCC events in the Near Detector, after fiducial cuts (*top row*) and after the preselection (*bottom row*). The MRCC quality cuts are applied. Most MRCC events have very high completeness and purity, especially those that pass the preselection, in which case completeness and purity are 98% and 96% respectively

6.4.2.3 Data-MC Discrepancy with Muon-Removed Events

As just mentioned, the same muon-removal procedure is applied both on the data and the MC. The same cuts utilized in the ν_e appearance search are then applied to the MRCC sample in order to find out how many ν_e candidates they contain.

Figure 6.17 shows the LEM pid distribution in the standard and the muon-removed (MR, or MRCC) samples, in both data and MC. There are significant differences in the LEM pid shape between the two samples, particularly in the high-PID, where the standard data shows an increase of ν_e-like events that is not reproduced in the MRCC sample. These differences are not unexpected, as the muon-removal process is not perfect and the standard sample includes a number of events that are not present in the standard sample (such as coherent π^0's or beam ν_e's). Despite these differences, the data-MC ratio is very similar between the standard and MRCC samples. The same can be said for the three variables that form the LEM pid, shown in Figure 6.18.

Figure 6.19 shows the reconstructed energy spectra of the standard and MRCC samples, for the LEM pid cut at 0.65 and at 0.80. Again, the ratio of data to MC is

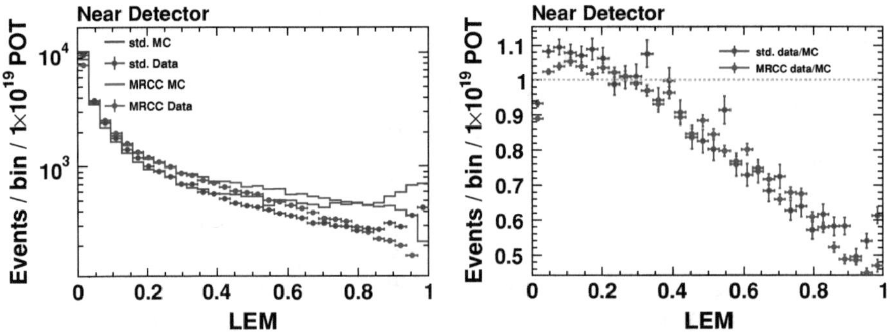

Fig. 6.17 LEM pid distribution in the standard and the MRCC samples (*left*), and the data/MC ratios for each sample (*right*). There are significant differences in the LEM distribution between the two samples (particularly in the high pid region), but the data-MC ratios are very similar

very similar in the two samples, particularly at low energies which is where most of the events are. The remarkable similarity in the data-MC disagreement between the standard and MRCC samples, not only in shower topology but also in reconstructed energy, strongly suggests that the two samples are affected by the same simulation uncertainties. In particular, since the basis for the MRCC sample is ν_μ CC events which are well understood and modeled in MINOS, this is strong confirmation that the hadronic shower simulation is the primary contributor to the data-MC disagreement.

6.4.3 Decomposing the Near Detector Spectrum

6.4.3.1 Description of the Method

The MRCC sample thus constitutes an independent, highly pure sample of quasi-NC events whose data-MC disagreement closely mimics the data-MC disagreement in the standard sample. As such, it can be used to correct the MC expectation for NC events in order to match the data. The correction is provided by the MRCC ratio of data to MC,

$$N_{\mathrm{NC}}^{pred} = N_{\mathrm{NC}}^{MC} \frac{N_{MRCC}^{data}}{N_{MRCC}^{MC}}, \tag{6.8}$$

where N_{NC}^{pred} is the number of NC events predicted by the MRCC method in the ND, N_{NC}^{MC} is the MC prediction, N_{MRCC}^{data} is the number of events selected in the MRCC data, and N_{MRCC}^{MC} is the data-MC ratio in the MRCC sample, shown on the right-hand side of Fig. 6.19.

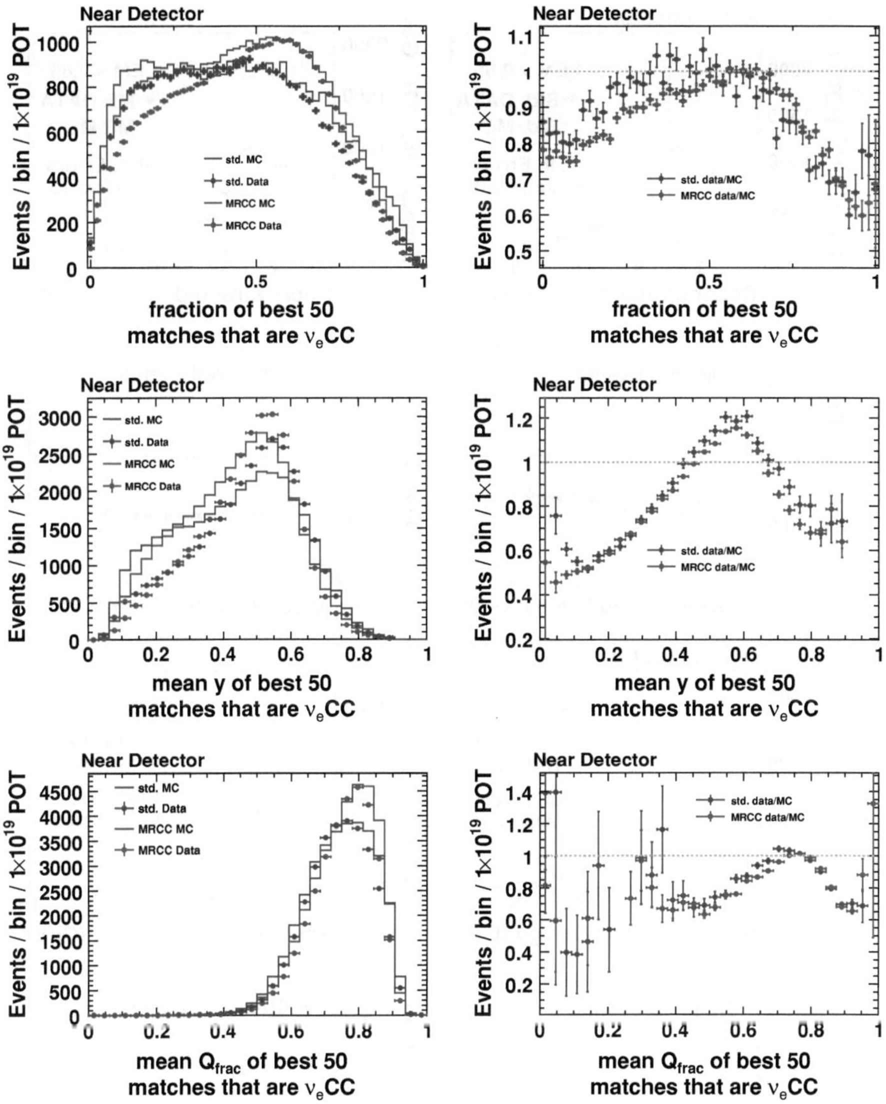

Fig. 6.18 The three variables that form the LEM pid in the standard and the MRCC samples, and the data/MC ratios for each sample (*right*). There are significant differences in the distributions between the two samples, but the data-MC ratios are similar

As in the HOO method, the beam ν_e component is obtained from the simulation:

$$N^{pred}_{beam\ \nu_e} = N^{MC}_{beam\ \nu_e} \tag{6.9}$$

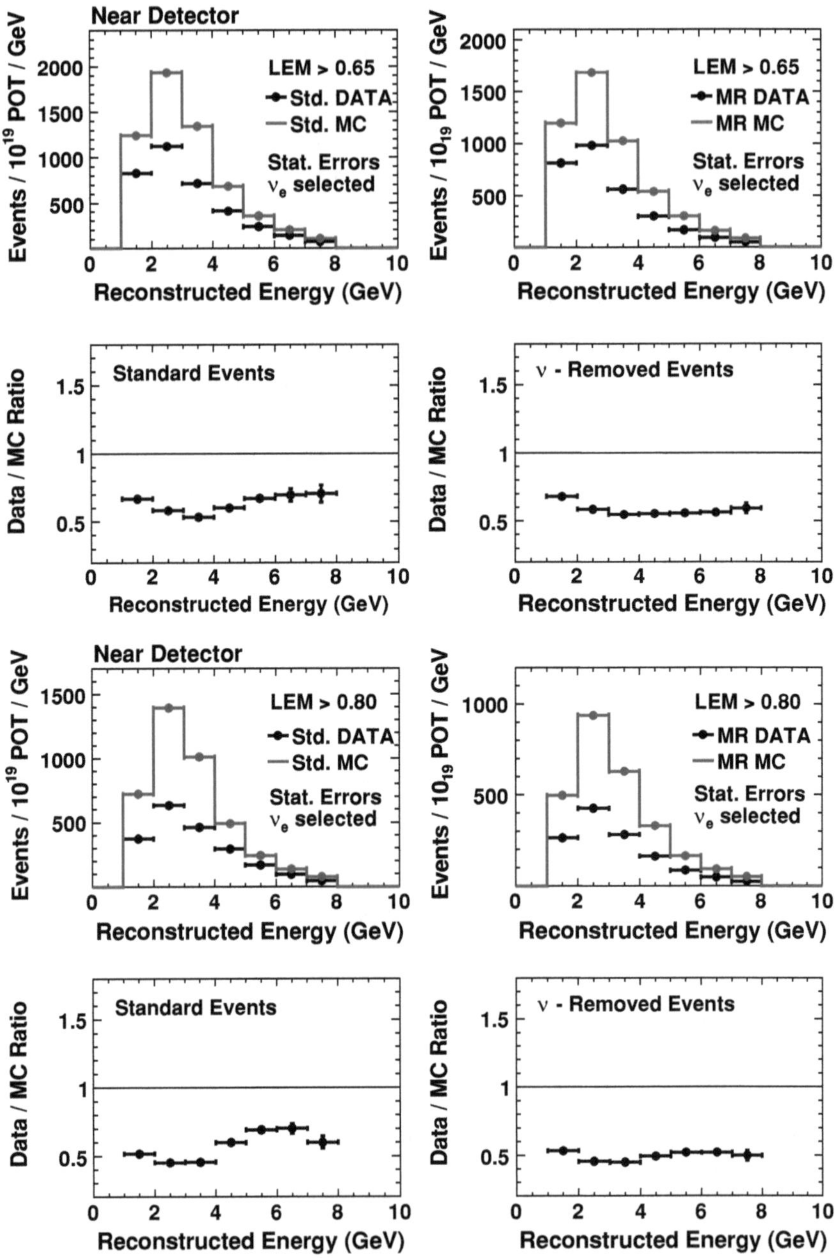

Fig. 6.19 Reconstructed energy spectra of LEM selected events with a cut at 0.65 (*first row*) and at 0.80 (*third row*), in the standard sample (*left*) and in the MRCC sample (*right*). The corresponding ratios of data to MC are also shown in the second and fourth rows for the cut at 0.65 and at 0.80 respectively. The data-MC disagreement is similar between the two samples as a function of reconstructed energy, for the two LEM cut values

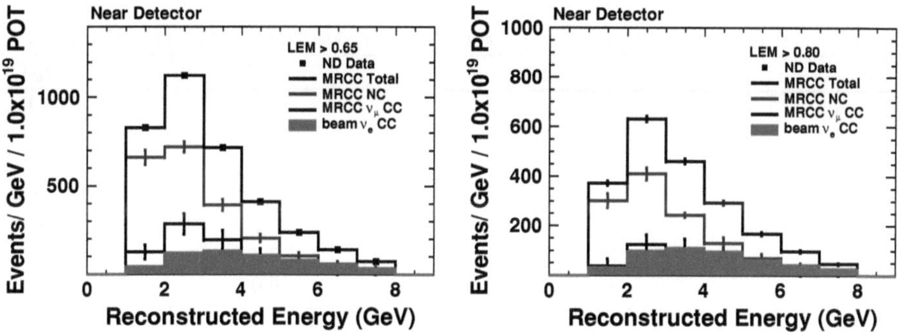

Fig. 6.20 Energy spectra of LEM selected events decomposed by MRCC with a cut at 0.65 (*left*) and at 0.80 (*right*)

The ν_μ CC component is obtained by requiring the sum of the components to equal the total number of events observed in the data N^{data}:

$$N^{pred}_{\nu_\mu\,CC} = N^{data} - N^{pred}_{\rm NC} - N^{pred}_{{\rm beam}\,\nu_e}. \tag{6.10}$$

Equations 6.8–6.10 are applied for every bin of reconstructed energy, of width 1 GeV.

6.4.3.2 Systematic Uncertainties

The sources of systematic error affecting the MRCC decomposition fall into two different categories:

- Category 1: uncertainties in the simulation.
- Category 2: differences between the MRCC and the standard samples.

The uncertainties in the simulation that affect the MRCC decomposition are those in the cross-section model, in the beam flux and in the hadronic shower energy scale. The error introduced by these uncertainties is addressed by varying each source of error within its uncertainty and redoing the decomposition. The change in the predicted background components is then taken as a measure of the systematic error. The systematic error contributions due to these uncertainties are shown in Table 6.4. The error introduced by the uncertainty in the beam ν_e component is also part of the category 1 systematics. As in the HOO method, the total error on the beam ν_e component is conservatively taken to be 30%, which has an impact of comparable size in the determination of the ν_μ CC component.

The systematic errors under category 2 involve all differences between the MRCC events and the standard NC events. Such differences can arise from imperfections in the muon-removal process, such as muon remnants left in the events or the removal of hits that are not really related to the muon track. It is

Table 6.4 Breakdown of systematic errors in the MRCC decomposition, for each background component. The values shown here correspond to the LEM selection with a cut value at 0.65

Type	Description	NC	ν_μ CC
Category 1	beam flux	0.5%	6.1%
	cross-sections	0.9%	8.3%
	shower energy	3.5%	10.8%
	beam ν_e	0.0%	21.7%
Category 2	shower charge	3.7%	10.3%
	MR vs. Standard	3.6%	9.9%
	Total	6.0%	30.3%

precisely to mitigate the impact of these imperfections that the NC component is corrected using a ratio of MRCC events, since in this way the biases in the muon-removal process cancel to first order.

The differences between the MRCC and the standard samples can also be physics related, given that the hadronic showers generated during ν_μ CC interactions are different from NC showers. In particular, NC showers originate in the exchange of a Z boson and ν_μ CC events in the exchange of a W^+ boson. This means that the net charge of NC showers is zero, while that of ν_μ CC events is $+1e$.[8] In order to assess this difference, the decomposition is done using $\overline{\nu}_\mu$ CC muon-removed events, and the results are compared to those obtained with ν_μ CC muon-removed events. Since antineutrino showers are mediated by the W^- boson and thus have a difference in net charge of $2e$ with respect to ν_μ CC showers, the "shower charge" systematic error is estimated by taking half of the difference in the decomposition as obtained with neutrinos and antineutrinos.

Unfortunately, there can be more sources of differences between the two samples. The kinematics of CC showers are simply different from those of NC showers, and it is possible that the ν_e selection is sensitive to these differences. The simulation indicates that the two samples are relatively close in terms of their kinematic distributions, but the simulation can only be trusted to a certain point. In order to estimate the impact of the remaining differences between the two samples, the double ratio of data to MC in the MRCC and the standard samples is used. The justification behind this procedure is that if the MRCC and the standard samples were identical, this double ratio would be unity.[9]

The decomposition uncertainties for the errors pertaining to category 2 are shown in Table 6.4. The fact that there is no direct and data-based approach for assessing the impact of the differences between the MRCC and the standard samples in the decomposition is the main reason why HOO is chosen as the primary decomposition method. Notwithstanding, the MRCC method constitutes a very useful cross-check and also provides, as will be shown later, an independent sideband on which to test the analysis.

[8] e denotes the electron charge.

[9] This is not strictly true, as there is a small amount of beam ν_e events in the background.

Table 6.5 Near Detector decomposition as obtained from the MRCC decomposition method, for an exposure of 1×10^{19} POT. The errors quoted for the MC values are all statistical. The total quoted in the MRCC row is the total number of selected events in the Near Detector data, and the error quoted in that case is just statistical. For the NC and ν_μ CC components the error is the one resulting from the MRCC decomposition. The beam ν_e component is obtained from the simulation and the error quoted in the MRCC row corresponds to 30% of the expectation

		Total	NC	ν_μ CC	Beam ν_e
LEM > 0.65	MC	5860 ± 57	3640 ± 45	1650 ± 30	570 ± 18
	MRCC	3528 ± 28	2170 ± 135	789 ± 238	570 ± 172
	1-MRCC/MC	40.0%	40.3%	52.2%	–
LEM > 0.80	MC	4073 ± 16	2520 ± 13	1089 ± 8	464 ± 5
	MRCC	2068 ± 21	1210 ± 106	394 ± 189	464 ± 140
	1-MRCC/MC	49.2%	52.0%	63.8%	–

6.4.3.3 Results

The results of the MRCC decomposition in the Near Detector can be seen in Table 6.5, and they are also illustrated in Fig. 6.20. MRCC predicts the NC and ν_μ CC components respectively to be 6% and 30% in the case of the cut at 0.65, and to 8% and 48% for the cut at 0.80. As in the HOO method, the error in the MRCC decomposition is higher for the tighter cut. It is also interesting to note that in the MRCC decomposition the correction that is applied to the ν_μ CC component is consistently ∼10% higher than the one for the NC component, whereas this is not the case in the HOO method.

6.5 Comparison of Near Detector Decomposition Results

Given that the HOO and MRCC methods are independent from each other, it is instructive to compare their results. This comparison is done graphically in

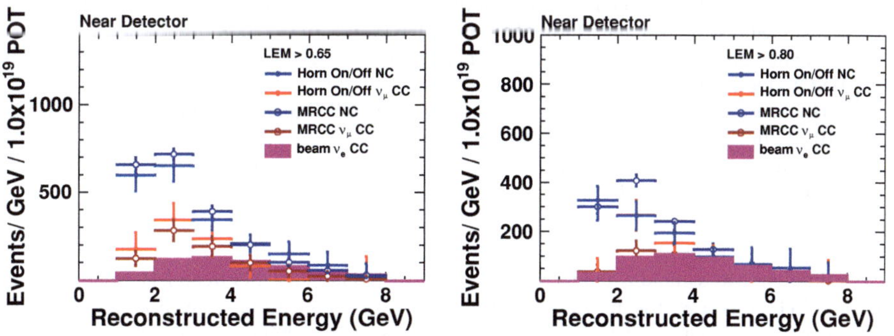

Fig. 6.21 Background decomposition of LEM selected events as given by HOO and by MRCC with a cut at 0.65 (*left*) and at 0.80 (*right*). The two decomposition methods are consistent to each other, and the differences are within 1σ for most bins of reconstructed energy

Table 6.6 Near Detector decomposition as obtained from the HOO, MRCC, and PORP decomposition methods, for an exposure of 1×10^{19} POT. The totals are by construction equal to the number of events observed in the data, and the error shown in that case is just statistical. For the NC and ν_μ CC components the error is the one resulting from the decomposition, except for PORP for which no systematic errors are calculated. The beam ν_e component is obtained from the simulation and a total uncertainty corresponding to 30% of the expectation is imputed

		Total	NC	ν_μ CC	Beam ν_e
LEM > 0.65	HOO	3528 ± 28	2073^{+260}_{-258}	865^{+351}_{-216}	590 ± 177
	MRCC	3528 ± 28	2170 ± 135	789 ± 238	570 ± 172
	PORP	3528 ± 28	2060	898	570
LEM > 0.80	HOO	2068 ± 21	1021^{+213}_{-205}	582^{+284}_{-176}	464 ± 139
	MRCC	2068 ± 21	1210 ± 106	394 ± 189	464 ± 140
	PORP	2068 ± 21	1153	479	436

Fig. 6.21. Also, a compilation of the results from each method is shown in Table 6.6. The MRCC and HOO results are consistent with each other, and in most energy bins the differences are well within 1σ.

Also shown in Table 6.6 are the results from the PORP method, described in Sect. 5.4.3. No systematics are determined for the PORP decomposition, given that this method is not used for the final Far Detector predictions but only for optimization and other studies where a prediction needs to be obtained for each cut value. Nevertheless, it can be seen that the PORP results are completely consistent[10] with those of MRCC and HOO.

These results constitute the basis for the next chapter, where the Far Detector background predictions are obtained.

References

1. P. Adamson et al., A study of muon neutrino disappearance using the fermilab main injector neutrino beam. Phys. Rev. **D77**, 072002 (2008)
2. G. Pawloski, L. Whitehead, Estimation of the neat detector NC and CC spectra for the $\nu_\mu \to \nu_e$ Analysis with Horn-On and Horn-Off Data, MINOS-doc-3935, 2008
3. R.B. Patterson, J.P. Ochoa-Ricoux, The impact of background composition on nue sensitivity, MINOS-doc-3760, 2007
4. A. Holin, Electron Neutrino Appearance in the MINOS Experiment. FERMILAB-THESIS-2010-41. In: Ph.D. Thesis University College London
5. P. Adamson et al., Measurement of neutrino oscillations with the MINOS detectors in the NuMI beam. Phys. Rev. Lett. **101**, 131802 (2008)

[10] A different MC sample with less statistics and which makes a better treatment of non-linearity in the PMTs was used for the LEM > 0.65 MRCC decomposition, as well as for the PORP decomposition. Because of this the values of the beam ν_e component do not match exactly between each row. It is also because of this that the raw MC values in Table 6.5 do not correspond exactly to those in Table 6.3. The impact of using a different MC sample in the actual decomposition is negligible however, as the methods are data based and the differences are well within the total errors.

Chapter 7
Predicting the Far Detector Rates

Now that the Near Detector background rates have been measured it is necessary to convert them into Far Detector rates. The extraction of the Far-Near conversions from the MC used to predict (extrapolate) the NC and ν_μ CC components, as well as the details on the prediction of the beam ν_e and ν_τ CC components, are shown in Sect. 7.1. In our analysis it is not only necessary to predict the backgrounds but also the amount of signal that would be observed in the case of a nonzero θ_{13}. Section 7.2 shows how the MC portrays the signal well and how, after applying a small correction due to the mismodeling of hadronic showers, it can be used to obtain the ν_e CC selection efficiency at the Far Detector. Section 7.3 includes a summary of all the predictions.

7.1 Far Detector Background Rates

7.1.1 Flux Differences between the Two Detectors

In the NuMI beamline the hadrons produced at the target are focused in the direction of the two detectors. Given that the neutrinos are not emitted colinearly with their parent hadrons and that there is no way to affect them after they have been produced, the neutrino beam inevitably widens as it travels under the earth from Illinois to Minnesota. This yields a difference in neutrino flux in the order of 10^6 between the two detectors, as shown in Fig. 7.1. This number can be reproduced in a first principles calculation if we assume that the beam flux density diminishes with the square of the distance to the source, as expected in the case of isotropic emission (in the rest frame of the neutrino parent). Knowing that the average neutrino is produced at a distance of 0.84 km upstream of the Near Detector and that the Near-Far distance is 735 km, an expected reduction in the

J. P. Ochoa Ricoux, *A Search for Muon Neutrino to Electron Neutrino Oscillations in the MINOS Experiment*, Springer Theses, DOI: 10.1007/978-1-4419-7949-0_7,

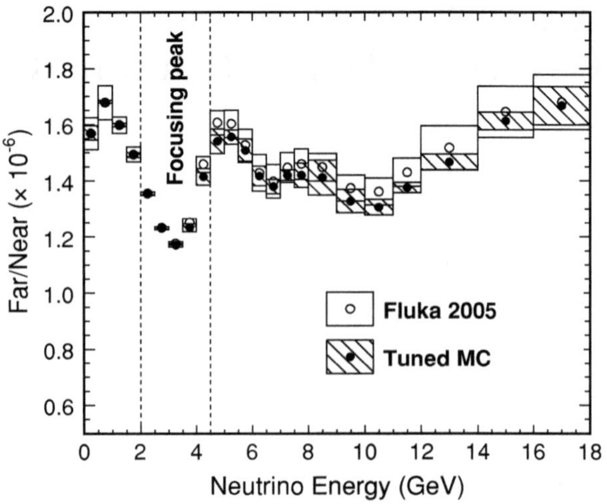

Fig. 7.1 ν_μ CC flux ratio between the Far and Near detectors in the absence of oscillations, from [1]. The ratio is shown before and after the beam fit described in Sect. 3.4.2. The fact that the ratio is not strongly affected by the beam fitting shows that the uncertainity in the flux difference between the two detectors is small.

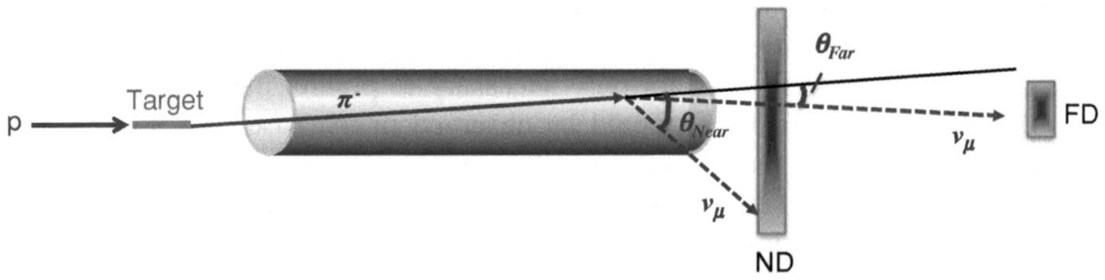

Fig. 7.2 Schematic of the NuMI beamline and the two detectors. Although the Far Detector is in reality larger than the Near Detector, here we display it with a smaller size in order to illustrate how, due to the widening of the beam as it travels, the two detectors subtend different solid angle regions. The Near Detector sees an extended neutrino source (the decay pipe), while the Far Detector essentially sees the source as a point. Also, when neutrinos are created at a finite distance from the beam axis the decay angles θ_{Near} and θ_{Far} required for the respective interception of the Near and Far detectors are different

flux between the Far and Near detectors of 1.3×10^{-6} is obtained, which is remarkably close to the simulation's prediction.

The Far/Near differences in ν_μ CC flux are not limited to a constant scale factor, but also display an energy dependence. As inferred from Fig. 7.1, the shapes of the Far and Near energy spectra differ by up to 30%. This energy dependence originates primarily from the fact that the neutrinos are produced with some transverse momentum from the decay of the parent particle. Consequently, the Near and Far detectors sample different decay angle distributions of the neutrino parents, as illustrated in Fig. 7.2. For the two-body decay of relativistic mesons $\pi/K \to \mu\nu$, the neutrino energy is a function of the decay angle, as given by

$$E_\nu \approx \frac{(1 - m_\mu^2/M^2)E}{1 + \gamma^2 \tan^2\theta_\nu}, \tag{7.1}$$

where m_μ and M are the muon and parent hadron masses, E is the parent hadron energy, $\gamma = E/M$ is the parent's Lorentz boost, and θ_ν is the angle (in the lab) between the neutrino and parent hadron directions. Due to its proximity to the beam source, the Near Detector is in general able to subtend larger decay angles than the Far Detector, particularly for those neutrinos that are produced near the end of the decay pipe. Given the small size of the decay pipe when compared to the Far-Near separation of 735 km, the neutrino source looks like a point to the Far Detector, for which the solid angle acceptance is essentially constant and significantly smaller than the one of the Near Detector. Also, when a neutrino is produced at a finite radius from the beam axis, different angular ranges are covered if the neutrino is to intercept the Near and Far detectors. The fact that the neutrinos intercepting the detectors originate from different decay angle distributions generate, according to Eq. (7.1), the differences in the energy spectra.

The main differences in flux between the two detectors are thus ultimately due to decay kinematics, focusing, angular acceptance, and beam geometry. These factors are well modeled in the MINOS simulation and are incorporated in the beam fit described in Sect. 3.4.2. As shown in Fig. 7.1, the Far/Near flux ratio changes very little when the beam fit is applied, which indicates that the uncertainty on the predicted flux ratio is small.

7.1.2 Choice of Extrapolation Method

While the differences in neutrino flux are the most significant factors, there are two more factors that affect the relative event rate between the detectors:

- Oscillations,
- Detector-specific differences.

Assuming that no oscillations to sterile neutrinos occur, NC events are unaffected by oscillations. This is not the case for ν_μ CC events,[1] of which 40% of the events below 5 GeV disappear. Also, while the detectors are functionally identical in terms of the environment where the neutrino interactions occur (strip size and steel width), there are differences in their readout, in neutrino intensity and in their energy calibration, which result in small disparities in their selection efficiencies. Their fiducial masses are also significantly different.

All of these differences must be taken into account when extrapolating (predicting) the Far Detector spectrum from the Near Detector spectrum. We extract all of these corrections from the full MINOS simulation, in the form of the

[1] Beam ν_e events are also affected by oscillations, although to a small degree. Using the same terminology as in Sect. 4.1.2, we have that $P(\nu_e \rightarrow \nu_e) = 1 - \alpha^2 \sin^2 2\theta_{12} \frac{\sin^2 A\Delta}{A^2} - 4\sin^2(\theta_{13}) \frac{\sin^2(A-1)\Delta}{(A-1)^2}$ [2]. This means that, for a value of θ_{13} at the CHOOZ limit, $< 6\%$ of beam ν_e's disappear.

Far-Near ratio of selected MC events $\mathrm{FD}^{\mathrm{MC}}/\mathrm{ND}^{\mathrm{MC}}$. The observed Near Detector rate of selected events is multiplied by the expected Far-Near ratio for every background component i at every bin of reconstructed energy E_{reco}:

$$\mathrm{FD}_i^{\mathrm{pred}}(\mathrm{E}_{\mathrm{reco}}) = \mathrm{ND}_i^{\mathrm{data}}(\mathrm{E}_{\mathrm{reco}}) \frac{\mathrm{FD}_{\mathrm{i}}^{\mathrm{MC}}(\mathrm{E}_{\mathrm{reco}})}{\mathrm{ND}_i^{\mathrm{MC}}(\mathrm{E}_{\mathrm{reco}})}. \tag{7.2}$$

In this way, the backgrounds in the Far Detector are actually based on measurements in the Near Detector. The simulation is only used to provide the necessary corrections in event rates between the detectors, which are very robust against modeling uncertainties. Section 6.1.2 showed how the largest systematic errors in the simulation (such as the one in the hadronic model) cancel to first order when taking the Far-Near ratio. The small residual uncertainty in the extrapolation is assessed later in this chapter.

In reality the events in a particular true energy bin at the Near Detector originate from decays that belong to different energy bins at the Far Detector, as illustrated in Fig. 7.3. This indicates that a better treatment of the relationship between the Far and Near energy spectra can be obtained with a two-dimensional matrix rather than with the one-dimensional ratio of Eq. (7.2). The "beam-matrix" extrapolation method, described in [1], follows this two-dimensional approach in the context of the ν_μ CC disappearance analysis.

In the ν_μ disappearance analysis the Far Detector unoscillated energy spectrum is compared with the Far Detector data in order to extract the oscillation parameters. Since it is precisely the location of the first oscillation minimum and its depth that yield Δm^2_{32} and $\sin^2 2\theta_{23}$ respectively, it is important for the shape of the unoscillated energy spectrum to be as accurate as possible. For the ν_e appearance analysis described in this thesis however, studies [3] show that there is practically no gain in sensitivity to θ_{13} when an energy-dependent fit is done to the observed

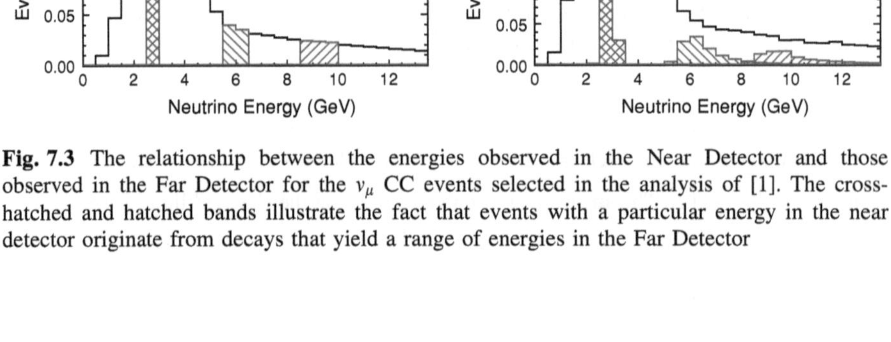

Fig. 7.3 The relationship between the energies observed in the Near Detector and those observed in the Far Detector for the ν_μ CC events selected in the analysis of [1]. The cross-hatched and hatched bands illustrate the fact that events with a particular energy in the near detector originate from decays that yield a range of energies in the Far Detector

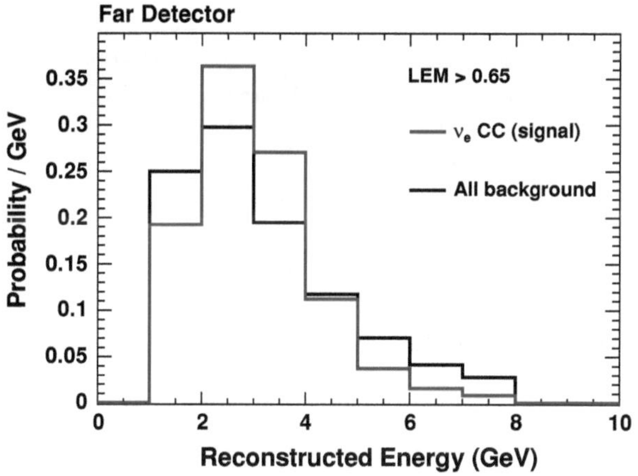

Fig. 7.4 Predicted background and expected signal in the Far Detector for LEM > 0.65, normalized to unity. The background is predicted using the PORP method (described in Sect. 5.4.3), which yields a total background of 21.6 events spread over the $1-8$ GeV energy range as shown by the black distribution. The shapes of the background and the potential signal are too similar, and the statistics too low, for an energy-dependent fit to improve the sensitivity obtained when the total number of events observed in the data is compared to the total background expectation integrated over all energies

Far Detector energy spectrum in order to separate the background from a potential signal. This is due to the small statistics involved in the analysis (which uses 3.14×10^{20} POT), as well as the fact that the signal and background energy spectra are similar in shape, as shown in Fig. 7.4. Because of this, the bounds on θ_{13} derived in our analysis are obtained through what is commonly referred to as a counting experiment, where the total number of selected events in the data is compared with the total predicted background integrated over all energies (1–8 GeV). For our purposes then, the extrapolation method of Eq. (7.2) is completely acceptable and presents the significant advantage that it is more straightforward in its application, which significantly simplifies the task of systematic error evaluation.

7.1.2.1 The Mechanics of the Extrapolation

The extrapolation, as described by Eq. (7.2), is carried out in 1 GeV bins. Even if only the total number of predicted background events integrated over the entire energy range is involved in the final derivation of the θ_{13} limits, the Far-Near ratios are energy-dependent as shown below. Also, it is useful to have a prediction for every energy bin in order to ensure that the observation has an energy distribution that is consistent with what is expected for the background and the potential signal (for instance, if an excess is observed, it is important to see that it is not coming one energy bin alone).

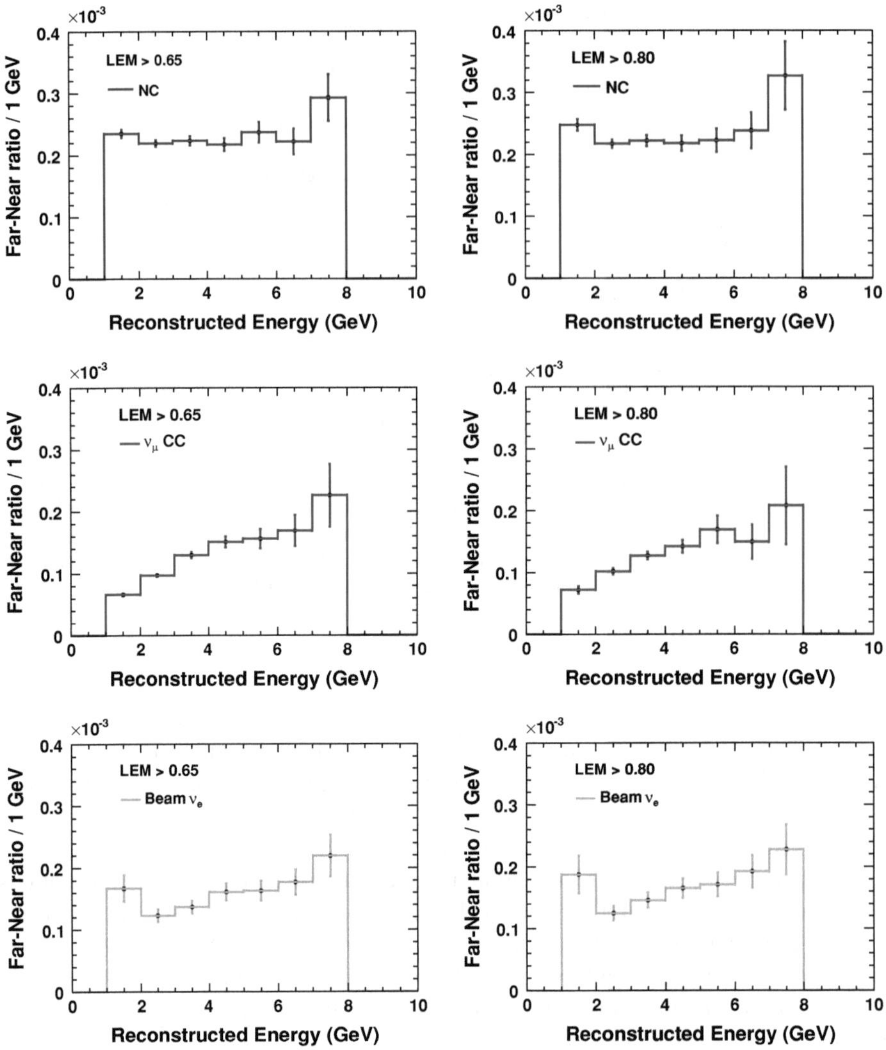

Fig. 7.5 Simulated Far-Near ratios from a LEM cut at 0.65 (*left*) and at 0.80 (*right*), for NC (*top*), ν_μ CC (*middle*) and beam ν_e (*bottom*) events. Oscillations are included with standard oscillation parameters $\Delta m^2_{32} = 2.4 \times 10^{-3}\text{eV}^2$ and $\theta_{23} = \pi/4$. The error shown is statistical only. The ratios do not change when the cut value is tightened, but are significantly different depending on the event type considered

The Far-Near ratios are shown in Fig. 7.5 for the LEM cuts at 0.65 and at 0.80. These ratios encapsulate our knowledge of meson decay kinematics, beamline geometry, angular acceptance, oscillations and detector differences. As such, they do not vary significantly between cut values. All the physics effects that have a strong effect on the selection and that do depend on the cut value (such as the uncertainties in the hadronic model) cancel to first order in the ratio.

The ratios do change significantly depending on the event type considered. The Far-Near ratio for NC events is mostly flat, whereas the ratios for ν_μ CC and beam ν_e events are not. This is the main reason why the extrapolation must be done for each component separately and the Near Detector spectrum decomposed. In the case of ν_μ CC events the drop at low energies is caused by oscillations. As a matter of fact, if it were not for ν_μ CC disappearance, the ν_μ CC Far-Near ratio would be very similar to the NC one. In the case of beam ν_e events the distinctive shape of the Far-Near ratio is due to the fact that they are mostly produced through secondary muon decay in the beam pipe, as explained in Sect. 6.2. The increase in the Far-Near ratio with energy suggests that the extra decay in the beam ν_e production chain, which gives these neutrinos a wider distribution of transverse momentum, also makes them more sensitive to the different angular acceptances of the detectors, particularly at low energies (with the exception of the 1–2 GeV energy bin perhaps).

It is important to note that the beam ν_e Far-Near ratios are shown just for comparison purposes. The ratios are not actually used in the extrapolation, given that this component is obtained directly from the simulation. Rather than taking this component from the Near Detector simulation and extrapolating it to the Far Detector, it is just obtained directly from the Far Detector simulation. Even if both procedures yield identical results by definition, it is simpler to follow the second approach.

The reader may have also noticed that the ratios of Fig. 7.5 are significantly different in magnitude from the flux ratio shown in Fig. 7.1. The main reason is that the flux ratio, by definition, does not account for the fact that the fiducial masses of the detectors are vastly different. As mentioned in Sect. 5.2.1, the fiducial volume amounts to 29 t in the case of the Near Detector, and to 3.9 kt in the case of the Far Detector. This means a correction in the order of ~ 130 must be applied to the flux ratio if it is to be converted into an approximate Far-Near ratio. When this is done an approximate Far-Near ratio in the order of 2.0×10^{-4} is obtained, which is very close to the values seen in Fig. 7.5.

7.1.2.2 Oscillated ν_τ CC and ν_e CC Events

The ν_τ CC background and the potential signal ν_e CC cannot be measured in the Near Detector and then extrapolated to the Far Detector, given that they originate through oscillations. The rate of their ν_μ CC parents can however be directly measured in the Near Detector, which makes it the best handle to predict the rate of the oscillated ν_τ CC and ν_e CC events.

The details of the method used to predict the expected rates of ν_τ CC and ν_e CC events in the Far Detector based on the observed ν_μ CC Near Detector rate can be found in [4]. The first step is to get a Far Detector prediction of ν_μ CC-like events as a function of true energy. This is achieved by using the same ν_μ CC selection as in the most recently published MINOS disappearance analysis [5], and following

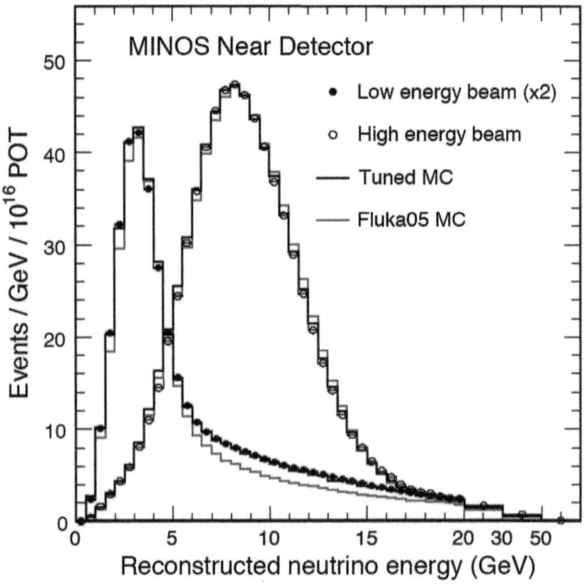

Fig. 7.6 Energy spectra in the Near Detector for ν_μ CC-like events in the low-energy (LE) and high-energy (HE) beam configurations, as obtained from [5]. The tuned simulation (which is the one used throughout the ν_e analysis) agrees very well with the data

the same approach of Eq. 7.2, but expressing the simulated Far Detector rate as a function of true energy rather than reconstructed energy:

$$\mathrm{FD}^{\mathrm{pred}}_{\nu_\mu CC-\mathrm{like}}(\mathrm{E}_{\mathrm{true}}) = \mathrm{ND}^{\mathrm{data}}_{\nu_\mu CC-\mathrm{like}}(\mathrm{E}_{\mathrm{reco}}) \frac{\mathrm{FD}^{\mathrm{MC}}_{\nu_\mu \mathrm{CC}-\mathrm{like}}(\mathrm{E}_{\mathrm{true}})}{\mathrm{ND}^{\mathrm{MC}}_{\nu_\mu CC-\mathrm{like}}(\mathrm{E}_{\mathrm{reco}})}. \tag{7.3}$$

An alternative approach is to use a two-dimensional matrix to convert true to reconstructed energy. Both approaches give similar results and are used in different stages of the analysis. The purity and the efficiency of the selection as obtained from the simulation are then used to convert $\mathrm{FD}^{\mathrm{pred}}_{\nu_\mu CC-\mathrm{like}}(\mathrm{E}_{\mathrm{true}})$ into the total distribution of unoscillated ν_μ CC events in true energy $\mathrm{FD}^{\mathrm{pred}}_{\mathrm{All}-\nu_\mu CC}(\mathrm{E}_{\mathrm{true}})$. The ratio of cross-sections between ν_μ CC and ν_τ CC or ν_e CC events, as well as the corresponding oscillation probabilities, are then applied to this distribution as a function of true energy to obtain $\mathrm{FD}^{\mathrm{pred}}_{\mathrm{All}-\nu_\tau CC}(\mathrm{E}_{\mathrm{true}})$ and $\mathrm{FD}^{\mathrm{pred}}_{\mathrm{All}-\nu_e CC}(\mathrm{E}_{\mathrm{true}})$. Finally, knowing the energy resolution for ν_τ CC and ν_e CC events, these distributions are expressed in terms of reconstructed energy, at which point the selection efficiencies are applied. In the case of ν_τ CC events, the selection efficiency is taken directly from the MC. In the case of ν_e CC events, the selection efficiency is estimated as explained in Sect. 7.2.

Although more complex in its implementation, a simple way of viewing this data-based method is that the observed ratio of data to MC in the Near Detector for ν_μ CC-like events is used to correct the MC expectation for ν_τ CC and ν_e CC appearance. Figure 7.6 shows that the data and the MC agree very well for the

Table 7.1 Predicted Far Detector backgrounds with the PORP, HOO and MRCC methods, for a LEM cut at 0.65 and at 0.80. The predictions are scaled to 3.14×10^{20} POT. Despite the simplicity of PORP, it yields results that are consistent with HOO and MRCC. The total HOO and MRCC predictions differ by only 2.7 and 3.8% in the case of the cut at 0.65 and at 0.80, respectively

	Method	NC	ν_μ CC	Beam ν_e	ν_τ CC	Total
LEM > 0.65	PORP	14.7	3.1	2.7	1.1	21.6
	HOO	14.8	2.9	2.7	1.1	21.4
	MRCC	15.4	2.8	2.7	1.1	22.0
	$1 - \frac{\text{HOO}}{\text{MRCC}}$	3.9%	−3.6%	–	–	2.7%
LEM > 0.80	PORP	8.3	1.8	2.2	0.7	12.9
	HOO	7.4	2.2	2.2	0.7	12.5
	MRCC	8.7	1.5	2.2	0.7	13.1
	$1 - \frac{\text{HOO}}{\text{MRCC}}$	14.9%	−46.7%	–	–	3.8%

sample of selected ν_μ CC-like events.[2] Because of this, the results of this data-based appearance method are very similar to what is obtained if the ν_τ CC and ν_e CC appearance rates are obtained directly from the simulation.[3] Nevertheless, it will be shown in the subsequent chapter that this appearance method allows for a considerable reduction in the systematic error of expected ν_e CC events, which is why it is used in the analysis.

7.1.2.3 Predicted Background Rates

The results of the PORP, Horn-on/off (HOO) and Muon-Removed (MRCC) Near Detector decomposition methods shown in Sect. 6.5 are the input to the extrapolation process. In particular, the values for the NC and ν_μ CC components from these methods are plugged into Eq. (7.2), where they are multiplied by the Far-Near ratios of Figure 7.5. The ν_τ CC component is obtained from the appearance procedure just described in the previous section, and the beam ν_e component from the simulation, which is why these two share the same values for all methods. The extrapolation results for each of the methods can be seen in Table 7.1. For the reasons explained in Sect. 6.4.3, HOO is the default decomposition method in our analysis.

The energy spectra of the predicted background can be seen in Fig. 7.7 for the HOO and MRCC decomposition methods. The corresponding ratios of HOO to MRCC as a function of reconstructed energy can be seen in Fig. 7.8. Given that the Far-Near ratios used in the extrapolation are the same for both methods, the

[2] Please note that only the low-energy configuration of the beam concerns us in the ν_e appearance analysis.

[3] For simplicity, the simulation's prediction is sometimes used in studies throughout this thesis, when indicated.

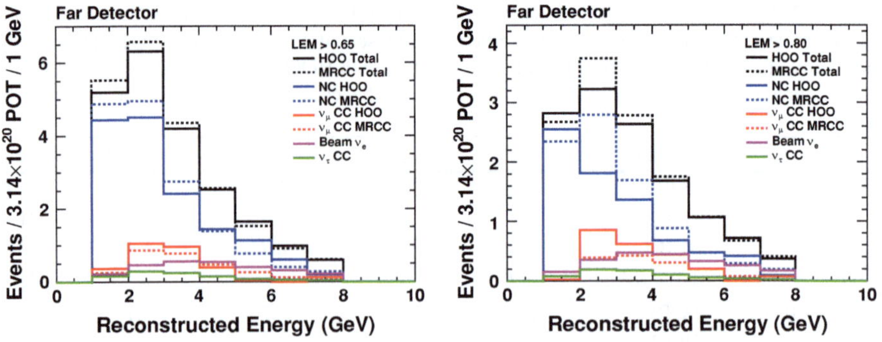

Fig. 7.7 Predicted Far Detector background spectra with the HOO and MRCC methods, for a LEM cut at 0.65 (*left*) and at 0.80 (*right*). Predictions are scaled to 3.14×10^{20} POT

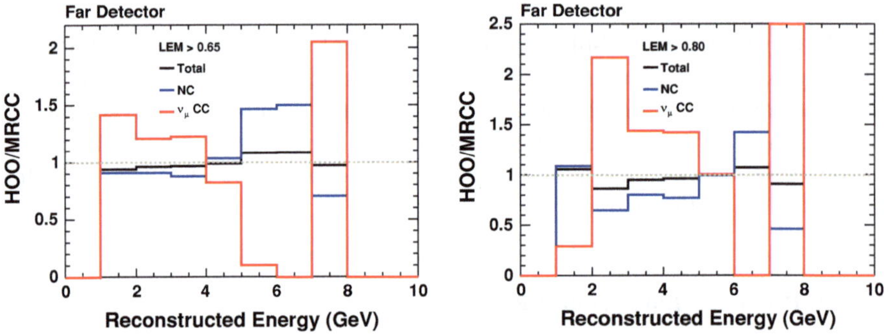

Fig. 7.8 Comparison between the HOO and MRCC Far Detector predictions as a function of energy, for a LEM cut at 0.65 (*left*) and at 0.80 (*right*). Even if the NC (blue) or ν_μ CC (red) components disagree somewhat between the HOO and MRCC methods, the total predicted background (black) agrees very well

differences between methods present in the Near Detector remain in the Far Detector, when the NC and ν_μ CC components are looked at individually. As seen in Fig. 7.8, these differences can be large, particularly for the ν_μ CC component where the HOO and MRCC methods disagree by more than a factor of two in some energy bins. Nevertheless, due to the constraint that the sum of the Near Detector components must equal the total Near Detector data, whenever one method predicts more than the other for one component, the opposite happens for the other component in order to compensate. Visually, this means than the red and blue curves of Fig. 7.8 are always in opposite sides of the central grey line at unity. Because of this correlation between the two main components, the total HOO and MRCC Far Detector predictions agree very well with each other, to 2.7% for the LEM cut at 0.65 and to 3.8% for the cut at 0.80. This is a point to which we will come back subsequently in this chapter, when combining the systematic errors for the total prediction.

7.2 Expected Far Detector Signal Rate

7.2.1 Overview

Now that the background has been predicted, it remains to accurately estimate the number of signal events expected at the Far Detector in the case of a nonzero θ_{13}. While the Near Detector allows us to make a direct measurement of the backgrounds, as explained in the previous chapter, it cannot be used to measure the oscillated signal. Because of this, the expected ν_e CC rate is predicted from the measured ν_μ CC rate in the Near Detector, using the appearance method described in Sect. 7.1.2. This method however requires the selection efficiency of ν_e CC events as an input.

Unfortunately the Near Detector cannot be used to directly study the selection efficiency of ν_e CC events, as it is practically impossible to isolate a significant sample of beam ν_e events. Therefore another way of assessing the selection efficiency at the Far Detector must be found in the case of signal events.

In addition to the electron resulting from the neutrino interaction, oscillated ν_e CC signal events have a hadronic shower component to them. As shown below, electron showers are well modeled in the MINOS simulation, while we have seen that hadronic showers are not. However, the impact of the hadronic shower mismodeling is small for the ν_e CC events selected by LEM, as shown by Table 8.8 in the next chapter. This allows us to estimate the signal efficiency by relying on the simulation, but applying a small correction obtained by using events where simulated electrons are combined with muon-removed (MRCC) showers. The details of the signal efficiency estimation procedure can be found in the next two sections (7.2.2 and 7.2.3).

7.2.2 Electrons in CalDet

While it is nearly impossible to isolate single electron events in MINOS it is straightforward in the CalDet [6], which was exposed to an electron test beam in CERN. Being a scaled-down but functionally equivalent version of the MINOS detectors, the same LEM selection can be applied to the CalDet electron data with a few minor modifications to the LEM code. Such modifications involve, for instance, changing the standard position of the detector to which compact events are moved to, given that the CalDet is smaller than the Near and Far Detectors.

The agreement between data and MC is thus found to be very good for LEM selected CalDet electrons, for the LEM variables that make up the pid and also for the LEM pid itself [7]. The relevant quantity for this study is the difference between data and MC in the selection efficiency of electrons. Figure 7.9 shows the selection efficiencies of CalDet electrons as a function of electron momentum in the data and in the MC, as well as their corresponding ratio. The data/MC ratio is

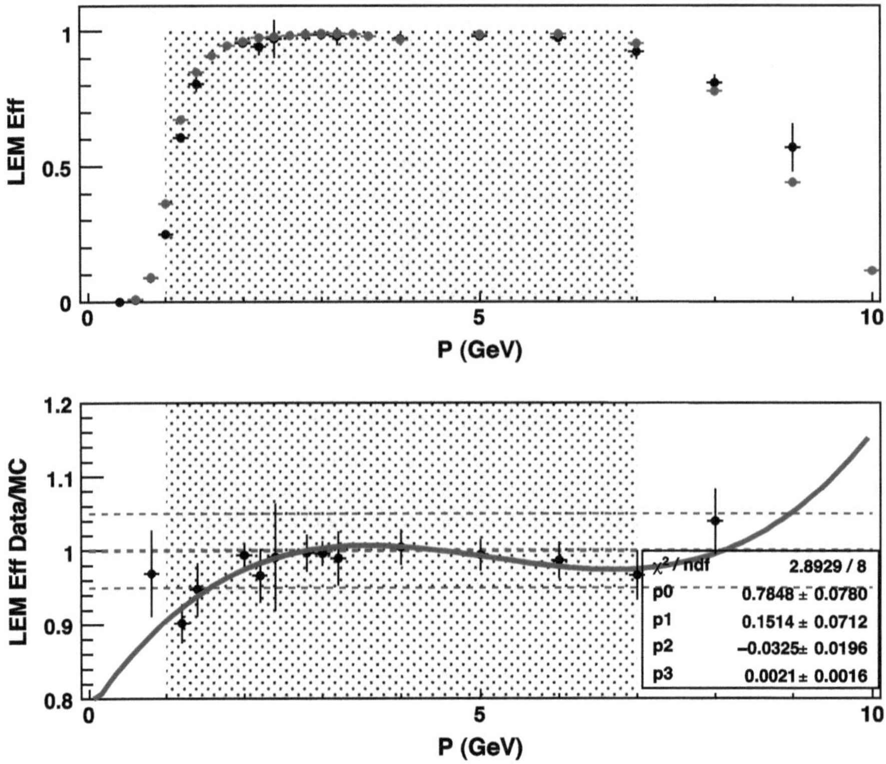

Fig. 7.9 (*top*) Selection efficiencies of CalDet electrons in the data (points) and the MC (solid curve), and (*bottom*) the corresponding data to MC ratio. The data to MC ratio is fit to a third-order degree polynomial. The shaded region is where the fit is done, which is also the energy region where most of the signal is

well within ±5% in the energy region of interest, and is consistent with being unity given the statistics. The fact that the selection efficiency decreases at higher energies is not surprising, given that LEM library events lie primarily in the 1–4 GeV energy region, where most of the signal is expected.

In order to estimate a systematic error on the MC efficiency, the data-MC discrepancy in CalDet is parameterized with a third degree polynomial, also shown in Fig. 7.9. This function provides a weighting function that is applied to the standard MC as a function of electron momentum, resulting in a 2.2% change in the number of Far Detector LEM selected signal events.[4] Given that the CalDet electron simulation employs the same models as the standard MINOS simulation, this constitutes a systematic error on the signal selection efficiency due to the

[4] This study was done using the LEM cut at 0.80, and we use the results also for the cut at 0.65 as we do not expect any significant changes.

mismodeling of electrons. This small systematic is combined with the error on the MRE efficiency prediction, as shown next.

7.2.3 *The Impact of Hadronic Shower Mismodeling*

7.2.3.1 Uncertainties in the Hadronic Model

Even if electrons are well modeled in the MINOS simulation, ν_e CC events have a hadronic shower component to them that, as we saw in the previous chapter, is not. It is thus necessary to estimate the impact of the hadronic shower mismodeling on the selection of ν_e CC events.

This can be achieved by varying the hadronic model within its uncertainties and recording the change in the number of selected ν_e CC events. This procedure is followed while estimating all the systematic uncertainties involved in the analysis, as described in Sect. 8.1, and the results can be found in Table 8.4. As seen there, the uncertainty in the number of expected ν_e CC events from uncertainties in the hadronic model amounts only to 2.2%, while it exceeds 50% for NC and ν_μ CC events. If we include the other aspects of the simulation that affect hadronic showers, such as the uncertainties in hadron multiplicity and in intranuclear rescattering (addressed, respectively, in Sect. 8.1.2), the uncertainty in the number of expected ν_e CC events rises only to 2.9%. The mismodeling of hadronic showers thus has a very small impact on the selection of signal ν_e CC events. This suggests that the ν_e CC events selected by LEM are, as expected, largely dominated by the electromagnetic component.

One possible approach is to estimate the number of expected ν_e CC events at the Far Detector entirely from the simulation. In that case the total systematic error on the signal prediction would be approximately 14%, as shown in Table 8.4. An even better approach is to use the measured ν_μ CC-like spectrum at the Near Detector, as explained in Sect. 7.1.2. In that case the total systematic error is reduced to approximately 8%, as shown in Table 8.8. This reduction in error is due to a better estimation of the flux and to a cancellation of some of the systematics, but the signal selection efficiency at the Far Detector still needs to be obtained from the MC.

7.2.3.2 A Data-Based Correction from MRE Events

Even if, as just shown, the signal selection efficiency from the simulation is only a few percent off due to hadronic shower mismodeling, it is possible to go the extra mile and obtain a data-based correction for it. This way an independent confirmation of the size of the effect from the uncertainties in the hadronic model can be obtained, as well as a slightly more accurate prediction of the expected signal.

This data-based correction can be obtained through a data-MC comparison of "Muon-Removed with Electron addition" (MRE) events in the Near Detector. An MRE event consists of a golden ν_μ CC with a well-identified track that is removed and replaced by a simulated electron with the same 4-momentum. The event is then re-reconstructed as a shower dominated event. This process is illustrated in Fig. 7.10.

The details of the MRE algorithm can be found in Chap. 6 of [4]. The electron that is added to the muon-removed shower is always obtained from the simulation. The muon-removed shower however can be obtained from the data or the MC, in which case MRE data or MC is obtained. Figure 7.11 shows a data-MC comparison for MRE events passing the preselection as a function of the three variables that compose the LEM pid. As expected, the distributions peak toward the signal-like regions (i.e., high ν_e fraction of best matches, low mean y, and high Q_{frac}). Despite the fact that the MRE data and MC samples have very different hadronic showers in them, the agreement between the data and the MC remains

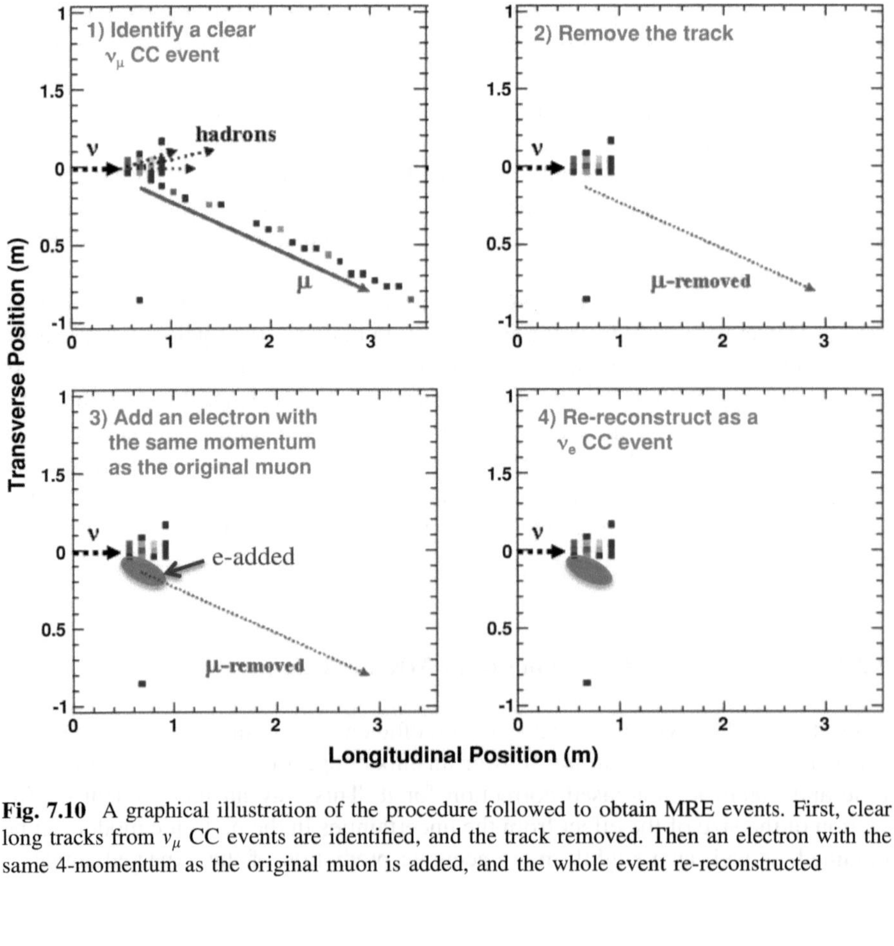

Fig. 7.10 A graphical illustration of the procedure followed to obtain MRE events. First, clear long tracks from ν_μ CC events are identified, and the track removed. Then an electron with the same 4-momentum as the original muon is added, and the whole event re-reconstructed

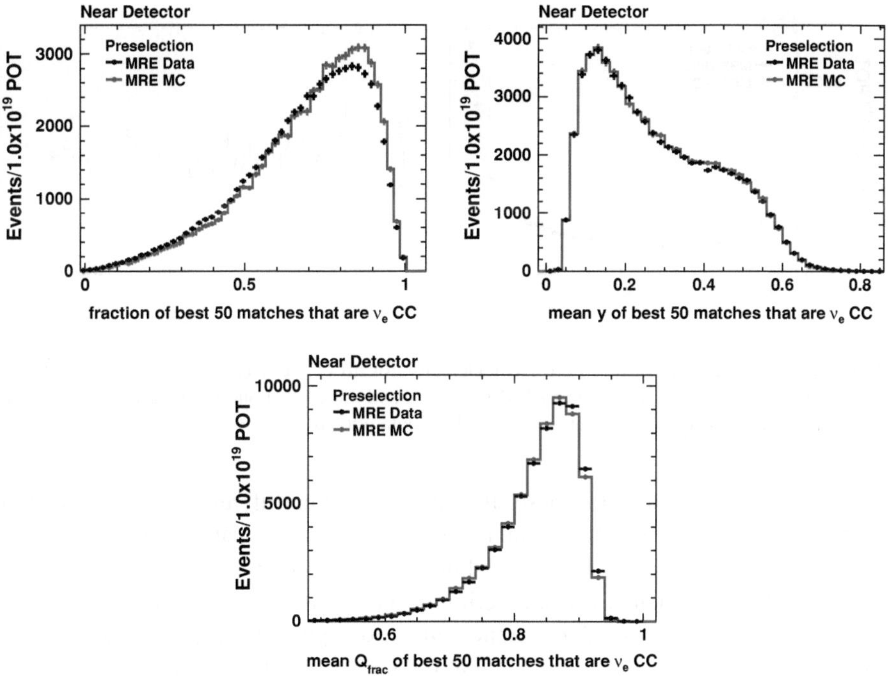

Fig. 7.11 MRE data and MC in the Near Detector for all events passing the MRE quality cuts and the preselection, as a function of the three variables that make up the LEM pid. There is very good agreement between the MRE data and MC

very good, and in general better than 5%. The same can be said of the LEM pid distribution, shown in Fig. 7.12. These results thus constitute an independent confirmation that the impact of the hadronic shower mismodeling is small when it comes to selecting signal events.

Beyond confirming this fact, the MRE data can also be used to extract a correction to the MC signal efficiency due to hadronic shower mismodeling. Figure 7.13 shows the selection efficiencies in the MRE MC, MRE data, and in the Far Detector MC. While the efficiencies are very similar in the MRE data and MC, they differ significantly with the one from the Far Detector MC, most likely due to imperfections in the MRE process. Because of this, and in an analogous way to what is done in the MRCC decomposition, we do not use the MRE events directly to estimate the signal efficiency. Rather, the ratio of efficiencies between the MRE data and MC in the Near Detector is used to correct the MC expectation,

$$\epsilon_i^{\text{pred}} = \epsilon_i^{\text{MC}} \frac{\epsilon_i^{\text{MRE-Data}}}{\epsilon_i^{\text{MRE-MC}}}, \tag{7.4}$$

where ϵ_i denotes the signal selection efficiency at energy bin i. In this way, the imperfections and biases associated with the muon removal and electron addition

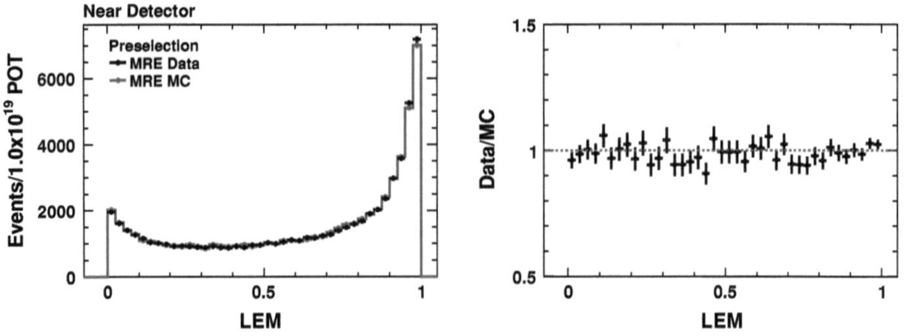

Fig. 7.12 MRE data and MC in the Near Detector for all events passing the MRE quality cuts and the preselection as a function of LEM pid (*left*), and the corresponding data/MC ratio (*right*). There is very good agreement between the MRE data and MC

process cancel to first order. Also, this allows for the differences in cross-section and interaction type between ν_μ and ν_e CC events to be treated in the ratio.

The right-hand side of Fig. 7.13 shows the corrections $\epsilon_i^{\text{MRE-Data}}/\epsilon_i^{\text{MRE-MC}}$ to the MC selection efficiency as a function of reconstructed energy. This results in a change of -4.0% and of -2.7% in the number of signal events predicted at the Far Detector, for the cut value at 0.65 and at 0.80, respectively. It is interesting to note that the correction goes in the direction that is naively expected. Data-MC comparisons in the standard sample show that the MINOS hadronic showers are less electromagnetic than predicted by the simulation. This means that when the hadronic showers in ν_e CC events are replaced by the ones from the data, fewer events should be selected on average, as observed.

It is necessary to obtain a systematic error on the MRE predicted selection efficiency ϵ_i^{pred}. In order to achieve this, uncertainties in the physics model are considered, such as in the beam flux and in the cross-sections. In order to assess the impact of these systematics, the MC samples are reweighted in the same manner as described in the next section, where the evaluation of systematics is done for the entire analysis. Also considered are systematic errors due to differences in the data and the MC that may not originate from the difference in hadronic showers. In particular, the track-related quantities used in the preselection show some disagreement between MRE data and MC, and could introduce some of the discrepancy between MRE data and MC that is attributed to the difference in hadronic showers and used in the efficiency prediction. These types of systematic errors are addressed by varying the preselection cuts independently between the two samples. Detector modeling uncertainties such as energy shifts between the data and the MC are also considered. Finally, the slight mismodeling of electrons could contribute to the difference in MRE data and MC. This was estimated to be a 2.2% effect in the previous section, and is also incorporated in the efficiency prediction uncertainty.

The details on the MRE efficiency prediction uncertainty determination can be found in Chap. 6 of [4]. When put together, all of the uncertainties on ϵ_i^{pred}

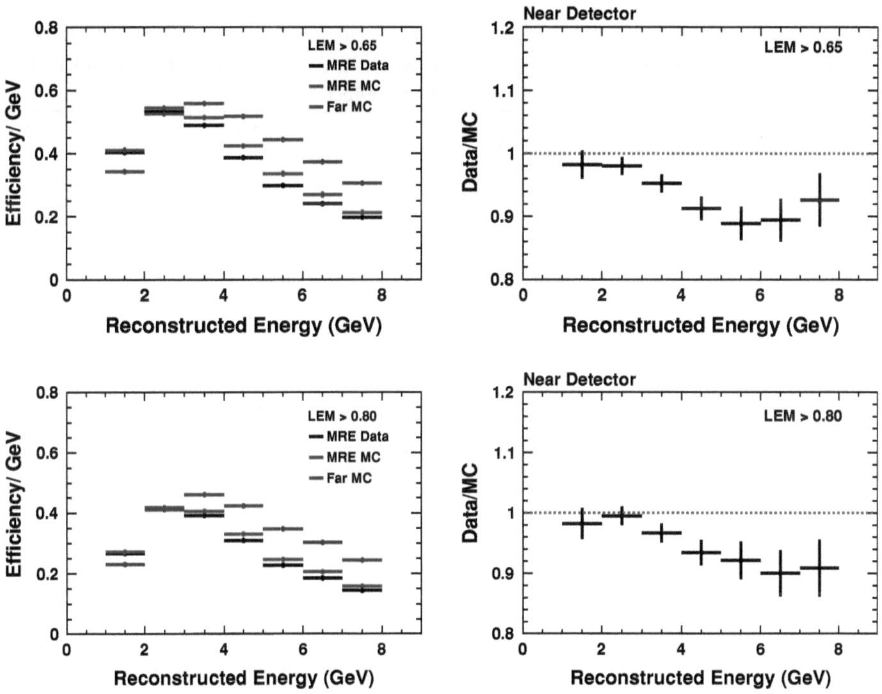

Fig. 7.13 ν_e CC selection efficiencies in the MRE MC, MRE Data, and Far Detector MC (*left*), and the ratio of data/MC MRE efficiencies (*right*), for the LEM cut at 0.65 (*top*) and at 0.80 (*bottom*)

considered amount to $^{-3.1\%}_{+3.4\%}$ for the LEM cut at 0.65, and to $^{-3.1\%}_{+3.5\%}$ for the cut at 0.80 [8]. The selection efficiency predicted by the MRE procedure accounts for the fact that hadronic showers in the simulation are not correctly modeled. Therefore, when using the MRE efficiency in the appearance method that yields the ν_e CC events at the Far Detector, the systematic errors due to the mismodeling of the hadronic showers are removed and replaced by the error in the MRE efficiency. This yields a total error of approximately 8.5% on the ν_e CC prediction, as shown in the next chapter (Table 8.9 in particular).

7.3 Summary of the Far Detector Predictions

The predicted Far Detector background and signal, as obtained through the procedures described in this chapter, are summarized in Table 7.2. The signal shown corresponds to a value of θ_{13} at the CHOOZ limit (i.e., $\sin^2(\theta_{13}) = 0.15$), with no matter effects or CP violation. With the cut value at 0.65 and with the HOO decomposition a total of 21.4 background events are expected. If θ_{13} is at the

Table 7.2 Signal and background (broken down by components) predictions at the Far Detector for the LEM > 0.65 and LEM > 0.80 selections. The predictions are normalized for an exposure of 3.14×10^{20} POTs. The signal corresponds to a value of θ_{13} at the CHOOZ limit ($\sin^2(2\theta_{13}) = 0.15$), ignoring CP violation and matter effects

	Method	NC	ν_μ CC	Beam ν_e	ν_τ CC	Total bg.	Signal
LEM > 0.65	HOO	14.8	2.9	2.7	1.1	21.4	11.1
	MRCC	15.4	2.8	2.7	1.1	22.0	11.1
LEM > 0.80	HOO	7.4	2.2	2.2	0.7	12.5	8.7
	MRCC	8.7	1.5	2.2	0.7	13.1	8.7

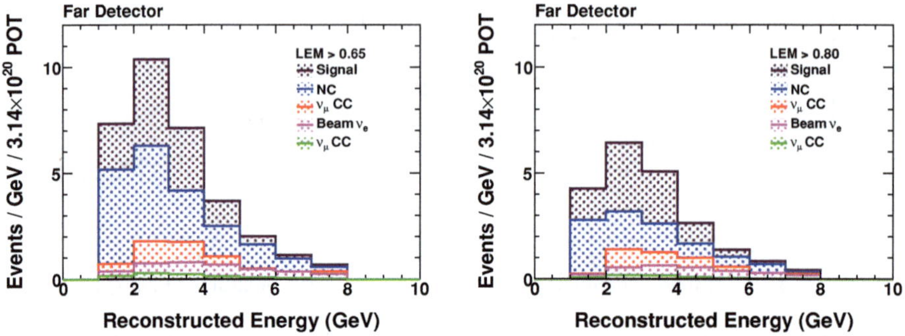

Fig. 7.14 Stacked energy spectra of the possible signal and the predicted background at the Far Detector, for the LEM selection with a cut value at 0.65 (*left*) and at 0.80 (*right*), using the HOO decomposition. The predictions are normalized for an exposure of 3.14×10^{20} POTs. The signal corresponds to a value of θ_{13} at the CHOOZ limit ($\sin^2(2\theta_{13}) = 0.15$), ignoring CP violation and matter effects

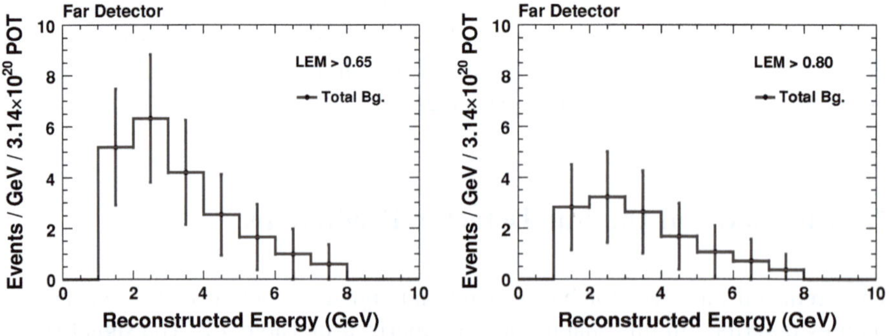

Fig. 7.15 Total background prediction for the LEM selection with a cut value at 0.65 (*left*) and at 0.80 (*right*), using the HOO decomposition. The statistical error at each bin assuming Gaussian statistics is shown

CHOOZ limit then an excess of 11.1 signal events will be observed over the predicted background. With the cut at 0.80 the background is strongly reduced. Only 12.5 background events are expected versus 8.7 signal events. This puts the signal-to-background ratio at ~ 0.5 for the cut at 0.65, and at ~ 0.70 for the cut at 0.80.

Figure 7.14 shows the energy spectrum of the background predicted with the HOO decomposition and the possible signal at the Far Detector. The results are very similar for the MRCC decomposition and are consequently not shown. The signal and background energy spectra are very similar in shape. Also, the statistical error at each bin is very large, as shown for the background in Fig. 7.15. These are the reasons why, as previously explained, no energy-dependent fit is done in the analysis to extract the possible signal. Only the total number of selected events in the data is compared to the background expectation in order to derive the limits on θ_{13}. Before this is done, however, a systematic uncertainty envelope must be derived for both the background prediction and the signal expectation. This is done as explained in the next chapter.

References

1. P. Adamson et al., A study of muon neutrino disappearance using the fermilab main injector neutrino beam. Phys. Rev. D **77**, 072002 (2008)
2. E.K. Akhmedov et al., Series expansions for three-flavor neutrino oscillation probabilities in matter. J. High Energy Phys. **2004**(04), 078 (2004)
3. R.B. Patterson, Energy fit in nue analysis. MINOS-doc-4834 (2008)
4. J. Boehm, Measurement of electron neutrino appearance with the MINOS experiment. FERMILAB-Thesis-2009-17
5. P. Adamson et al., Measurement of neutrino oscillations with the MINOS detectors in the NuMI beam. Phys. Rev. Lett. **101**, 131802 (2008)
6. P. Adamson et al., The MINOS calibration detector. Nucl. Instrum. Meth. Phys. Res. Sect. A: Accelerators. Spectrometers, Detectors and Associated Equipment. **556**(1) 119–133 (2006)
7. T. Vahle, Nue PID in Caldet, MINOS-doc-5546 (2009)
8. J. Boehm, Measuring ν_e CC selection efficiency and systematic error using the muon removal with electron addition process. MINOS-doc-5385 (2009)

Chapter 8
Systematic Errors and θ_{13} Sensitivity

This chapter begins with an estimation of all the systematic uncertainties affecting the background and signal predictions. These are divided into uncertainties in the physics models and in the detector models, and are addressed respectively in Sects. 8.1.2 and 8.1.3. The contributions from each source of systematic error, as well as the total uncertainty in each component and in the totals, are calculated in Sect. 8.1.4. From this information it is possible to study the sensitivity to θ_{13}, as done in Sect. 8.2.

8.1 Systematic Uncertainties

8.1.1 Overview

A systematic error estimate must be obtained for all the predictions that have been derived so far.

The largest background components (ν_μ CC and NC) are extrapolated from the Near Detector via Eq. 7.2:

$$\mathrm{FD}_i^{\mathrm{pred}}(\mathrm{E_{reco}}) = \mathrm{ND}_i^{\mathrm{data}}(\mathrm{E_{reco}})\frac{\mathrm{FD}_i^{\mathrm{MC}}(\mathrm{E_{reco}})}{\mathrm{ND}_i^{\mathrm{MC}}(\mathrm{E_{reco}})}.$$

In this equation, each Near Detector component is acted on by the corresponding Far–Near ratio in order to yield the corresponding Far Detector prediction. Consequently, there are two types of error associated with this process:

- The error on the decomposition,
- The error in the Far–Near ratios.

J. P. Ochoa Ricoux, *A Search for Muon Neutrino to Electron Neutrino Oscillations in the MINOS Experiment*, Springer Theses, DOI: 10.1007/978-1-4419-7949-0_8,

The error on the decomposition is already estimated for each background component, as shown in Chap. 6. The combination of the errors of each component, as well as the decomposition error's contribution to the total uncertainty, can be found in Sect. 8.1.4.

The error in the Far–Near ratios is also referred to as the error in the extrapolation throughout this thesis. Given that these ratios involve the number of selected events as given by the simulation in the two detectors, it follows that anything that is not well modeled in the MC and that affects the two detectors differently becomes a source of systematic error.

As discussed previously, most physics modeling related systematics cancel to first order when taking the ratio, although there are some residual effects that must be quantified. In addition, there are some physical differences between the detectors:

- Readout patterns: The readout is double-ended in the Far Detector and single-ended in the Near Detector, which causes differences in the light level. Furthermore, there is multiplexing (8 fibers per pixel) in the Far Detector.
- PMTs: The PMTs are different in the two detectors, which causes different gains and cross-talk patterns.
- Intensity: The Near Detector is subject to a much higher neutrino event rate, which causes many (~ 8) events to occur in the same spill, while all Far Detector events occur alone.

Because of these differences, most of the uncertainties in the detector models do not cancel in the Far–Near ratio. Both the physics modeling and detector modeling systematics are addressed respectively in Sects. 8.1.2 and 8.1.3.

The beam ν_e component is obtained directly from the simulation and not by extrapolation from the Near Detector. Nevertheless, the systematic error evaluation for this component is done in parallel with the assessment of the uncertainties in the extrapolation, given that this assessment involves studying every component that is not properly modeled. The same applies for the ν_τ CC component which, although derived through the more sophisticated appearance approach of Sect. 7.1.2, at the end is affected by the same uncertainties as the extrapolation.

Finally it should be mentioned that, given the large amount of information that is presented in the subsequent assessment of all the systematic errors, the resulting numbers are only presented for the LEM cut at 0.65. The systematic evaluation procedure is, however, the same for the cut at 0.80, and the corresponding systematic errors are very similar to the ones for the cut at 0.65 both in magnitude and in behavior.

8.1.2 Uncertainties in the Physics Models

8.1.2.1 Uncertainties in the Hadronic Model

As explained in Sect. 3.4.2, the AGKY model represents an improvement over previous hadronic models. AGKY's improved performance is achieved by dividing the kinematic range into regions where other models are applicable and verified experimentally. This way, the KNO model is used in the low-invariant mass region ($W < 2.1$ GeV), and the PYTHIA/JETSET model in the high-W region ($W > 3.0$ GeV). In the medium-W region, which is the regime that is of most interest to our analysis, AGKY performs a phenomenological extrapolation between the KNO-based model and the PYTHIA/JETSET model. Unfortunately, there is a lack of external data to constrain such an extrapolation. Consequently, there are uncertainties in many of the model parameters that can very strongly impact the LEM selection.

In order to assess the error in the hadronic model, six uncertainties of the model, labeled T1–T6, are considered. Each of these uncertainties is varied within its range, and the effects on LEM are recorded. A more detailed account on each of these uncertainties can be found in [1].

T1: baryon x_f selection. As mentioned previously, one of the main improvements over the KNO based model in the AGKY implementation was the correction of the parton distribution. In the T1 uncertainty treatment this correction is undone, in order to conservatively estimate the uncertainties in the selection of the baryon four-momentum.

T2: π^0 probability. In this treatment the π^0 creation probability is altered by $\pm 20\%$. This is the only alteration to the hadronic model that is done in two directions, and it is consequently termed T2+ or T2− accordingly.

T3: charged-neutral particle multiplicity correlation. While some measurements indicate that the charged and neutral pion multiplicities are independent, the current implementation of the AGKY model does not reflect this. As a consequence, the charged particle topological cross sections are not properly reproduced by the current model. A new multiplicity algorithm exists in an updated version of the AGKY model, described in [2], which agrees much better with the bubble chamber data in the kinematic region dominated by the KNO model. In the T3 treatment this new multiplicity algorithm is deployed within the standard version of AGKY.

T4: implementation ambiguities. The AGKY model has many implementations, such as in the NEUGEN [3] and GENIE [4] packages. For example, if an event is rejected when the p_T squeezing step is applied to a phase space decay, in one code the phase space decay is regenerated, while in the other the algorithm returned to an earlier step and regenerated the baryon as well. In order to estimate the impact of all implementation ambiguities, the T4 treatment considers the AGKY implementation in GENIE as opposed to the standard one in NEUGEN.

T5: p_T squeezing. During a hadronic remnant decay, the average summed transverse momentum p_T of the showers is constrained through a weighting factor. In the T5 treatment the p_T squeezing parameter in remnant system decay is changed from its default value, resulting in broader showers.

T6: isotropic two-body decays. The AGKY model uses the assumption that all two-body decays are isotropic. The T6 treatment involves a similarly strong assumption in another direction, namely that all two-body decays occur at right angles with respect to the momentum transfer direction.

Table 8.1 Changes in the number of selected Near Detector NC events and in the HOO Far Detector NC prediction when the hadronic model modifications T1–T6 are applied

	T1(%)	T2+(%)	T2−(%)	T3(%)	T4(%)	T5(%)	T6(%)	Total(%)
FD	−36.6	11.0	−10.6	−21.2	−16.3	−15.0	−16.1	51.6
ND	−37.4	10.8	−10.4	−19.6	−15.5	−14.1	−14.5	50.5
Pred.	0.3	0.3	−0.3	−2.3	−1.0	−1.2	−1.9	3.4

The total hadronic model uncertainty in the simulation is larger than 50% in both the Near Detector and the Far Detector. The changes however are very similar in the two detectors and consequently cancel to first order when taking the Far–Near ratio, yielding a total uncertainty in the Far Detector prediction of only 3.4%

The T1–T6 modifications are applied by reweighting the simulation as a function of W^2, summed transverse momentum p_T and electromagnetic fraction. The effects on the LEM selection are summarized in Table 8.1 for NC events. The systematic error in the simulation due to hadronic model uncertainties is very large and exceeds 50% in both detectors. But the error that is relevant for the analysis is the error in the Far Detector prediction, which is obtained by redoing the extrapolation (the Far–Near ratio in particular) after the application of each of the six uncertainty treatments T1–T6. As seen in Table 8.1, the effects of the T1–T6 treatments on LEM are very similar and practically cancel between the two detectors.[1] This is to be expected given that the two simulations use the same hadronic model. Due to detector differences in readout, calibration and intensity, there remains a small residual systematic that amounts to 3.4%.

There is one more variation of the hadronic model that is considered. The average hadron multiplicity is determined from bubble chamber experiments, which are not all in agreement. Therefore the multiplicity is varied in order to cover the range of the different measurements. Also, the Levy function which determines the KNO scaling is varied within its uncertainties. The uncertainties associated with this study are labeled as Hadron Multiplicity and impact the Far Detector prediction to less than 0.7%, as shown in Table 8.2.

8.1.2.2 Uncertainties in the Cross-Sections

In order to assess the uncertainties in the cross-sections we follow the same approach as in the ν_μ disappearance analysis [5], where the three parameters used by NEUGEN to model the cross-sections in the few-GeV region (M_A(QE), M_A(RES) and KNO) are varied within their respective uncertainties of 15, 15 and 50%. This is an appropriate approach for all of the interactions involved in the ν_e appearance analysis, except for ν_τ CC events. For these events an overall 10% uncertainty is applied to the quasi-elastic and resonant events, and a 50%

[1] Note that the Far–Near ratios are calculated as a function of energy. Consequently the prediction uncertainty in the last row of Table 8.1 cannot be obtained from the middle two rows, although their algebraic difference usually provides a reasonable estimate.

Table 8.2 Changes in the number of selected Near Detector NC events and in the HOO Far Detector NC prediction under the hadronic multiplicity study

Shift	FD (%)	ND (%)	Pred. (%)
Hadron multiplicity			
$+1\sigma$	+1.3	+1.9	+0.7
-1σ	−1.9	−2.6	−0.6

uncertainty for the deep inelastic events, as recommended by the MINOS physics simulation group. This extra uncertainty is labeled $\sigma_{\nu_\tau CC}$.

The resulting cross-section uncertainties are shown in the summary tables of Sect. 8.1.4. For the NC and ν_μ CC components the systematic error in the simulation is close to 3%. Furthermore, there is a large cancellation between the two detectors, thus reducing the systematic error on the Far Detector prediction to less than 0.5%. For beam ν_e events the situation is different, as the error amounts to approximately 10% in each detector and there is no cancellation, since this component is obtained directly from the simulation. As a matter of fact, the uncertainties in the cross-sections and in the beam flux are the dominant ones and alone account for 95% of the total beam ν_e uncertainty. Similarly, the total systematic error on the ν_τ CC events is dominated by the error in the cross-sections. Due to the fact that most ν_τ CC events are deep inelastic, their total systematic error is close to 50%. It should be noted that while the summary tables show the effect of varying M_A(QE), M_A(RES) and KNO in the number of selected ν_τ CC events, these uncertainties are not considered independently from $\sigma_{\nu_\tau CC}$, as that would result in double counting.

8.1.2.3 Uncertainties in the Beam Flux

Due mainly to the scarcity of external data to constrain hadron production at the MINOS target, there are uncertainties in the beam flux. Hadron production can however be constrained with the ν_μ CC data in the Near Detector, as the position of the horn and the target determines the region of transverse and longitudinal pion momentum that contributes to the ν_μ spectrum. This is done in the beam fit described in Sect. 3.4.2 and which is incorporated in the simulation. In addition to the hadron production parameters, the fit includes other parameters pertaining to the beam focusing elements (such as the horn current, target position and baffle scraping, among others) which also affect the beam flux. In the course of the fit, a one-sigma error envelope is determined for each parameter involved. In our analysis, we estimate the uncertainty in the beam flux by varying these parameters within their errors.[2]

[2] For the beam ν_e component the one-sigma error envelope only includes the error due to hadron production at the target. Following the recommendation from the beam systematics group, an additional 1.77% is added in quadrature to account for uncertainties in the beam-focusing elements.

The beam flux uncertainty in LEM selected events is on the order of 10% for all components in both detectors. However, due to the cancellation occurring in the Far–Near ratio, the uncertainties in the NC and ν_μ CC predictions are less than 1%. As with the cross-section uncertainties, the fact that there is no cancellation for the beam ν_e component makes the beam flux one of the largest uncertainties. It is interesting to note that the beam ν_e error due to beam flux uncertainties is larger in the Far Detector than in the Near Detector. This is because in the Far Detector a larger fraction of events originate from the decay of kaons, which are not as well constrained by the fit as pions by are.

8.1.2.4 Uncertainties in the Intranuclear Rescattering Model

The AGKY model describes the final state of particles during a neutrino–nucleon interaction. However, most of the neutrino interactions occur in steel, where the nucleons are bound to at least 50 other nucleons in the nuclei. Therefore, some of the final state hadrons continue to interact in the nucleus, which alters the visible energy and the kinematic distributions of the shower. These intranuclear rescattering processes are handled by the INTRANUKE package. It was recently discovered [6] that the hadron absorption probability in INTRANUKE was being underestimated by roughly 10%. It was impossible but also unnecessary to regenerate all of the simulation files in time for the analysis, given that this effect cancels to first order when taking the Far–Near ratio. In order to estimate the small residual systematic the simulation with the corrected version of INTRANUKE was compared to the one used in the analysis.

The INTRANUKE uncertainties have a large ($\sim 15\%$) impact on those components that consist mostly of hadronic showers, such as NC and ν_μ CC events. For beam ν_e's the effect is only of $\sim 3\%$, due to the dominance of the electron shower in these events. Nevertheless, due to the fact that the two detectors are affected identically, the residual error in the prediction is less than 1% for the NC and ν_μ CC predicted events.

8.1.3 Uncertainties in the Detector Models

8.1.3.1 Uncertainties in the Low Pulse Height Hits

When constructing the NC/CC discriminant used in the ν_μ CC disappearance analysis, it was discovered by R. Ospanov that low pulse-height hits are incorrectly modeled in the MINOS simulation [7]. Figure 8.1 illustrates this for the small showers that pass the preselection. In general, the strip charge distribution is very well modeled. Below ~ 1 PE however, there is a large excess in the data with respect to the simulation, the causes of which are not fully understood. As seen in Fig. 8.1, crosstalk is very important at low pulse-heights, and the number of

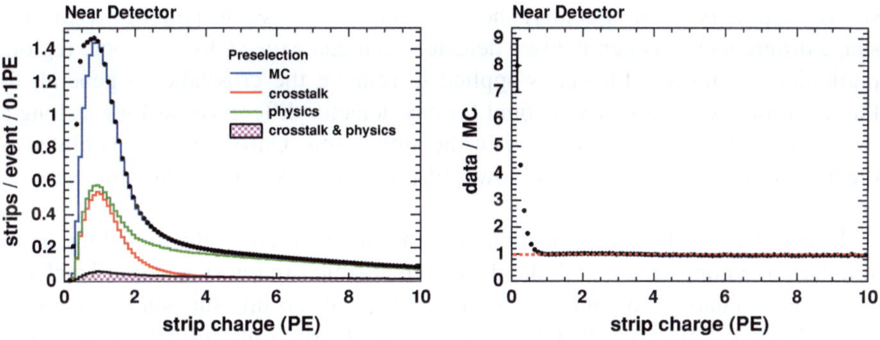

Fig. 8.1 Near Detector strip charge distribution in photoelectrons for events passing the preselection broken (*left*), and the corresponding data to MC ratio (*right*). The sum of the red, green and purple curves does not equal the blue one given that there are other types of hits not mentioned (for instance, hits leaking from other buckets). The data and the MC agree very well, except below ~1 PE where there is a very large excess in the data with respect to the simulation

crosstalk hits per event is actually comparable to the number of physics hits below 2 PE. This makes the crosstalk simulation one of the main suspects. Studies show however that even a very aggressive tuning of the crosstalk model cannot, by itself, remove the discrepancy. Therefore there seem to be other causes, such as detector noise mismodeling or uncertainties in the PMT gains (in particular the single PE widths).

Because of this discrepancy, together with the uncertainties in the crosstalk model (discussed in the next section), the LEM algorithm ignores all hits with a pulse-height below 3.0 PEs when doing the matching, as explained in Sect. 5.3.3. While this PE cut certainly mitigates the issue, it does not fully remove it as the low pulse-height mismodeling might still create a bias at the reconstruction stage. For instance, it could be that the purity and the completeness of the event builder is affected by the excess of low pulse-height hits. In order to assess the size of such an effect, a special MC sample is produced where the hits below 2 PE are removed before the reconstruction occurs. The difference between the number of LEM selected events in this special sample with respect to the standard sample is taken as a measure of the systematic error.

The low-pulse height systematic error is found to be on the order of a few percent in each detector, being more important in the Near Detector. Due to the relative cancellation between the Far and Near detectors, the uncertainty in the NC + ν_μ CC prediction is 1.6%.

8.1.3.2 Uncertainties in the Crosstalk Model

The fact that the two detectors use different PMTs results in different crosstalk patterns. In order to separate the signal from the background, the MINOS ν_e selections, and LEM in particular, have to be very sensitive to differences in the shower topology. Consequently, the differences in crosstalk patterns create PID shape differences between the two detectors that can exceed 30% in some regions, particularly if no low-PE cut is applied to remove the crosstalk. In general, the Far–Near ratio can correctly account for any detector differences as long as long as their causes are properly modeled in the simulation. Unfortunately, as hinted by the previous discussion on the low pulse-height hits, this is not the case for crosstalk.

In order to understand and quantify the state of crosstalk modeling in the simulation, a comparison with the data from cosmic ray tracks is done. Tracks are the cleanest channel to study crosstalk, as most of the hits surrounding them are crosstalk hits. Figures 8.2 and 8.3 show the distributions of hits around long cosmic ray tracks, for the Near Detector and the Far Detector respectively. Most of the hits in those distributions are crosstalk hits, as other processes yielding hits around tracks such as delta-rays and bremsstrahlung radiation are mitigated by rejecting tracks with reconstructed showers in them.

The key to understanding the patterns of the distributions on the right of Figs. 8.2 and 8.3 are the pixel vs. strip number mappings such as the ones shown in Fig. 8.4. For the Near Detector, the mapping is such that neighboring pixels frequently lie ±13 and ±20 strips away. The default simulation underestimates crosstalk by about 30% at those positions. For the Far Detector, the two peaks at around ±10 are crosstalk hits in the adjacent pixels, while the peaks at ±1, ±2 and ±3 strips are crosstalk hits in the diagonal neighboring pixels. The adjacent crosstalk is slightly overestimated by the default MC, while the diagonal crosstalk is underestimated.

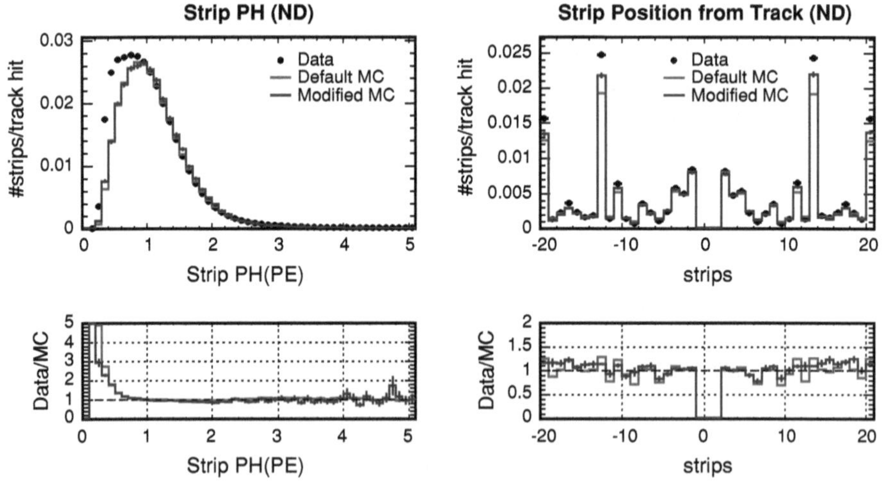

Fig. 8.2 Near Detector strip charge distribution of crosstalk hits around tracks (*left*), and distribution of transverse strip position from track center (*right*). The corresponding ratios of data to MC are shown in the *bottom*. The modified (retuned) crosstalk simulation improves the data to MC agreement, particularly for the strip position distribution. Images obtained from [8]

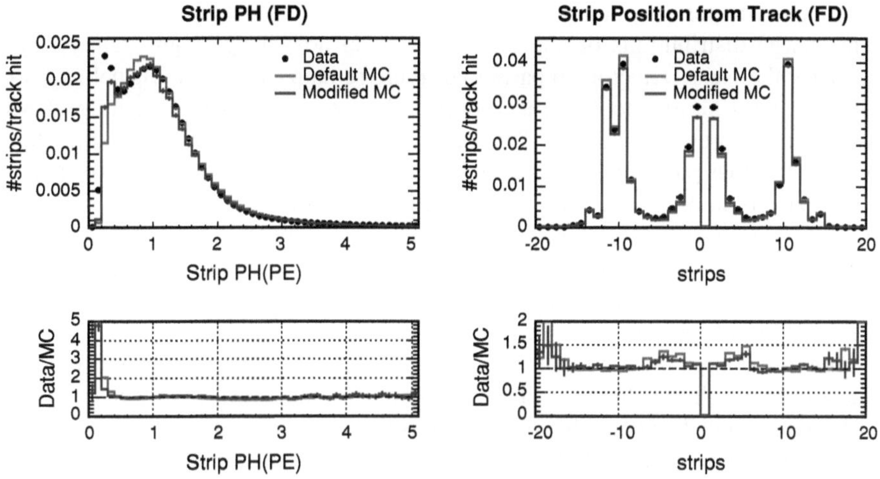

Fig. 8.3 Far Detector strip charge distribution of crosstalk hits around tracks (*left*), and distribution of transverse strip position from track center (*right*). The corresponding ratios of data to MC are shown in the *bottom*. The modified (retuned) crosstalk simulation improves the data to MC agreement, although the agreement there was already good. The strip charge distribution in the data has a double peak structure, corresponding most likely to the contributions from the electrical (charge) and optical crosstalk. Images obtained from [8]

M64 PMT, Alner Box Type 1

Pixel 0 23	Pixel 1 54	Pixel 2 41	Pixel 3 11	Pixel 4 35	Pixel 5 22	Pixel 6 53	Pixel 7 40
Pixel 8 43	Pixel 9 9	Pixel 10 37	Pixel 11 24	Pixel 12 55	Pixel 13 42	Pixel 14 10	Pixel 15 36
Pixel 16 39	Pixel 17 26	Pixel 18 57	Pixel 19 44	Pixel 20 8	Pixel 21 38	Pixel 22 25	Pixel 23 56
Pixel 24 59	Pixel 25 46	Pixel 26 6	Pixel 27 19	Pixel 28 27	Pixel 29 58	Pixel 30 45	Pixel 31 7
Pixel 32 4	Pixel 33 17	Pixel 34 29	Pixel 35 60	Pixel 36 47	Pixel 37 5	Pixel 38 18	Pixel 39 28
Pixel 40 31	Pixel 41 62	Pixel 42 49	Pixel 43 3	Pixel 44 16	Pixel 45 30	Pixel 46 61	Pixel 47 48
Pixel 48 51	Pixel 49 1	Pixel 50 14	Pixel 51 32	Pixel 52 63	Pixel 53 50	Pixel 54 2	Pixel 55 15
Pixel 56 12	Pixel 57 34	Pixel 58 21	Pixel 59 52	Pixel 60 0	Pixel 61 13	Pixel 62 33	Pixel 63 20

M16 PMT, one example

Pixel 0 50 4 166 27 96 73 189 143	Pixel 1 154 39 16 85 62 131 177 108	Pixel 2 190 167 51 97 5 74 120 28	Pixel 3 178 155 40 132 86 63 17 109
Pixel 4 156 87 41 133 18 64 110 179	Pixel 5 121 144 52 191 98 29 75 6	Pixel 6 157 65 134 180 42 111 88 19	Pixel 7 53 145 30 7 122 168 76 99
Pixel 8 8 146 31 54 100 77 123 169	Pixel 9 66 20 89 112 135 158 181 43	Pixel 10 101 55 32 78 9 147 124 170	Pixel 11 67 136 21 182 44 90 159 113
Pixel 12 22 91 137 68 160 45 114 183	Pixel 13 33 10 56 102 79 125 148 171	Pixel 14 46 115 69 184 138 23 92 161	Pixel 15 126 57 172 149 80 34 11 103

Fig. 8.4 An example of the relationship between pixel number and strip number for the M64 (*left*) and the M16 (*right*). The example on the left serves one partial U plane in the ND, while the one on the right serves 2/3 of a FD plane. Images obtained from [8]

By relating the charge of the crosstalk hits to the charge deposited by the muon in the original pixel as measured from cosmic ray events, the crosstalk fractions are calculated for each pixel and a tuned version of the crosstalk model is obtained [8]. As seen in Figs. 8.2 and 8.3, the retuned crosstalk model does significantly better, particularly for the strip position distributions on the right. The strip charge distributions on the left however do not change significantly, especially for the Near Detector. In particular, there remains a data excess below 1 PE, which is very similar to the one observed between the standard data and MC (shown in the previous section). This suggests that crosstalk mismodeling explains most of the low pulse-height disagreement in the standard sample. Nevertheless, electrical crosstalk (where charge in the PMT amplification process spreads into the neighboring channels) dominates at those pulse-heights, and it would have to be unrealistically amplified in order to account for the discrepancy. This is why, as mentioned in the previous section, it is believed that other factors contribute to the mismodeling of the low pulse-height hits.

Due to the large uncertainties inherent in the crosstalk and the low pulse-height hits simulation, hits below 3.0 PEs are ignored by the LEM algorithm, as previously mentioned. The cut value of 3.0 PEs is not arbitrary, and its determination is entirely driven by the need to keep the crosstalk systematic under reasonable limits (i.e., within a few percent) in the Near Detector, which is where the uncertainties are the largest. As a measure of the crosstalk systematic error, the default simulation is compared to the one that incorporates the retuned crosstalk model. Studies show that if a cut of 0.5 PEs is applied, the crosstalk systematic error in the Near Detector is larger than 40%, which is unacceptable. For a cut at 2.5 PEs, it falls down to approximately 6%, and for the cut at 3.0 PEs it becomes approximately 2%. Since the effect is smaller at the Far Detector, the total error in the NC + ν_μ CC prediction is approximately 4%. This makes the crosstalk systematic the third in magnitude.

As the PE cut is increased the low pulse-height hits and crosstalk systematics are greatly reduced. Unfortunately, so is the sensitivity of the algorithm, as physics hits that are part of the shower are also thrown away. Figure 8.5 shows the example of a ν_e CC event before and after the 3.0 PE cut. As seen there, practically all of the crosstalk hits are removed after the application of the cut, but so are some hits that are part of the shower. In the next analysis with the full dataset, where a

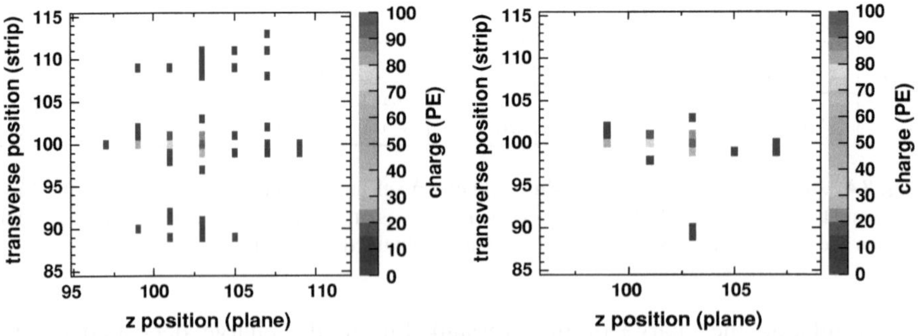

Fig. 8.5 A simulated ν_e CC event before (*left*) and after (*right*) the 3.0 PE cut applied by the LEM algorithm. The cross-talk hits form a cluster approximately ten strips below and above the shower hits. Practically all of them are removed by the 3.0 PE cut

better crosstalk model will be used, it might be possible to lower this PE cut and thus regain some sensitivity. Studies show that the sensitivity to θ_{13} is degraded by approximately 3–4% for every PE in the cut. This means that if the cut were reduced to 2.0 PE then an additional 3–4% in sensitivity might be gained.

8.1.3.3 Uncertainties in the Calibration

As explained in Sect. 3.5, the task of the MINOS calibration chain is to remove the variations within each detector (attenuation, strip to strip, etc.), to relate the energy depositions from the two detectors (relative calibration) and to translate the response into energy (absolute calibration). In this section we concern ourselves with those portions of the calibration chain that affect the determination of the energy deposition at each individual strip. The four items described in this section can have important contributions to the total systematic uncertainties given that changes in the relative energy deposition between the strips change the topology of the shower. In Sect. 8.1.3 we address the uncertainties that affect the strips as a group.

Attenuation correction uncertainty. As explained in Sect. B.1, light attenuation in the fiber makes for events that have a shorter fiber pathlength to the PMTs to appear more energetic. While this effect is mostly accounted for by the attenuation corrections, a residual variation in the stopping muon response as a function of position along the strip on the order of a few percent is observed in the data [9]. In order to estimate the attenuation systematic error, an MC sample is produced where the attenuation response as a function of position is rescaled to match the data distributions. The corresponding change in the number of LEM selected events is less than 0.3% for all background components. This effect goes in opposite directions in the two detectors and between the NC and CC components, which causes the systematic error on the NC + ν_μ CC predicted background to be practically zero.

Strip-to-strip correction uncertainty. The strip-to-strip correction removes the differences in strip response within a detector, which can be as large as 30% [9]. According to studies done by the MINOS Calibration Group, after the corrections the mean variation in strip-to-strip response is 0.5%. An MC study sample where the response of each strip is independently scaled by using a Gaussian distribution with a width of 0.5% is produced. The effects on the number of LEM selected events are very small. The beam ν_e component is the most affected, and changes by -0.5%. The error in the NC + ν_μ CC predicted background is only of 0.1%.

Linearity correction uncertainty. The linearity correction removes the non-linearity effects caused by PMT saturation at high pulse-heights. In the linearity portion of the calibration an error is calculated that reflects the uncertainty in the correction that was applied. The linearity correction uncertainty is estimated by producing an MC study sample where the linearity corrections are independently varied within their errors. Once again, the impact in the LEM selection is small, yielding a total uncertainty in the NC + ν_μ CC predicted events of only 0.3%.

Gains uncertainty. The PMT gains are only used in the conversion between ADCs and photoelectrons (PEs). As such, they do not play a role in the energy determination of an event. They however play a very important role in the LEM algorithm, where photoelectrons are used in the following two processes:

- the comparisons between input and library events (explained in Sect. 5.3.4).
- the 3.0 PE cut that is applied to remove cross-talk hits.

There is a 7% channel-to-channel uncertainty in the gains. However, due to the way they are determined from the light injection data, there can also be an overall systematic shift as high as ±5% [10]. In order to estimate the gains systematic error, a MC sample with all the gains systematically shifted by ±5% and then individually varied within 7% is produced. The effects of varying each channel individually within 7% is very small. However, systematically scaling all the gains by ±5% produces an effect of similar magnitude in the number of LEM selected events. As is the case with most of the other detector modeling uncertainties, the uncertainty in the gains is independent in the two detectors, which means that there is no cancellation in the Far–Near ratio. This translates into a $\sim$8% error in the NC + ν_μ CC prediction at the Far Detector.

The gains uncertainty is the largest source of systematic error in the LEM background prediction at the Far Detector. Section 10.1.1 includes a brief discussion on how this systematic error can be reduced in future versions of LEM.

8.1.3.4 Uncertainties in the Energy Scale

Relative energy scale uncertainty. The previous section described those effects that change the response at the individual strip level within the detector. As seen in Sect. 3.5, once the drift, attenuation, strip-to-strip and linearity calibrations have been applied, the resulting corrected number of ADCs is converted to an energy unit that is universal between the detectors. These units are termed MIPs.

Because of differences in light level, the conversion from corrected ADCs to MIPs is different for each detector. The uncertainty in this conversion factor is 2.3 and 2.4% in the Near Detector and the Far Detector respectively [9]. Modified MC samples where these factors are scaled up and down by their respective uncertainties are produced and the effects on the LEM selection recorded. Energy is a direct input in the LEM pid calculation (see Sect. 5.4.3), which is why this systematic is the second largest one in magnitude after gains. The resulting uncertainty in the predicted number of NC + ν_μ CC events from the relative energy scale uncertainty is approximately 5%.

Absolute energy scale uncertainty. Once the energy deposition has been converted to MIPs, there is one more conversion factor needed to convert this number to GeV. The uncertainty in this conversion factor is 5.7% [9].

The effect of varying the MIP to GeV conversion factor in the reconstruction by its uncertainty is quite large. The change in the number of LEM selected events in each detector surpasses 10% for some components. However, this conversion factor is used for both detectors, and thus the effects cancel to first order in the Far–Near ratio. The resulting uncertainty in the NC + ν_μ CC prediction is less than 1%.

Hadronic/EM shower energy scale uncertainty. So far, we have estimated the systematic error due to uncertainties in the event's total energy determination. But it is also possible for the hadronic and electromagnetic (EM) shower energy scales to be differently mismodeled, even if the total energy in the event remains correct. Such an effect could have an impact on the selection. For instance, if EM showers deposit more energy than expected in the detector models, then more events would be selected, as the EM component would have more importance over the hadronic one.

In order to assess the size of this systematic error, the selection efficiency as a function of true EM and hadronic energy is determined from the simulation. Then a variation of $\pm 5\%$ in the hadronic energy is considered (as motivated by Caldet studies), and the corresponding selection efficiency calculated. The ratio of new-to-old selection efficiencies provides a weight for each event in the simulation. Due to the fact that changing the hadronic energy scale while leaving the EM energy scale untouched modifies the total energy of the event, the total energy is also shifted accordingly. This process results in a change of less than 3% in the number of selected events in each detector. When it comes down to the predicted number of NC + ν_μ CC events, the effect is on the order of 1%.

8.1.3.5 Uncertainties in Neutrino Intensity

One more difference between the Far and Near detectors is the neutrino intensity at which they are exposed. For normal intensity, events in the Near Detector are reconstructed among another ~ 8 events, while Far Detector events occur one per spill. This can create a bias in the reconstruction and in the selection.

Studies reveal that the selection efficiency of LEM in the Near Detector is indeed intensity dependent [11], being approximately 15% higher in the low intensity regime ($\sim 2 \times 10^{12}$ protons-per-pulse) than in the normal intensity regime ($> 20 \times 10^{12}$ protons-per-pulse). This difference is not a good estimate of the systematic error however, as the simulation incorporates this pileup effect to some accuracy. It is observed that the selection efficiency is different for the first event in the spill. This is somewhat expected as events that occur late in the spill can be affected by residues of earlier events (PMT afterpulsing, neutrons, etc.). Thus a conservative approach to estimating the intensity systematic is to take the difference in selection efficiency for the first event in the spill between the data and MC. This yields an estimate of 4.0% for LEM, which applies only for those components that are extrapolated from the Near Detector (as opposed to ν_τ CC or oscillated ν_e CC events).

8.1.3.6 Uncertainties in the Preselection

If the reconstruction variables employed at the preselection are mismodeled differently in the two detectors, then this would introduce a bias in the Far–Near ratio. In order to evaluate a systematic error for such an effect, the selection efficiencies for the four major cuts involved in the preselection were calculated [12]. A comparison of the efficiencies between the two detectors showed differences smaller than 1%. Consequently a 1% systematic error is conservatively assigned to the predicted Far Detector background due to preselection uncertainties.

8.1.3.7 Uncertainties in the Normalization

There is a systematic error in the total exposure recorded, originating from uncertainties in the detector's live time as well as in the fiducial mass of the detectors. Both of these uncertainties affect the number of predicted events in the form of a scale factor independent of energy. This normalization systematic is evaluated using the estimates made for the ν_μ CC disappearance analysis [13]. In particular, the Far Detector live time is considered to have an uncertainty of 1.0%.[3] Also, there is an uncertainty in the steel and scintillator thicknesses of 0.2%, and an uncertainty in the fiducial volume of 2.1% as determined from data/MC comparisons. These effects amount to 2.4%.

8.1.3.8 Additional Uncertainties for Oscillated ν_τ CC and ν_e CC Events

The oscillated ν_τ CC and ν_e CC events are determined from the measured ν_μ CC flux using the appearance method described in Sect. 7.1.2. Therefore the selection uncertainties for ν_μ CC events have to be included as well. For our analysis, only the main three systematic errors in the ν_μ CC disappearance analysis [14] are considered. In addition to a normalization error of 4%, an error on the NC background and on the ν_μ CC shower energy scale of 50 and 10% are included respectively. The normalization error is larger for the ν_μ CC selection than for the ν_e selection given that a selection bias between the detectors of 3.0% is also included [13]. Given the small amount of NC background events that are selected in the ν_μ CC disappearance analysis, the error on the NC background amounts to less than 1% uncertainty in the predicted ν_τ CC and ν_e CC events. The uncertainty on the CC shower energy scale yields an uncertainty of 1% for the ν_τ CC prediction but of $\sim$3.5% for the ν_e CC prediction.

---

[3] The Near Detector live time uncertainty is negligible. It is also important to note that while the POT counting process in the beamline is also considered to have a 1% uncertainty, it affects the two detectors identically and thus cancels out.

8.1.4 Summary of Systematic Uncertainties

8.1.4.1 Simulation Uncertainties

The combination of all the systematic errors is done using the same software as developed in [15]. Each systematic shift is evaluated at each energy bin. The relevant quantity for the analysis, however, is the error on the total number of predicted events. Furthermore, there are very large bin-to-bin correlations. Because of this, we calculate and present the systematic errors on the total components (summed over all energies) throughout the rest of this chapter.

While assessing the uncertainties in the extrapolation we analyzed every aspect of the simulation that does not accurately represent the data. During this work the shifts for each systematic were obtained, allowing us to calculate the overall uncertainty in the simulation at both detectors. Tables 8.3 and 8.4 show the systematic uncertainties in the Near and Far MC events selected by LEM. Note that some of the uncertainties previously described originate in the extrapolation process and thus do not apply when considering the simulation independently at each detector. Such systematics are Normalization, Preselection, Intensity and Relative Energy.

As seen in Tables 8.3 and 8.4, the total systematic error in the simulation is very large. The error in the Far Detector is smaller only because the signal events, being less dependent on the hadronic model, have a smaller uncertainty than the hadronic backgrounds (NC and ν_μ CC). But if no signal is considered at the Far Detector then the total error is larger than 50%, as in the Near Detector.

As previously explained, the error in the Far Detector prediction as obtained in our analysis is much smaller, due to the fact that the backgrounds are measured in the Near Detector. This error is the one relevant for our analysis, and is calculated in the following sections.

8.1.4.2 Combining all the Uncertainties on the Prediction

The total number of background predicted events F^{BG} is given by

$$F_i^{BG} = F_i^{NC} + F_i^{\nu_\mu} + F_i^{\text{Beam}\,\nu_e} + F_i^{\nu_\tau}, \tag{8.1}$$

where F_i^{NC}, $F_i^{\nu_\mu}$, $F_i^{\text{Beam}\,\nu_e}$ and $F_i^{\nu_\tau}$ are the number of predicted NC, ν_μ CC, beam ν_e and ν_τ CC events respectively at each energy bin i. The NC and ν_μ CC components are extrapolated from the Near Detector, the beam ν_e component is obtained from the MC, and the ν_τ CC component is obtained from the ν_μ CC measured flux. Consequently, we have

$$F_i^{NC} = R_i^{NC} N_i^{NC}, \tag{8.2}$$

$$F_i^{\nu_\mu} = R_i^{\nu_\mu} N_i^{\nu_\mu}, \tag{8.3}$$

Table 8.3 Systematic errors on the events selected at the Near Detector with LEM > 0.65

Systematic	Shift	NC (%)	ν_μ CC (%)	Beam ν_e (%)	Total (%)
Hadronic model	T1	−37.4	−51.7	−1.6	−35.8
	T2+	10.8	12.4	0.3	9.7
	T2−	−10.4	−12.1	−0.3	−9.4
	T3	−19.6	−13.9	−0.7	−15.5
	T4	−15.5	−21.2	−0.7	−14.8
	T5	−14.1	−19.7	−0.7	−13.6
	T6	−14.5	−23.1	−1.2	−14.8
Hadron multiplicity	$+1\sigma$	1.9	0.4	−1.7	1.1
	-1σ	−2.6	−4.1	−0.8	−2.9
M_A(QE)	−15%	−0.1	−0.1	−4.4	−0.5
	+15%	0.1	0.1	4.5	0.5
M_A(RES)	−15%	−2.0	−1.5	−6.3	−2.3
	+15%	2.9	2.0	6.5	3.0
KNO	±50%	±2.3	±2.2	±6.4	±2.6
Beam flux	$\pm1\sigma$	±9.8	±9.0	±9.0	±9.5
Intranuke	1σ	−17.3	−19.0	3.3	−15.9
Low pulse-height	1σ	−3.5	−6.4	−1.9	−5.0
Crosstalk	1σ	−3.8	−0.1	1.1	−2.3
Attenuation	1σ	−0.3	0.1	0.1	−0.2
Strip-to-strip	1σ	0.1	0.3	−0.5	0.1
Linearity	1σ	−0.6	0.3	0.5	−0.2
Gains	-1σ	−5.3	−3.2	−1.2	−4.3
	$+1\sigma$	5.4	9.3	−0.8	6.0
Abs. energy	$+1\sigma$	7.5	10.0	8.4	8.3
	-1σ	−8.4	−10.8	−11.4	−9.3
Had/EM energy	$+1\sigma$	−1.7	−1.6	−1.8	−1.7
	-1σ	1.9	2.7	3.1	2.3
Total		−55.5	−70.8	−18.5	−52.3
		55.4	71.3	16.9	51.3

The total error exceeds 50%. The double line in the middle separates the physics modeling systematics from the detector modeling systematics

$$F_i^{\text{Beam}\,\nu_e} = R_i^{\text{Beam}\,\nu_e} N_i^{\text{Beam}\,\nu_e} \text{ (from MC)}, \tag{8.4}$$

$$F_i^{\nu_\tau} = RCC_i^{\nu_\tau} \epsilon_i^{\nu_\tau} N_i^{\nu_\mu CC\text{-like}}, \tag{8.5}$$

where R^j denotes the Far–Near ratio of background component j, RCC^k denotes the conversion factor required to turn a Near Detector ν_μ CC-like spectrum into a Far Detector spectrum of oscillated k events,[4] and ϵ^j stands for the selection efficiency of j events in the Far Detector.

4 As such, RCC^k includes the extrapolation to the Far Detector, the conversion from reconstructed to true energy, the spectral corrections due to the purity and efficiency of the ν_μ CC selection, the $\nu_\mu \rightarrow k$ oscillation probability, and the ratio of ν_μ CC to k cross-sections. The details are explained in Sect. 7.1.2.

Table 8.4 Systematic errors on the simulated background events selected at the Far Detector with LEM > 0.65

Systematic	Shift	NC (%)	ν_μ CC (%)	Beam ν_e (%)	ν_τ CC (%)	Osc. ν_e CC (%)	Total (%)
Hadronic model	T1	−36.6	−52.6	−2.0	NA	−1.7	−23.2
	T2+	11.0	12.6	0.2	NA	0.4	6.6
	T2−	−10.6	−12.2	−0.2	NA	−0.4	−6.3
	T3	−21.2	−13.3	0.4	NA	−0.5	−11.4
	T4	−16.3	−22.7	−0.8	NA	−0.7	−10.2
	T5	−15.0	−21.6	−0.8	NA	−0.6	−9.5
	T6	−16.1	−23.8	−0.9	NA	−0.7	−10.3
Hadron multiplicity	$+1\sigma$	1.3	−0.5	−1.5	−0.7	−1.5	0.2
	-1σ	−1.9	−2.8	0.1	−0.2	−0.7	−1.6
M_A(QE)	−15%	−0.1	−0.1	−4.0	−10.3	−5.0	−1.7
	+15%	0.2	0.1	4.2	11.7	5.1	1.8
M_A(RES)	−15%	−2.1	−0.8	−6.3	−7.3	−6.2	−3.3
	+15%	2.9	1.2	6.6	8.9	6.1	3.8
KNO	±50%	±2.5	±1.5	±6.9	±4.4	±6.2	±3.6
Beam flux	$\pm1\sigma$	±9.0	±8.9	±13.8	±10.7	±7.1	±8.8
Intranuke	1σ	−16.8	−19.5	−2.2	−2.9	−2.0	−12.4
Low pulse-height	1σ	−1.7	−4.0	−0.9	−1.3	−1.8	−2.4
Crosstalk	1σ	0.2	0.1	0.1	−1.8	−0.6	−0.0
Attenuation	1σ	−0.2	−0.1	0.3	−0.2	−0.2	−0.2
Strip-to-strip	1σ	−0.0	0.0	0.1	0.0	0.0	0.0
Linearity	1σ	−0.1	−0.2	−0.4	−0.1	−0.0	−0.1
Gains	-1σ	−4.5	−6.0	−2.2	−3.6	−2.7	−4.1
	$+1\sigma$	6.1	8.9	0.4	3.5	3.1	5.3
Abs. energy	$+1\sigma$	7.9	10.0	7.3	9.0	3.6	7.1
	-1σ	−7.9	−12.2	−6.7	−7.6	−4.3	−7.5
Had/EM energy	$+1\sigma$	−1.9	−2.6	−1.9	−0.9	−0.7	−1.5
	-1σ	3.8	7.1	1.9	1.3	0.7	2.9
Total		−55.9	−72.9	−19.0	−48.3	−13.9	−36.8
		56.2	73.2	19.2	49.1	13.6	35.5

The signal is calculated assuming θ_{13} at the CHOOZ limit and ignoring CP violation and matter effects. The total error exceeds 35%. The double line in the middle separates the physics modeling systematics from the detector modeling systematics

When calculating the variations of F_i^{BG} it is advantageous to consider the correlations that exist in the Near Detector decomposition. In particular, all decomposition methods are subject to the constraint that the total number of events must equal the data N^{data}:

$$N_i^{\text{data}} = N_i^{NC} + N_i^{\nu_\mu} + N_i^{\text{Beam}\,\nu_e}. \tag{8.6}$$

This is why even if the HOO and the MRCC decompositions yield different results in the Near Detector, their Far Detector predictions do not differ by more than a few percent, as explained in the end of Sect. 7.1.2. By expressing the number of NC events in the Near Detector as a function of N^{data}, which has only a small statistical error, we can take into account the correlation between the background components and thus get a smaller systematic error on the total Far Detector prediction:

$$N_i^{NC} = N_i^{\text{data}} - N_i^{\nu_\mu} - N_i^{\text{Beam}\ \nu_e}. \tag{8.7}$$

Rewriting Eq. 8.1 in more detail we get

$$\begin{aligned} F_i^{BG} = R_i^{NC}&(N_i^{\text{data}} - N_i^{\nu_\mu} - N_i^{\text{Beam}\ \nu_e}) \\ &+ R_i^{\nu_\mu} N_i^{\nu_\mu} \\ &+ R_i^{\text{Beam}\ \nu_e} N_i^{\text{Beam}\ \nu_e} \\ &+ RCC_i^{\nu_\tau} \epsilon_i^{\nu_\tau} N_i^{\nu_\mu CC-\text{like}}. \end{aligned} \tag{8.8}$$

The error in the total Far Detector prediction is thus given by

$$\begin{aligned} \delta F_i^{BG} = &\ \delta R_i^{NC} N_i^{NC} + \delta R_i^{\nu_\mu} N_i^{\nu_\mu} + \delta R_i^{\text{Beam}\ \nu_e} N_i^{\text{Beam}\ \nu_e} \\ &+ (\delta RCC_i^{\nu_\tau} \epsilon_i^{\nu_\tau} + RCC_i^{\nu_\tau} \delta\epsilon_i^{\nu_\tau}) N_i^{\nu_\mu CC-\text{like}} \\ &+ R_i^{NC} \delta N_i^{\text{data}} + (R_i^{\nu_\mu} - R_i^{NC}) \delta N_i^{\nu_\mu} \\ &+ (R_i^{\text{Beam}\ \nu_e} - R_i^{NC}) \delta N_i^{\text{Beam}\ \nu_e} + RCC_i^{\nu_\tau} \epsilon_i^{\nu_\tau} \delta N_i^{\nu_\mu CC-\text{like}}. \end{aligned} \tag{8.9}$$

The first four terms (equivalent to the first two lines) in the previous equation are the contributions to the total error from uncertainties in the extrapolation, while the last four terms (in the last two lines) give the contributions from uncertainties in the decomposition. These two contributions are respectively presented in the following two sections.

8.1.4.3 Contributions from the Extrapolation

The two main background components, NC and ν_μ CC, are extrapolated from the Near Detector. As such, they share the same set of systematic errors and are consequently presented together in Tables 8.5 and 8.6, for the HOO and MRCC decompositions respectively. The first two columns in these tables present the relative uncertainties for each of these two background predictions, that is $\left(\sum_i \delta R_i^{NC} N_i^{NC}\right)/F^{NC}$ and $\left(\sum_i \delta R_i^{\nu_\mu} N_i^{\nu_\mu}\right)/F^{\nu_\mu}$. In order to account for the fact that the systematics affect the NC and ν_μ CC components in a correlated way, the fractional error on their sum,

$$\frac{\delta F_{\text{extrap}}^{NC+\nu_\mu}}{F^{NC+\nu_\mu}} = \frac{\sum_i \left(\delta R_i^{NC} N_i^{NC} + \delta R_i^{\nu_\mu} N_i^{\nu_\mu}\right)}{F^{NC} + F^{\nu_\mu}}, \tag{8.10}$$

is obtained. This way the NC and ν_μ CC components are considered as a single one when it comes to calculating the systematic error on the total background prediction. The systematic uncertainty on the NC + ν_μ CC component is presented graphically in Fig. 8.6.

Table 8.5 Systematic errors for the NC and ν_μ CC Far Detector predictions with LEM > 0.65, based on the HOO decomposition

Systematic	Shift	NC		ν_μ CC		NC + ν_μ CC	
		Min (%)	Max (%)	Min (%)	Max (%)	Min (%)	Max (%)
Hadronic model	T1	−0.3	0.3	−2.5	2.5	−0.2	0.2
	T2	−0.3	0.3	0.2	−0.2	−0.2	0.3
	T3	−2.3	2.3	−0.4	0.4	−1.9	1.9
	T4	−1.0	1.0	−1.3	1.3	−1.1	1.1
	T5	−1.2	1.2	−1.5	1.5	−1.2	1.2
	T6	−1.9	1.9	−1.1	1.1	−1.7	1.7
Hadron multiplicity	$\pm 1\sigma$	−0.6	0.7	0.4	0.2	−0.4	0.6
M_A(QE)	±15%	−0.0	0.0	−0.0	0.1	−0.0	0.0
M_A(RES)	±15%	−0.0	0.0	−0.5	0.4	−0.1	0.1
KNO	±50%	−0.2	0.2	−0.3	0.4	−0.1	0.1
Beam flux	$\pm 1\sigma$	−0.7	0.8	−0.7	0.9	−0.7	0.9
Intranuke	1σ	−0.6	0.6	−0.5	0.5	−0.4	0.4
Low pulse-height	1σ	−1.5	1.5	−2.1	2.1	−1.6	1.6
Crosstalk	1σ	−4.2	4.2	−1.1	1.1	−3.7	3.7
Attenuation	1σ	−0.1	0.1	−0.3	0.3	−0.0	0.0
Strip-to-strip	1σ	−0.2	0.2	−0.2	0.2	−0.1	0.1
Linearity	1σ	−0.5	0.5	−0.7	0.7	−0.3	0.3
Gains	ND $\pm 1\sigma$	−5.0	5.8	−8.4	4.5	−5.5	5.6
	FD $\pm 1\sigma$	−4.2	5.9	−6.8	9.5	−4.6	6.5
Rel. Energy	ND $\pm 1\sigma$	−3.1	2.1	−4.7	3.0	−3.4	2.2
	FD $\pm 1\sigma$	−2.4	4.0	−7.3	6.4	−3.2	4.4
Abs. energy	$\pm 1\sigma$	0.5	0.3	−6.6	2.8	−0.6	0.7
Had/EM energy	$\pm 1\sigma$	−0.3	1.3	0.0	0.7	−0.2	1.2
Intensity	$\pm 1\sigma$	−4.0	4.0	−4.0	4.0	−4.0	4.0
Preselection	$\pm 1\sigma$	−1.0	1.0	−1.0	1.0	−1.0	1.0
Normalization	$\pm 1\sigma$	−2.4	2.4	−2.4	2.4	−2.4	2.4
Total		−10.7	12.1	−16.6	14.5	−11.1	12.2

The beam ν_e component is obtained from the simulation. The error in the Far Detector simulation was already presented in Table 8.4. To that estimate we just need to add the uncertainties in the Far Detector live time and fiducial mass (i.e., the normalization uncertainties). Also, the correction to the ν_e selection efficiency from the MRE procedure is applied to the beam ν_e component, which contributes an additional $^{-3.1\%}_{+3.4\%}$ uncertainty.[5] The fractional systematic error on the beam ν_e component is shown in Table 8.7.

[5] Given that the error from the simulation already accounts for the fact that the hadronic showers are mismodeled, there is some double counting by also including the error in the selection efficiency as determined from the MRE procedure, as the MRE correction accounts precisely for the same effect. Due however to the fact that the total error on the beam ν_e component on the order of 20%, it makes essentially no difference to remove it.

Table 8.6 Systematic errors for the NC and ν_μ CC Far Detector predictions with LEM > 0.65, based on the MRCC decomposition

Systematic	Shift	NC		ν_μ CC		NC+ν_μ CC	
		Min (%)	Max (%)	Min (%)	Max (%)	Min (%)	Max (%)
Hadronic model	T1	−0.6	0.6	−2.8	2.8	−0.1	0.1
	T2	−0.3	0.3	0.0	−0.0	−0.3	0.3
	T3	−2.4	2.4	−0.5	0.5	−1.9	1.9
	T4	−1.0	1.0	−1.9	1.9	−1.2	1.2
	T5	−1.1	1.1	−2.0	2.0	−1.3	1.3
	T6	−1.9	1.9	−1.2	1.2	−1.8	1.8
Hadron multiplicity	$\pm 1\sigma$	−0.6	0.8	0.3	0.2	−0.5	0.7
M_A(QE)	±15%	−0.0	0.0	−0.0	0.0	−0.0	0.0
M_A(RES)	±15%	−0.0	0.0	−0.4	0.4	−0.1	0.1
KNO	±50%	−0.2	0.2	−0.3	0.3	−0.2	0.1
Beam flux	$\pm 1\sigma$	−0.7	0.9	−0.6	0.8	−0.7	0.9
Intranuke	1σ	−0.4	0.4	−0.3	0.3	−0.3	0.3
Low pulse-height	1σ	−1.3	1.3	−2.5	2.5	−1.5	1.5
Crosstalk	1σ	−4.2	4.2	−0.5	0.5	−3.4	3.4
Attenuation	1σ	−0.1	0.1	−0.2	0.2	−0.0	0.0
Strip-to-strip	1σ	−0.0	0.0	−0.4	0.4	−0.0	0.0
Linearity	1σ	−0.5	0.5	−0.3	0.3	−0.4	0.4
Gains	ND $\pm 1\sigma$	−5.0	5.8	−7.2	2.6	−5.4	5.3
	FD $\pm 1\sigma$	−4.5	6.1	−5.8	8.7	−4.7	6.5
Rel. Energy	ND $\pm 1\sigma$	−3.1	2.5	0.1	2.9	−2.6	2.6
	FD $\pm 1\sigma$	−2.6	4.1	−4.6	4.3	−3.0	4.1
Abs. energy	$\pm 1\sigma$	0.6	0.5	−5.8	2.2	−0.4	0.8
Had/EM energy	$\pm 1\sigma$	−0.1	1.5	−1.4	4.3	−0.3	1.9
Intensity	$\pm 1\sigma$	−4.0	4.0	−4.0	4.0	−4.0	4.0
Preselection	$\pm 1\sigma$	−1.0	1.0	−1.0	1.0	−1.0	1.0
Normalization	$\pm 1\sigma$	−2.4	2.4	−2.4	2.4	−2.4	2.4
Total		−10.8	12.3	−13.8	13.5	−10.7	12.1

The oscillated ν_τ CC and ν_e CC components are obtained from the measured ν_μ CC-like spectrum in the Near Detector and thus are affected by the additional uncertainties described in Sect. 8.1.3. Table 8.8 summarizes the fractional error on these two components, as obtained from Eq. 8.9:

$$\frac{\delta F^{\nu_\tau CC}_{\text{extrap}}}{F^{\nu_\tau CC}} = \frac{\sum_i (\delta RCC_i^{\nu_\tau} \epsilon_i^{\nu_\tau} + RCC_i^{\nu_\tau} \delta\epsilon_i^{\nu_\tau}) N_i^{\nu_\mu CC-\text{like}}}{F^{\nu_\tau CC}}, \tag{8.11}$$

$$\frac{\delta F^{\nu_e CC}_{\text{extrap}}}{F^{\nu_e CC}} = \frac{\sum_i (\delta RCC_i^{\nu_e} \epsilon_i^{\nu_e} + RCC_i^{\nu_e} \delta\epsilon_i^{\nu_e}) N_i^{\nu_\mu CC-\text{like}}}{F^{\nu_e CC}}. \tag{8.12}$$

Table 8.8 shows the total error on the ν_e CC component if the selection efficiency is taken from the MC, in which case a systematic for uncertainties in the hadronic model and in the intranuclear rescattering is necessary. As discussed in Sect. 7.2.3 however, the MRE correction is applied to the MC selection efficiency in order to adjust the expected number of signal events due to hadronic shower

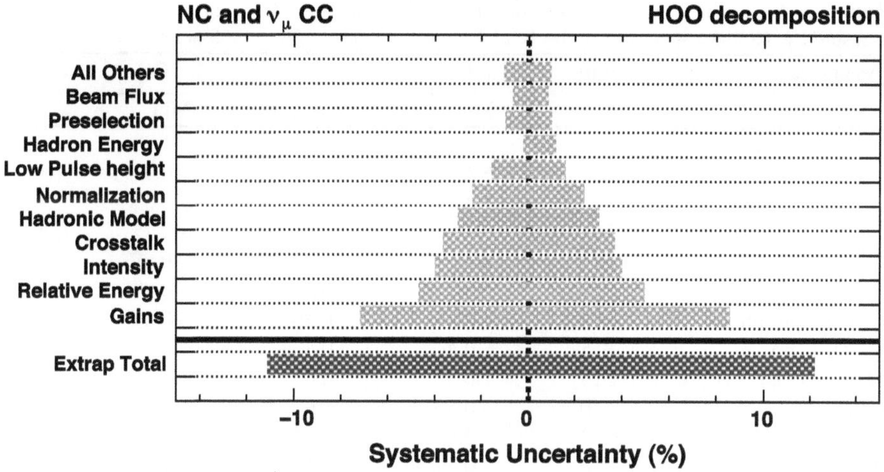

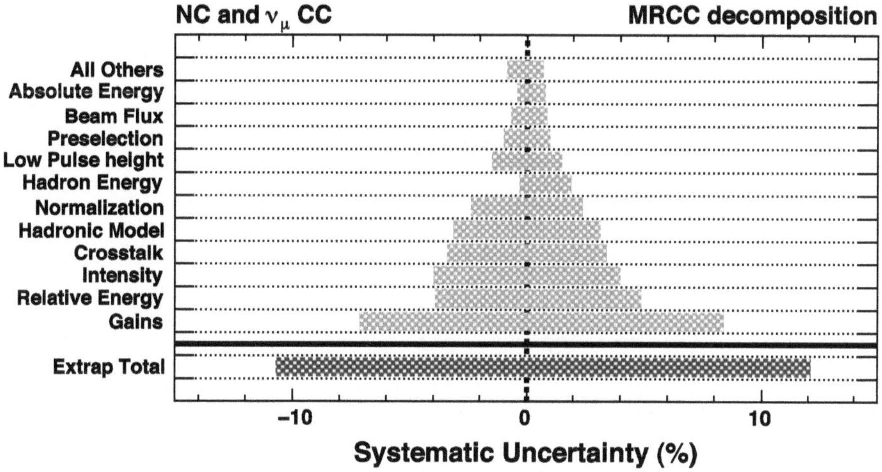

Fig. 8.6 Extrapolation systematic uncertainties in the NC + ν_μ CC prediction at the Far Detector, for the HOO decomposition (*top*) and the MRCC decomposition (*bottom*). The sources of systematic error are arranged in order of increasing importance. Systematic effects that contribute less than 0.7% are grouped into the all others category

Table 8.7 Systematic errors for the beam ν_e Far Detector events selected with LEM > 0.65

Systematic	Shift	Beam ν_e	
		Min (%)	Max (%)
Simulation	$\pm 1\sigma$	−19.0	19.2
Normalization	$\pm 1\sigma$	−2.4	2.4
Efficiency (MRE)	$\pm 1\sigma$	−3.1	3.4
Total		−19.4	19.6

The simulation error is obtained from Table 8.4

Table 8.8 Systematic errors for the oscillated ν_τ CC and ν_e CC events, predicted from the measured ν_μ CC-like spectrum in the Near Detector for the LEM > 0.65 selection

Systematic	Shift	ν_τ CC		ν_e CC	
		Min	Max	Min	Max
Hadronic model	T1	NA	NA	−1.7%	1.7%
	T2	NA	NA	−0.4%	0.4%
	T3	NA	NA	−0.5%	0.5%
	T4	NA	NA	−0.7%	0.7%
	T5	NA	NA	−0.6%	0.6%
	T6	NA	NA	−0.7%	0.7%
Hadron multiplicity	$\pm 1\sigma$	−0.4%	0.4%	−0.8%	0.9%
M_A(QE) (%)	±15	−0.4	0.4	−0.2	0.3
M_A(RES) (%)	±15	−0.3	0.4	−0.2	0.0
KNO (%)	±50	−0.5	0.6	−0.0	0.0
$\sigma_{\nu_\tau CC}$ (%)	±50	−49.0	49.2	NA	NA
Beam flux	$\pm 1\sigma$	−0.6%	0.5%	−1.1%	1.3%
Intranuke	1σ	−3.1%	3.1%	−1.7%	1.7%
Low pulse-height	1σ	−0.9%	0.9%	−1.7%	1.7%
Crosstalk	1σ	−1.9%	1.9%	−0.6%	0.6%
Attenuation	1σ	−0.1%	0.1%	−0.2%	0.2%
Strip-to-strip	1σ	−0.0%	0.0%	0.0%	0.0%
Linearity	1σ	−0.1%	0.1%	−0.1%	0.1%
Gains	$\pm 1\sigma$	−3.6%	3.5%	−2.7%	3.0%
Abs. energy	$\pm 1\sigma$	−5.6%	7.4%	−4.5%	4.0%
Preselection	$\pm 1\sigma$	−1.0%	1.0%	−1.0%	1.0%
Normalization	$\pm 1\sigma$	−4.0%	4.0%	−4.0%	4.0%
CC shower energy	$\pm 1\sigma$	−1.1%	1.1%	−3.7%	3.4%
Total		−49.8%	50.3%	−8.4%	8.2%

This table corresponds to the case where the selection efficiency at the Far Detector is taken from the simulation. As such, this table gives the final error on the ν_τ CC component but not for the signal, where the MRE correction is applied (see Table 8.9)

mismodeling. In that case the errors associated with the hadronic showers disappear and the error on the MRE efficiency correction takes their place, as shown in Table 8.9. This way we avoid double counting some of the systematic errors affecting the signal, to first order. The impact of adding or removing some of these systematics is very small anyway, given that the total error in the signal is dominated by the uncertainties in the absolute energy, normalization, and CC shower energy.

It is interesting to compare the magnitude of the total systematic errors on the oscillated ν_τ and ν_e CC events as obtained through the appearance method (Tables 8.8 and 8.9) and as obtained from the simulation (Table 8.4). There is practically no difference for the ν_τ CC component, because the overall systematic error in the two cases is dominated by the uncertainty in the deep-inelastic ν_τ cross-section. For the signal events however, by extrapolating the measured ν_μ CC-like spectrum and thus canceling some of the physics related systematics in the process (such as in the beam flux and in the cross-sections), the overall systematic error is reduced from 14 to 8.5% approximately.

Table 8.9 Systematic errors for the oscillated ν_e CC events, predicted from the measured ν_μ CC-like spectrum in the Near Detector for the LEM > 0.65 selection

Systematic	Shift	ν_e CC	
		Min (%)	Max (%)
M_A(QE)	±15%	−0.2	0.3
M_A(RES)	±15%	−0.2	0.0
KNO	±50%	−0.0	0.0
Beam flux	$\pm 1\sigma$	−1.1	1.3
Low pulse-height	1σ	−1.7	1.7
Crosstalk	1σ	−0.6	0.6
Attenuation	1σ	−0.2	0.2
Strip-to-strip	1σ	0.0	0.0
Linearity	1σ	−0.1	0.1
Gains	$\pm 1\sigma$	−2.7	3.0
Abs. energy	$\pm 1\sigma$	−4.5	4.0
Preselection	$\pm 1\sigma$	−1.0	1.0
Normalization	$\pm 1\sigma$	−4.0	4.0
CC shower energy	$\pm 1\sigma$	−3.7	3.4
Efficiency (MRE)	$\pm 1\sigma$	−3.1	3.4
Total		−8.5	8.4

This table corresponds to the case where the MRE correction is applied to the selection efficiency in order to account for the mismodeling of hadronic showers. The total error given here is the one used in the rest of the analysis for the signal ν_e CC events

Equation 8.9 gives the error on the background prediction as a function of each energy bin i. In order to obtain the error on the total background prediction, the error on the total of each component has to be summed in quadrature. Consequently, the error from uncertainties in the extrapolation is given by

$$\begin{aligned}\delta F^{BG}_{\text{extrap}} &= \sum_i \left(\delta R_i^{NC} N_i^{NC} + \delta R_i^{\nu_\mu} N_i^{\nu_\mu}\right) \\ &\oplus \sum_i \delta R_i^{\text{Beam }\nu_e} N_i^{\text{Beam }\nu_e} \\ &\oplus \sum_i \left(\delta RCC_i^{\nu_\tau} \epsilon_i^{\nu_\tau} + RCC_i^{\nu_\tau} \delta\epsilon_i^{\nu_\tau}\right) N_i^{\nu_\mu CC-\text{like}},\end{aligned} \tag{8.13}$$

where $\oplus$ denotes sum in quadrature, and where

$$\sum_i \delta R_i^{\text{Beam }\nu_e} N_i^{\text{Beam }\nu_e} = \sum_i \delta F_{MC,i}^{\text{Beam }\nu_e} \tag{8.14}$$

is the total error on the beam ν_e component as obtained from the simulation.

Table 8.10 combines the information on the number of predicted events (from Table 7.1) and on the extrapolation errors to calculate the extrapolation uncertainty on the total prediction F^{BG}. By adding the individual $\delta F^x_{\text{extrap}}$ contributions from each background component in quadrature as shown in Eq. 8.13, a fractional error

Table 8.10 Far Detector prediction F^x, systematic uncertainty from the extrapolation $\delta F^x_{\text{extrap}}$ and corresponding fractional error $\delta F^x_{\text{extrap}}/F^x$ for the three background components and the total background

	NC + ν_μ CC		Beam ν_e	ν_τ CC	Total	
	HOO	MRCC			HOO	MRCC
LEM > 0.65						
F^x	17.7	18.2	2.7	1.1	21.4	22.0
$\delta F^x_{\text{extrap}}$	2.16	2.20	0.53	0.55	2.29	2.33
$\delta F^x_{\text{extrap}}/F^x$ (%)	12.2	12.1	19.6	50.3	10.6	10.6
LEM > 0.80						
F^x	9.6	10.2	2.2	0.7	12.6	13.1
$\delta F^x_{\text{extrap}}$	1.32	1.31	0.48	0.36	1.45	1.44
$\delta F^x_{\text{extrap}}/F^x$ (%)	13.7	12.8	21.8	51.2	11.5	11.0

The fractional error on the total background prediction from uncertainties in the extrapolation amounts to approximately 11% in all cases

of 10.6% on the total background prediction at the Far Detector is obtained from uncertainties in the extrapolation for the LEM > 0.65 selection. Although the intermediate details are not shown, the same procedure as described so far is followed to estimate the extrapolation systematic uncertainties for the cut at 0.80. In that case the error is slightly larger than for the cut at 0.65, particularly for the HOO decomposition method.

8.1.4.4 Contributions from the Decomposition

The errors from the Near Detector decomposition have both a statistical and a systematic component to them, and these two are combined differently. The statistical errors are summed in quadrature over each energy bin, while the systematic errors are summed linearly, taking into account the sign of the deviation in each energy bin. Then the decomposition errors for each component are summed in quadrature. This is illustrated by the following equation, which is derived from Eq. 8.9:

$$\begin{aligned}\delta F^{BG}_{\text{decomp}} &= \sum_i^{\oplus} R_i^{NC}\delta N^{\text{data}}_{i,\text{STAT}} \\ &\oplus \sum_i^{\oplus}(R_i^{\nu_\mu} - R_i^{NC})\delta N^{\nu_\mu}_{i,\text{STAT}} \oplus \sum_i (R_i^{\nu_\mu} - R_i^{NC})\delta N^{\nu_\mu}_{i,\text{SYS}} \\ &\oplus \sum_i (R_i^{\text{Beam}\ \nu_e} - R_i^{NC})\delta N^{\text{Beam}\ \nu_e}_{i,\text{SYS}} \\ &\oplus \sum_i RCC_i^{\nu_\tau}\epsilon_i^{\nu_\tau}\delta N_i^{\nu_\mu CC\text{-like}}.\end{aligned} \tag{8.15}$$

Once again, $\oplus$ denotes sum in quadrature and $\sum_i^{\oplus}$ denotes the sum in quadrature over energy bins. The error in the LEM selected data is statistical only and is very

Table 8.11 Fractional error on the total background prediction from decomposition uncertainties for the two LEM cut values

Method	$\delta F^{BG}_{\mathrm{decomp}}/F^{BG}$ (stat) (%)	$\delta F^{BG}_{\mathrm{decomp}}/F^{BG}$ (syst) (%)	$\delta F^{BG}_{decomp}/F^{BG}$ (total) (%)
LEM > 0.65			
HOO	3.2	4.4	5.5
MRCC	1.2	4.1	4.2
LEM > 0.80			
HOO	5.0	6.8	8.4
MRCC	1.2	4.5	4.7

The error increases with the higher LEM cut, although in all cases it is significantly smaller than the error due to extrapolation uncertainties

small, which makes the first term practically negligible. Similarly, the error in the ν_μ CC-like spectra is even smaller and the last term in the previous equation is ignored altogether. During the decomposition an error of 30% was assumed at each energy bin on the beam ν_e component, which also covers the statistical uncertainties. This is why the fourth term in the previous equation has no statistical counterpart.

The results for $\delta F^{BG}_{\mathrm{decomp}}$ are shown in Table 8.11. In the case of HOO, the decomposition fractional error significantly increases with the more stringent cut. This is due in part to the fact that as the cut increases the less horn-off data statistics are available to do the decomposition. In both cases however the systematic error from the decomposition is significantly smaller than the one from the extrapolation.

8.1.4.5 Total Systematic Uncertainty on the Background

Having calculated the contributions to the error on the total background prediction from the decomposition and the extrapolation, these two are simply combined in quadrature to yield the overall uncertainty:

$$\delta F^{BG} = \delta F^{BG}_{\mathrm{extrap}} \oplus \delta F^{BG}_{\mathrm{decomp}}. \tag{8.16}$$

Based on the information from Tables 8.10 and 8.11, a total uncertainty in the background prediction of 12.0% is obtained for the LEM cut at 0.65 and of 14.2% for the LEM cut at 0.80, when using the preferred HOO decomposition method. This information is shown in Table 8.12.

8.2 Overall Physics Reach

8.2.1 Number of Selected Events

Now that all the predictions are finalized, together with an estimate of their systematic uncertainties, it is possible to study the LEM physics reach in θ_{13} for our analysis, which corresponds to an exposure of 3.14×10^{20} POT.

Table 8.12 Fractional systematic error on the total LEM background prediction due to uncertainties in the decomposition, in the extrapolation and in total

Method	Decomposition (%)	Extrapolation (%)	Tota (%)l
LEM > 0.65			
HOO	5.5	10.6	12.0
MRCC	4.2	10.6	11.4
LEM > 0.80			
HOO	8.4	11.5	14.2
MRCC	4.7	11.0	12.0

The total error in the number of predicted events is between 11.4 and 14.2%, and it is larger for the most stringent cut at 0.80

Table 8.13 Far Detector predictions for the ANN and LEM selections, as well as the corresponding systematic fractional errors

Method	Signal	σ_{sg}^{syst} (%)	Total bg.	$\sigma_{bg.}^{syst}$ (%)	Sig./bg.	Super-fom
ANN						
HOO	10.3	7.7	26.3	7.4	0.39	1.88
MRCC	10.3	7.7	27.8	7.1	0.37	1.83
LEM > 0.65						
HOO	11.1	8.5	21.4	12.0	0.52	2.10
MRCC	11.1	8.5	22.0	11.4	0.50	2.09
LEM > 0.80						
HOO	8.7	9.3	12.6	14.2	0.70	2.20
MRCC	8.7	9.3	13.1	12.0	0.66	2.20

The predictions are normalized to an exposure of 3.14×10^{20} POTs. The signal corresponds to a value of θ_{13} at the CHOOZ limit ($\sin^2(2\theta_{13}) = 0.15$), ignoring CP violation and matter effects. The signal to background ratio and the values of the super-fom for each case are also shown

Before deriving the expected exclusion contours, we summarize the results obtained so far in Table 8.13. There the amount of signal expected for a θ_{13} at the CHOOZ limit is shown, as well as its fractional systematic error σ_{sg}. Similarly, the expected background is shown together with its fractional systematic error σ_{bg}. The ratio of signal to background is calculated for each selection, as well as the "super-fom." The super-fom is the sensitivity estimator used to optimize the LEM selection, as described in Sect. 5.4.3. It is simply a measure of the expected signal with respect to the fluctuations of the background (systematic and statistical),

$$\text{super-fom} = \frac{F^{\text{signal}}}{\sqrt{\sigma_{\text{stat}}^2 + \sigma_{\text{syst}}^2}} = \frac{F^{\text{signal}}}{\sqrt{F^{\text{BG}} + (\sigma_{bg}^{\text{syst}} F^{\text{BG}})^2}}, \tag{8.17}$$

where F^{BG} and F^{signal} represent the total background and signal predictions respectively, and $\sigma_{bg}^{\text{syst}}$ represents the systematic uncertainty on the background.

The Far Detector signal and background predictions for the LEM selection, based on the HOO decomposition and assuming a value of θ_{13} at the CHOOZ limit are thus given by

$$F_{\text{LEM} > 0.65}^{BG} = 21.4 \pm 4.6 \text{ (stat)} \pm 2.6 \text{ (syst)}, \tag{8.18}$$

$$F^{\text{signal}}_{\text{LEM}>0.65} = 11.1 \pm 3.3\ (\text{stat}) \pm 0.9\ (\text{syst}), \tag{8.19}$$

and

$$F^{BG}_{\text{LEM}>0.80} = 12.6 \pm 3.5\ (\text{stat}) \pm 1.8\ (\text{syst}), \tag{8.20}$$

$$F^{\text{signal}}_{\text{LEM}>0.80} = 8.7 \pm 2.9\ (\text{stat}) \pm 0.8\ (\text{syst}). \tag{8.21}$$

For the amount of data currently available for this analysis (3.14×10^{20} POT), the measurement is thus statistics dominated. At least $\sim 10 \times 10^{20}$ POT of data are required for this not to be the case, assuming that the fractional systematic errors remain unchanged.

It is instructive to compare these results with the ones of the ANN selection, whose total predictions are given by

$$F^{BG}_{\text{ANN}>0.70} = 26.3 \pm 5.1\ (\text{stat}) \pm 1.9\ (\text{syst}), \tag{8.22}$$

$$F^{\text{signal}}_{\text{LEM}>0.70} = 10.3 \pm 3.2\ (\text{stat}) \pm 0.8\ (\text{syst}). \tag{8.23}$$

As previously mentioned, the ANN selection is a neural network consisting of 11 variables that describe the shower topology. It is the second-best selection method in MINOS (in terms of sensitivity to θ_{13}), after LEM. The details of how the ANN algorithm works can be found in [16].

In general, ANN selects more background than LEM. Studies show that LEM is able to perform a better separation between the signal and the background, thus keeping only those events that look very similar to the signal. This is one of the reasons why LEM has a higher fractional systematic error on the background prediction, as it is precisely those events that have the largest modeling uncertainties.

Despite the larger fractional systematic error on the background, LEM's sensitivity to θ_{13} is significantly better than ANN's, due mainly to the significant reduction in the selected background and the corresponding reduction in the statistical error. As seen in Table 8.13, even with the lower cut at 0.65, LEM selects $\sim 20\%$ less background with $\sim 8\%$ more signal. With the tighter cut at 0.80, LEM can reach a signal to background ratio that is approximately 80% higher than ANN's. When it comes to sensitivity to θ_{13}, LEM's is at least 16% higher than ANN's, as indicated by the super-fom difference. This makes LEM the best ν_e selection currently available in MINOS.

8.2.2 Expected Sensitivity Contours

8.2.2.1 A Feldman–Cousins Approach

From all the information available until this point it is possible to calculate the potential exclusion contours that LEM can establish. For the remainder of this

Table 8.14 Oscillation parameters and their uncertainties, as incorporated in the calculation of the Feldman–Cousins contours

Δm^2_{32}	θ_{23}	Δm^2_{21}	θ_{12}
$(2.43 \pm 0.13) \times 10^{-3}\,\mathrm{eV}^2$	$\frac{\pi}{4} \pm 0.122$	$(8.0 \pm 0.6) \times 10^{-5}\,\mathrm{eV}^2$	0.59365 ± 0.041

chapter we use the HOO decomposition as it is the primary method. As explained in Sect. 4.1.2, the probability for ν_e appearance depends not only on θ_{13}, but also on the CP violating phase δ_{CP} and the hierarchy of neutrino masses. The last two parameters are unknown, and therefore a different limit on θ_{13} must be obtained for each value of δ_{CP} and for each mass hierarchy. We achieve this by exploring the $\sin^2(2\theta_{13})$ and δ_{CP} parameter space for each mass hierarchy.

Our analysis probes a region of parameter space that is close to the $\sin^2(2\theta_{13}) = 0$ physical boundary. Moreover, the expectation at the Far Detector is on the order of tens of events (even in the presence of a signal), which leads to non-Gaussian behavior. Because of these reasons, a Feldman–Cousins approach [17] is used to derive the final physics result. The strategy is basically that, for each point in parameter space, a large number of fake experiments is generated and the cutoff $\Delta\chi^2_{x\%}$ which encloses $x\%$ of the fake experiments is recorded. The points at which $\Delta\chi^2_{x\%} = \Delta\chi^2_{\mathrm{DATA}}$ delineate the $x\%$ confidence limit contours. The details of the Feldman–Cousins implementation can be found in Appendix D. As shown there, the contours obtained have a certain thickness to them due to the fact that only an integer number of events may be observed in real life, combined with the low number of expected events. For aesthetic and practical purposes, all of the contours shown outside of Appendix D have been smoothed.

Furthermore, when generating the Feldman–Cousins surfaces it is necessary to incorporate the error in the solar and atmospheric oscillation parameters. This is done by varying them at each fake experiment, according to the values shown in Table 8.14. These uncertainties can have a very significant impact in the interpretation of an excess as a signal. In particular, the $\sim 15\%$ uncertainty in θ_{23} translates into a $\sim 25\%$ uncertainty in the $\nu_\mu \to \nu_e$ probability. In order to minimize the impact the parameter space is parameterized in terms of $2\sin^2(2\theta_{13})$ $\sin^2(\theta_{23})$ vs. δ_{CP}. All of the details are explained in Appendix D.

Lastly, it should be mentioned that the 90% CHOOZ limit on θ_{13} corresponding to the current best fit value of the atmospheric mass splitting is included in all the contour plots for comparison purposes. Such a limit is given by $\sin^2(2\theta_{13}) < 0.15$ and does not involve the atmospheric mixing angle θ_{23}, while our limits do. Consequently, the CHOOZ limit as drawn is only valid in the case of maximal mixing ($\theta_{23} = \pi/4$).

8.2.2.2 Expected Exclusion Limits

In order to get an idea of the average sensitivity of LEM we need to consider the case where the observation matches the expectation. This is strictly not possible,

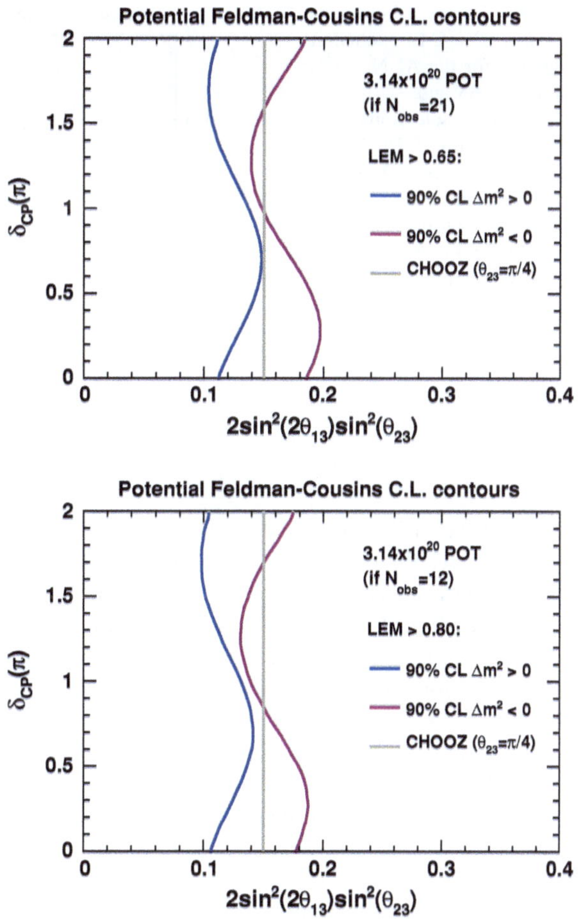

Fig. 8.7 Potential Feldman–Cousins 90% C.L. exclusion contours for the LEM > 0.65 (*top*) and LEM > 0.80 (*bottom*) selections, for an observation of 21 and 12 events respectively. The region of parameter space to the right of the blue and purple curves would be excluded at 90% C.L. for the normal and inverted mass hierarchy cases. In both situations the CHOOZ limit would be excluded for the case of maximal mixing and the normal hierarchy

however, due to the fact that only an integer number of events can be observed in the data. Because of this, we consider the case where the observation is the largest integer number for which the best fit signal corresponds to $\theta_{13} = 0$. In other words, we take the observed number of events to be the rounded down background expectation for $\theta_{13} = 0$.

The background predictions at $\theta_{13} = 0$ for LEM > 0.65 and LEM > 0.80 are approximately 21.7 and 12.7 events respectively,[6] for both mass hierarchies. We thus consider the case where we observe 21 and 12 events respectively. The

[6] The reader may notice that these numbers do not correspond exactly to the background predictions shown at Tables 7.2 and 8.13. This is because the background prediction is also affected by the value of θ_{13} considered, although to a very small degree. The official signal and background predictions are calculated for a value of θ_{13} at the CHOOZ limit. In the Feldman–Cousins contour generation however a new background and signal prediction is generated for each combination of the oscillation parameters.

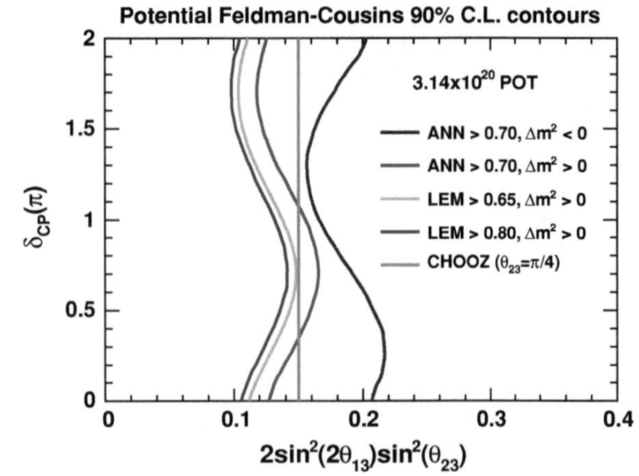

Fig. 8.8 Potential Feldman–Cousins 90% C.L. exclusion contours for the ANN, LEM > 0.65 and LEM > 0.80 selections, for an observation of 26, 21 and 12 events respectively. For LEM only the normal hierarchy contours are shown. LEM > 0.80 is the most sensitive selection, being approximately 15–20% more sensitive than ANN

resulting exclusion contours are shown in Fig. 8.7. If the normal mass hierarchy holds, our results could supersede the CHOOZ limit.

It is again instructive to compare with the average sensitivity achieved by the ANN selection. The $\theta_{13} = 0$ prediction for ANN > 0.70 is approximately 26.7 events, and consequently we consider the case where 26 events are observed. The 90% exclusion contours for ANN in the case of both mass hierarchies are shown in Fig. 8.8, alongside the ones for LEM > 0.80 and LEM > 0.65 in the case of the normal mass hierarchy. As expected, the LEM > 0.80 is the most sensitive, giving a limit that is 15–20% more restrictive than the one attained by ANN. LEM > 0.65 also achieves very good results, although not as good as LEM > 0.80.

8.2.2.3 Discovery Potential

If fewer events than the ones just considered are observed then the limits will be more restrictive. If, on the other hand, more events are observed, the contours will move to the right. If the number of observed events is large enough $\theta_{13} = 0$ will be excluded at some confidence level.

Table 8.15 Minimum number of observed events required to exclude $\theta_{13} = 0$ at 68% and 90% C.L.

	Prediction	68% C.L.		90% C.L.	
		N_{obs}	Excess	N_{obs}	Excess
LEM > 0.80	12.7	16	3.3	19	6.3
LEM > 0.65	21.7	26	4.3	30	8.3
ANN > 0.70	26.7	31	4.3	35	8.3

The excess at which the given number of observed events corresponds to is also shown, as well as the approximate background prediction at $\theta_{13} = 0$

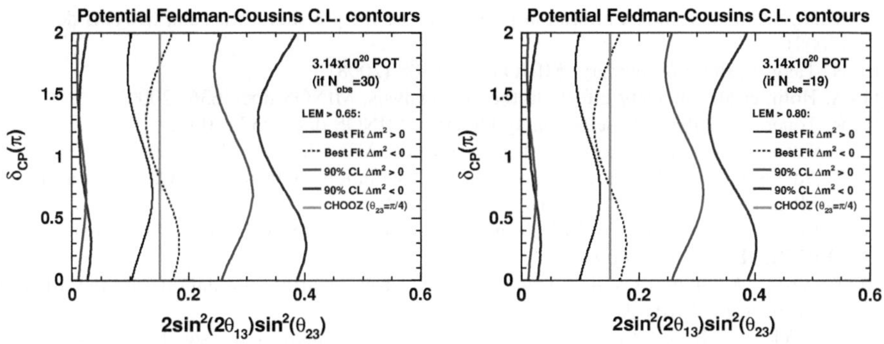

Fig. 8.9 Potential Feldman–Cousins 90% C.L. exclusion contours for the LEM > 0.65 (*left*) and LEM > 0.80 (*right*) selections, for an observation of 30 and 19 events respectively. If the number of observed events is equal to or higher than the values used here then $\theta_{13} = 0$ will be excluded at 90% C.L

Table 8.15 indicates how large the number of observed events must be in order for $\theta_{13} = 0$ to be excluded at 68 and 90% C.L. Only those values for which $\theta_{13} = 0$ is excluded at both mass hierarchies and for all values of δ_{CP} are recorded.

If 30 events or more are observed in the Far Detector with the LEM > 0.65 selection, then $\theta_{13} = 0$ will be excluded with 90% confidence. This corresponds to an excess of 8.3 events, and is equivalent to a value of θ_{13} slightly below the CHOOZ limit on average (it depends on the value of δ_{CP} and the mass hierarchy considered). Similarly, for the LEM > 0.80 selection 19 events or more must be observed to make a nonzero observation of θ_{13} to 90% confidence. In that case the excess needed is 6.3 events and the corresponding best fit θ_{13} is again slightly below the CHOOZ limit on average. The contours that would be obtained in both of these situations are shown in Fig. 8.9.

References

1. J. Boehm, H. Gallagher, T. Yang, Hadronization model uncertainties for the ν_e analysis, MINOS-doc-5392 (2008)
2. H. Gallagher, Hadronization model issues for GENIE 2.6.0, MINOS-doc-5317 (2008)
3. H. Gallagher, The NEUGEN neutrino event generator. Nucl. Phys. Proc. Suppl. **112**, 188–194 (2002)
4. C. Andreopoulos, The GENIE universal, object-oriented neutrino generator. Nucl. Phys. Proc. Suppl. **159**, 217–222 (2006)
5. C. Andreopoulos et al., Updated cross-section model uncertainties for the charged current analysis, MINOS-doc-2989 (2007)
6. S. Dytman, H. Gallagher, Changes to the determination of the mean free path of hadrons in nuclear matter in INTRANUKE, MINOS-doc-4358 (2008)
7. R. Ospanov, A measurement of muon neutrino disappearance with the MINOS detectors and NuMI beam, FERMILAB-THESIS-2008-04
8. G. Pawloski, T. Yang, Crosstalk tuning, MINOS-doc-5319 (2008)

9. MINOS Collaboration, Calibration position paper for pre-shutdown data, MINOS-doc-3941 (2008)
10. N. Tagg, Calibration plenary, MINOS-doc-5508 (2008)
11. A. Holin et al., Intensity effects in the nue analysis, MINOS-doc-5636 (2009)
12. R. Toner, ν_e analysis preselection systematics, MINOS-doc-5257 (2008)
13. J. Nelson, Notes on the normalization systematic, MINOS-doc-2106 (2006)
14. P. Adamson et al., Measurement of neutrino oscillations with the MINOS detectors in the NuMI beam, Phys. Rev. Lett. **101**, 131802 (2008)
15. J. Boehm, Measurement of electron neutrino appearance with the MINOS experiment, FERMILAB-THESIS-2009-17
16. T. Yang, A study of muon neutrino to electron neutrino oscillations in the MINOS experiment, FERMILAB-THESIS-2009-04
17. G.J. Feldman, R.D. Cousins, Unified approach to the classical statistical analysis of small signals, Phys. Rev. D **57**(7), 3873–3889 (1998)

Chapter 9
Far Detector Data Analysis and ν_e Appearance Results

At this point all the predictions are finalized and the systematic errors estimated. The Far Detector data is ready to be analyzed, but before doing that it is useful to test the decomposition and the extrapolation processes in a sample that is devoid of a significant amount of signal events, in order to make one last check for weaknesses or failures in the analysis. Three particular samples are used as Far Detector sidebands. Before looking at these samples in the context of the Far Detector it was determined that a discrepancy of 2σ or higher in any of them would cause us to avoid proceeding with the main analysis until a reasonable explanation was found or until the given problem was solved. One particular sideband that fails to satisfy the 2σ requirement and is thus analyzed in Sect. 9.2. After this study a decision to proceed with the analysis is reached. The final results of the ν_e appearance analysis are shown in Sect. 9.3.

9.1 Far Detector Sidebands

9.1.1 The Anti-LEM Sideband

The search for ν_e appearance is done by looking at the high LEM pid region (by cutting at either at 0.65 or at 0.80) and seeing if there is an excess over the predicted backgrounds. Before doing that however, it is possible to apply the same analysis to the subset of events that are preselected but that are rejected by the high LEM cut. This sample of events is referred to as the anti-LEM sideband.

As such, the anti-LEM sideband consists of small showers that are primarily hadronic in nature. Consequently the Far Detector data should be consistent with a background only prediction, within statistical errors. A few signal events could be present in the sideband however. In order to ensure that there is less than a 10% chance of observing a 2σ discrepancy in the sideband due to the signal expected

J. P. Ochoa Ricoux, *A Search for Muon Neutrino to Electron Neutrino Oscillations in the MINOS Experiment*, Springer Theses, DOI: 10.1007/978-1-4419-7949-0_9,

Table 9.1 Breakdown of Near Detector selected events in the anti-LEM sideband, for the default MC and also for the three decomposition methods. All numbers correspond to an exposure of 1×10^{19} POT. For the PORP, HOO and MRCC methods the total on the right matches the total number of events observed in the Near Detector data, and the error is just statistical in that case. An overall uncertainty of 30% is applied to the number of beam ν_e events. For the other cases the error shown corresponds to the error as obtained from the decomposition

Separation	NC	ν_μ CC	Beam ν_e	Total
Default MC	12809	14139	479	27427
PORP	12910	14140	479	27529 ± 166
HOO	12594^{+1990}_{-1968}	14460^{+2048}_{-1945}	479 ± 144	27533 ± 78
MRCC	12432 ± 350	14622 ± 365	479 ± 144	27533 ± 78

for a θ_{13} at the CHOOZ limit, a upper limit at LEM <0.55 is enforced. Also, those very few events that make it in the preselection sample but that do not find any matches in the LEM libraries (and consequently do not get a value of LEM pid assigned to them) are removed from the sideband, as they are in the main analysis.

Table 9.1 shows the breakdown of Near Detector selected events in the anti-LEM sideband, as predicted by the simulation as well as by the PORP, HOO and MRCC decomposition methods. As in the case of the standard sample, the three decomposition methods give consistent results. By construction, the total of each method is equal to the total number of events observed in the Near Detector data, and the error shown in that case is just statistical.[1] Thus it can be observed that the data-MC disagreement for the showers that constitute the anti-LEM sideband is very small, being less than 1% overall. This stands in sharp contrast with the standard sample, which contains the more EM-like background.

Another difference is that the anti-LEM sideband presents a higher proportion of ν_μ CC events, as can be seen by comparing with Sect. 6.5. While the standard sample contains roughly one ν_μ CC per two NC events, the anti-LEM sideband contains these two types of events in approximately equal amounts. This makes the sideband a good test of the Near Detector decomposition methods and the subsequent extrapolation. Since oscillations affect ν_μ CC events very strongly at low energies while they leave NC events intact, a higher proportion of ν_μ CC events means that the Far-Near spectral differences are stronger for the anti-LEM sample than they are for the standard sample.

Table 9.2 shows the number of Far Detector predicted events for the anti-LEM sideband. For simplicity, the number of expected ν_e CC and ν_τ CC events is obtained directly from the simulation, and the corresponding systematic uncertainties are taken from Table 8.4. In order to assess the systematic error in the NC and ν_μ CC components, the main sources of systematic error that do not cancel

[1] The PORP and HOO/MRCC totals in Table 9.1 disagree slightly because the PORP decomposition was done with a smaller data sample. The results are, however, consistent within errors.

Table 9.2 Breakdown of Far Detector predicted events in the anti-LEM sideband, for the default MC and the three decomposition methods. All numbers correspond to an exposure of 3.14×10^{20} POT. When shown, errors are systematic only. The signal corresponds to a value of θ_{13} at the CHOOZ limit, while ignoring matter effects and CP violation

Separation	NC	ν_μ CC	Beam ν_e	ν_τ CC	Total bg.	Signal
Default MC	99.1	50.9	2.4	3.1	155.4	7.7
PORP	98.59	51.1	2.4	3.1	155.2	7.7
HOO	96.9 ± 7.4	53.0 ± 4.1	2.4 ± 0.5	3.1 ± 1.6	155.3 ± 12.7	7.7 ± 1.1
MRCC	95.3 ± 7.5	51.6 ± 3.7	2.4 ± 0.5	3.1 ± 1.6	152.4 ± 10.9	7.7 ± 1.1

Table 9.3 Results of the anti-LEM sideband. The σ difference is the number of standard deviations from the prediction, where the statistical and systematic errors of a measurement of exactly the prediction are considered. Only the HOO and MRCC predictions are shown as they are the only ones that incorporate a systematic error. The observation of 176 events is consistent with the background-only prediction to better than 1.5σ

Sideband	Method	Prediction	Observation	σ diff.
Anti-LEM	HOO	155.3 ± 12.7(syst) ± 12.5(stat)	176	1.16σ
	MRCC	152.4 ± 10.9(syst) ± 12.3(stat)	176	1.44σ

between the detectors are reevaluated in the context of the anti-LEM sideband [1]. These systematics include the error in the absolute and relative energy scale, gains, attenuation, strip-to-strip corrections, crosstalk, low pulse-height modeling and normalization. All of these amount to a fractional uncertainty of approximately 7% for the NC + ν_μ CC prediction, which is considerably lower than for the standard sample. The total systematic errors for each component are included in Table 9.2. Again we find that the total number of background events predicted by the three methods are in very good agreement, in this case to better than 2%.

Once the predictions of Table 9.2 are finalized the Far Detector data can be looked at. A total of 176 events are observed for the anti-LEM sideband, which is higher than the predicted background. As shown in Table 9.3 however, the data are consistent with the background-only prediction to better than 1.5σ, which means that the 2σ requirement is passed. The reconstructed energy spectra, as well as the LEM pid distribution for the anti-LEM sideband, can be seen in Fig. 9.1. The excess of events is not observed to come from any particular energy bin or even from the higher region of the PID. As a matter of fact, a considerable part of the excess is found in the lowest bin ($0 < \text{LEM} < 0.05$), suggesting that it is not due to signal, but possibly due to the fact that the fraction of events that do not get any matches in the LEM libraries is slightly different between the data and the MC. The three variables that compose the LEM pid are also shown in Fig. 9.2. As with the reconstructed energy spectra and the LEM pid distribution, the agreement between the prediction and the data is found to be very good in all regions of the distributions.

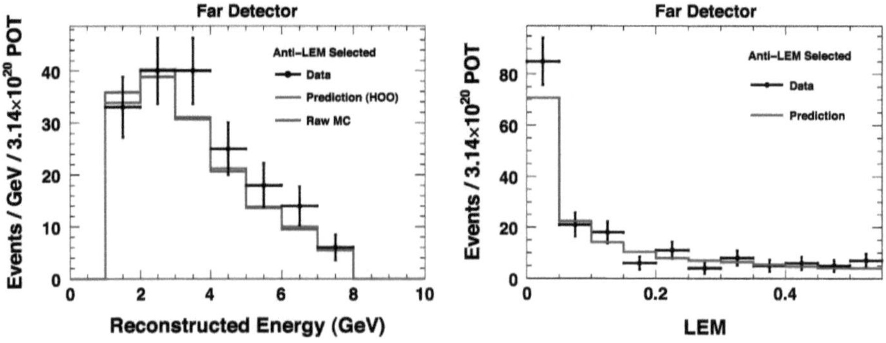

Fig. 9.1 Reconstructed energy spectra (*left*) and LEM pid distribution (*right*) for the events in the anti-LEM sideband. The prediction at each bin on the right plot is calculated using the PORP method. The data-MC agreement is very good in all regions of the distributions, given the available statistics

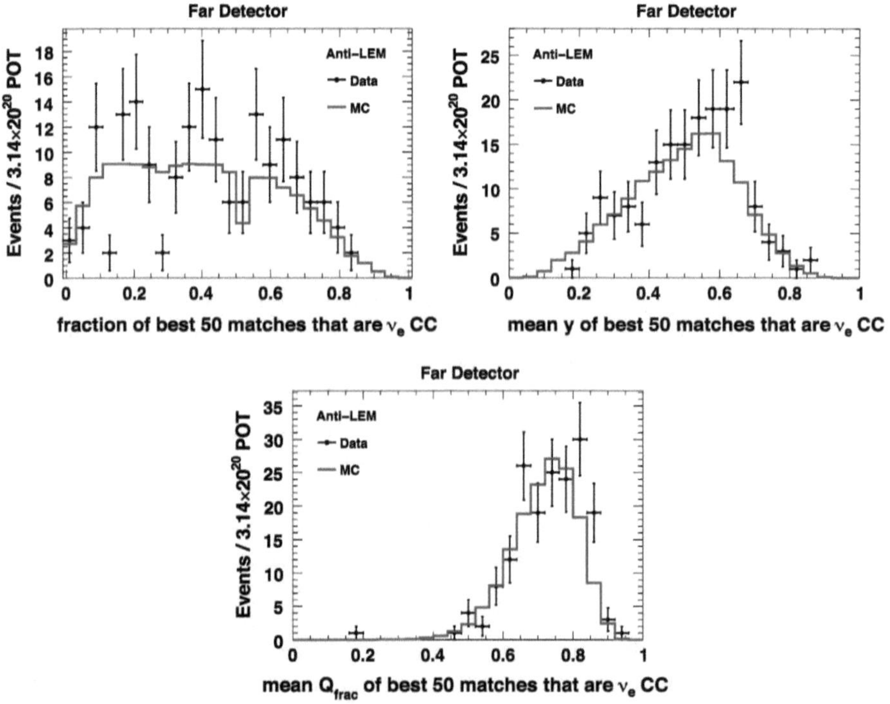

Fig. 9.2 The three variables that make up the LEM pid for events in the anti-LEM sideband. As expected, these events are mostly located in the background dominated regions, such as low f_{50}, high y_{50} and lower q_{50}. The predictions are calculated using the default MC, and are very close to those to what would be obtained with either of the decomposition methods given that the data-MC agreement is very good in the anti-LEM region

Table 9.4 Results for the near-LEM sideband, defined as those preselected events that satisfy $0.55 < \mathrm{LEM} < 0.65$. Only the HOO decomposition method was used to make the prediction. The observation is consistent with the background-only prediction to 1.79σ

Sideband	Prediction	Observation	σ diff.
near-LEM	7.2 ± 2.7	12	1.79σ

It should be mentioned that the near-LEM region, defined as the region not included by the LEM >0.65 selection but neither in the anti-LEM sideband was also looked at before performing the final analysis. Since this region could contain a significant amount of signal, it was determined that only a 3σ discrepancy or higher would be considered to indicate a problem with the analysis. Also, it was the last sideband to be analyzed before the final search for ν_e appearance was carried out. Table 9.4 shows the number of expected and observed events in this region. The data are consistent with the background-only prediction to 1.8σ.

9.1.2 The MRE Sideband

The MRE procedure, which is used to correct the selection efficiency of ν_e CC events as described in Sect. 7.2.3, can also be applied to Far Detector events. This provides us with another independent sample on which to test the machinery of the analysis. Unlike the anti-LEM sideband or to the MRCC sideband that is described in the next section, this sideband consists primarily of events that very closely resemble the expected signal.

The purity of the MRE selected sample is extremely good, and consequently no decomposition needs to be applied. Therefore, in an analogous way to what is done to NC and ν_μ CC background events, the MRE data is directly extrapolated to the Far Detector using the MRE Far-Near ratio as a function of reconstructed energy $\mathrm{E_{reco}}$:

$$\mathrm{FD^{pred}_{MRE}(E_{reco}) = ND^{data}_{MRE}(E_{reco})\frac{FD^{MC}_{MRE}(E_{reco})}{ND^{MC}_{MRE}(E_{reco})}}. \quad (9.1)$$

As shown in Sect. 7.2.3, the MRE data and MC agree very well in the ND. Consequently, the Far Detector prediction is very similar to the one obtained using the MRE Far Detector simulation. Contrary to the anti-LEM and near-LEM sidebands, the probability of a signal event leaking into this sample is extremely small. Also, when comparing with the data we consider only the statistical error in the prediction, as no systematic uncertainty was calculated for a Far Detector MRE prediction as obtained from an extrapolation process. The numbers in Table 8.9 that correspond to the expected signal are probably a good estimate, but the MRE procedure involves muon-removed events which could amplify some of the systematics considered there.

Table 9.5 Results of the MRE sideband. The σ difference is the number of standard deviations from the prediction, where only the statistical error on a measurement of exactly the prediction is considered. The observations are consistent with the background-only prediction to 0.2σ and 1.2σ for the LEM cuts at 0.65 and at 0.80 respectively

	Sideband	Prediction	Observation	σ diff. (stat)
LEM > 0.65	MRE	176 ± 13(stat)	180	0.3σ
LEM > 0.80	MRE	144 ± 12(stat)	158	1.2σ

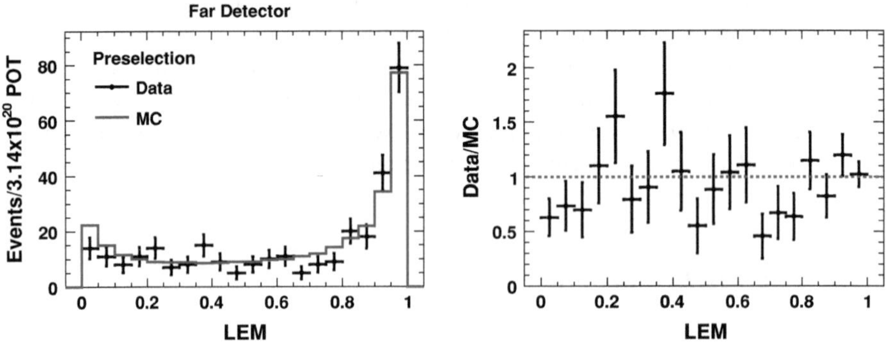

Fig. 9.3 LEM pid distribution of Far Detector MRE events (*left*) and the corresponding data to MC ratio (*right*). The agreement between the data and the MC is very good

In any case, Table 9.5 shows the results obtained when looking at the Far Detector MRE data. There it can be seen that the prediction is only 0.3σ and 1.2σ away from the observations obtained with cut values at 0.65 and at 0.80 respectively. The agreement is thus very good, even if the systematic error is not considered. Similarly, Fig. 9.3 shows the LEM pid distribution for Far Detector MRE events, and the corresponding data to MC ratio. The agreement between the data and the MC is very good all throughout the preselection sample. The same can be seen in Fig. 9.4, which shows the three variables that form the LEM pid.

9.1.3 The MRCC Sideband

The last two major sidebands that have been considered, anti-LEM and MRE, have allowed us to test the efficacy of the Near Detector decomposition and the Far Detector extrapolation for both background-like and signal-like events at the $\sim 10\%$ level, which is set primarily by the statistical error on these samples. Unfortunately on both of these samples the level of disagreement between the data and the MC is also less than 10%. Given that in the main analysis the data and the simulation disagree to more than 40%, it is important to find a similar sample where it can be tested that the disagreement seen in the Near Detector is indeed reproduced in the Far Detector.

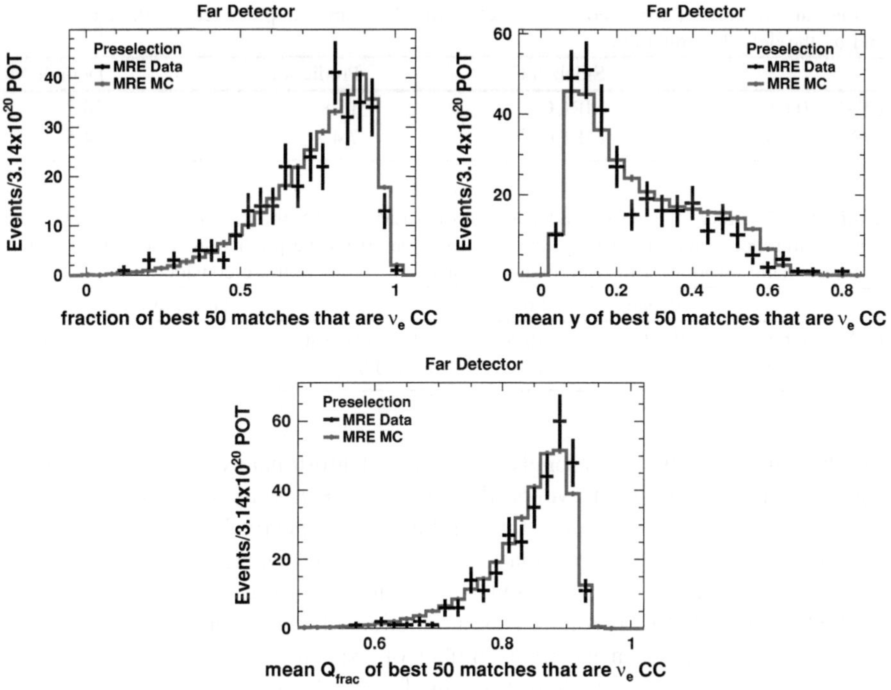

Fig. 9.4 The three variables that make up the LEM pid for events in the MRE sideband. As expected, events concentrate predominantly in the signal dominated regions, such as high f_{50}, low y_{50} and high q_{50}. The agreement between the data and the MC is very good in general

This can be achieved by applying the muon-removal (MRCC) process on Far Detector golden ν_μ CC events. The sample of Far Detector MRCC data events is kept blinded until a prediction is obtained, as is done in the main analysis. As in the case of the MRE sample, no decomposition needs to be applied to the MRCC Near Detector sample. Also, there is no need to worry about any possible signal appearing in this sample, given that it consists of true muon removed ν_μ CC events to better than 98%. The MRCC data from the Near Detector is thus directly extrapolated in the usual way to obtain the Far Detector prediction:

$$\mathrm{FD}^{\mathrm{pred}}_{\mathrm{MRCC}}(\mathrm{E_{reco}}) = \mathrm{ND}^{\mathrm{data}}_{\mathrm{MRCC}}(\mathrm{E_{reco}}) \frac{\mathrm{FD}^{\mathrm{MC}}_{\mathrm{MRCC}}(\mathrm{E_{reco}})}{\mathrm{ND}^{\mathrm{MC}}_{\mathrm{MRCC}}(\mathrm{E_{reco}})}. \tag{9.2}$$

The extrapolation is done in bins of 1 GeV. As shown in Sect. 6.4.2, the ratio of data to MC in the Near Detector for MRCC events is very similar to the one observed in the standard sample, being higher than 40%. Consequently, the Far Detector prediction by the extrapolation is significantly smaller than what is predicted by the simulation alone, differing by practically a factor of two in the case of the cut at 0.80. This is illustrated in Table 9.6. The MRCC sideband thus fulfills its purpose of providing a cross-check of the analysis with a sample where

Table 9.6 Number of expected Far Detector MRCC events, as predicted by the extrapolation and by the default simulation

	Sideband	Prediction	Default MC
LEM > 0.65	MRCC	16.8	28.7
LEM > 0.80	MRCC	7.4	14.5

Table 9.7 Results of the MRCC sideband for the LEM > 0.65 and the LEM > 0.80 selections. The σ difference is the number of standard deviations from the prediction, where the statistical and systematic errors of a measurement of exactly the prediction are considered

	Sideband	Prediction	Observation	σ diff.
LEM > 0.65	MRCC	16.8 ± 4.1(stat) $\pm$ 1.6(syst)	25	1.86σ
LEM > 0.80	MRCC	7.4 ± 2.7(stat) $\pm$ 0.8(syst)	16	3.03σ

the data and the MC are strongly discrepant. Unfortunately however, given the small number of expected events, this cross-check is only good to $\sim$24% and $\sim$37% in the case of the cut at 0.65 and at 0.80 respectively.

In order to evaluate the systematic error in the Far Detector prediction the same error that is obtained for NC events in the main analysis is used. This error is applied to the MRCC prediction as a function of energy in the up and down directions, the change in the total number of expected events recorded, and the largest of the two variations kept [2]. This follows from the previous observation that NC events and MRCC events look quite similar in the MINOS detectors, and must consequently have very similar extrapolation errors.

When the Far Detector MRCC data is looked at, an excess over the prediction is found for both cut values, as shown in Table 9.7. The excess obtained with the cut at 0.65 is still consistent with the prediction to better than 2σ. This is, however, not the case when the tighter cut at 0.80 is used. In that case a discrepancy of more than 3σ is obtained, which is too high to be a statistical fluctuation. The energy spectra of the data, the prediction and the default MC are shown in Fig. 9.5. The excess seems to be roughly spread out among all energies, given the available statistics.

As previously stated, the main purpose of the MRCC sideband is to check if the large data-MC disagreement seen in the Near Detector for the signal-like sample is reproduced in the Far Detector. It is thus instructive to compare the LEM pid distribution in both detectors, as done in Fig. 9.6. There is a large discrepancy between the data and the MC at high (> 0.5) LEM values in the Near Detector. The corresponding discrepancy exists between the prediction (in blue) and the default MC (in red) in the Far Detector, given that the prediction is obtained by extrapolating the Near Detector data. We find that, in general, the Far Detector data agrees better with the prediction than with the default simulation. At high LEM values however, a constant excess of events is observed in the data with respect to the prediction.

Figure 9.7 shows the distributions of the three variables that form LEM, also denoted as f_{50}, y_{50} and q_{50}. By using the data to MC ratio in the Near Detector for

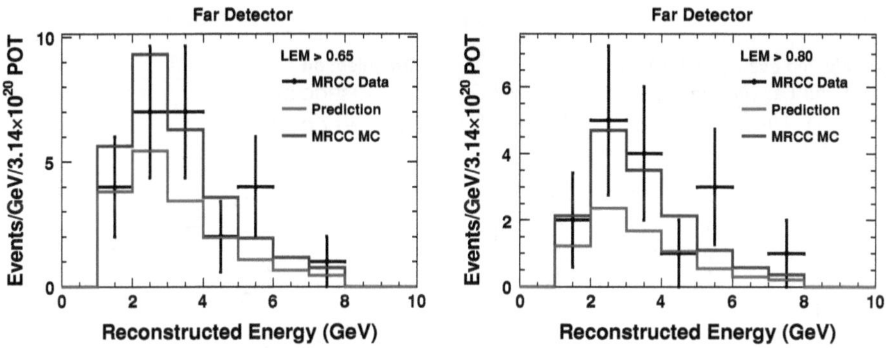

Fig. 9.5 The energy spectra of selected Far Detector LEM selected MRCC events, for a cut value of 0.65 (*left*) and of 0.80 (*right*). The data is shown alongside the prediction obtained from extrapolation from the Near Detector and the default simulation's prediction. These last two differ by almost a factor of two, particularly for the tighter cut at 0.80. In both cases there is an excess of events in the data with respect to the prediction

each of these variables it is possible to reweigh the default MC in order to obtain a prediction. Again we observe that, in most areas, the data seem to agree better with the prediction than with the default MC. The exception just seems to occur, as previously noted, for those few events at the high LEM region and thus with high f_{50} and, in particular, low y_{50}.

Given the requirement that all sidebands be within 2σ, these results force us to study the MRCC sideband and the excess observed in the high LEM region before proceeding with the main analysis. In particular, it is important to determine if the observed excess is due to a factor that affects the MRCC and standard samples equally, or not. Such a study is performed in the following section.

9.2 Addressing the MRCC Sideband Excess

9.2.1 Systematics in the MRCC Sample

As explained when assessing the systematics for the standard sample in Chap. 8, any effect that is not well modeled in the simulation and that acts differently in the two detectors is a source of systematic error. Such effects were properly evaluated in the context of the standard sample, although it is possible for some of them to have more pronounced effects in the MRCC sample. This would cause the systematic errors in the MRCC sideband to be underestimated.

One such effect could be crosstalk, which is one of the largest systematics in the standard sample and one of the biggest sources of differences between the two detectors. Different scenarios can be imagined where crosstalk plays a much more important role with the MRCC sample more than with the standard sample. For

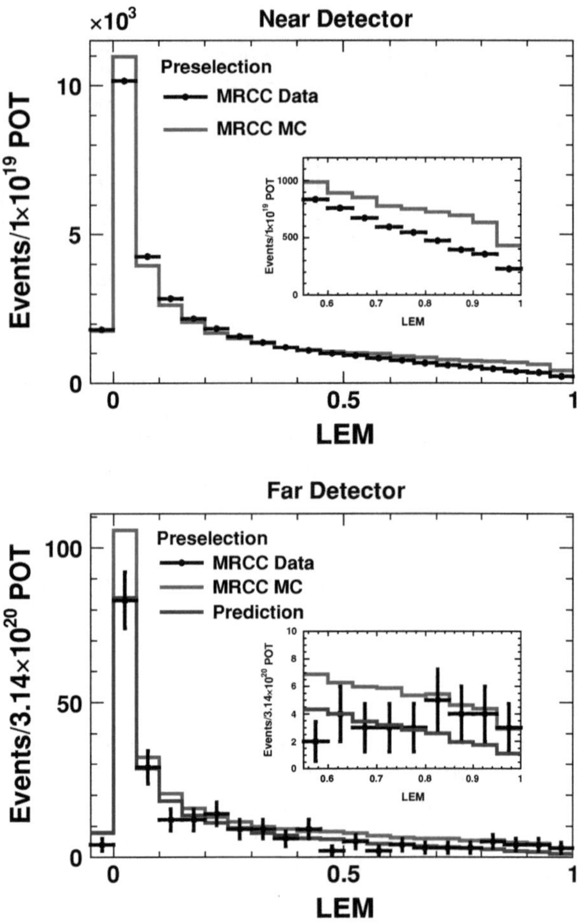

Fig. 9.6 The LEM pid distribution for MRCC preselected events in the Near (*top*) and Far (*bottom*) detectors. The Far Detector data agrees better with the prediction than with the default MC until about LEM ≈ 0.8, at which point a constant excess is observed. The events at LEM pid <0 are the preselected events that do not get any matches in the LEM library

instance, muon tracks leave a significant amount of crosstalk alongside their path, most of which is removed by the MRCC algorithm. Given the differences in crosstalk patterns, it is feasible that the crosstalk left by the MRCC algorithm is significantly different in the two detectors, thus altering the shower shape. While it is hard for such an effect to have gone unnoticed by an experienced scanner who checked the muon removal process [3] in the Far Detector, it could be that it is subtle enough to be invisible to the naked eye, or that it is stronger in the Near Detector.

In order to test the impact of this sytematic the MRCC algorithm is applied to MC events produced with the improved crosstalk simulation. The results are then compared to the normal MRCC events in the simulation. The Near Detector is where the differences between the standard and the improved crosstalk models differ the most, and is thus the best detector on which to perform this study. The results in the Near Detector show that the effects of changing the xtalk simulation

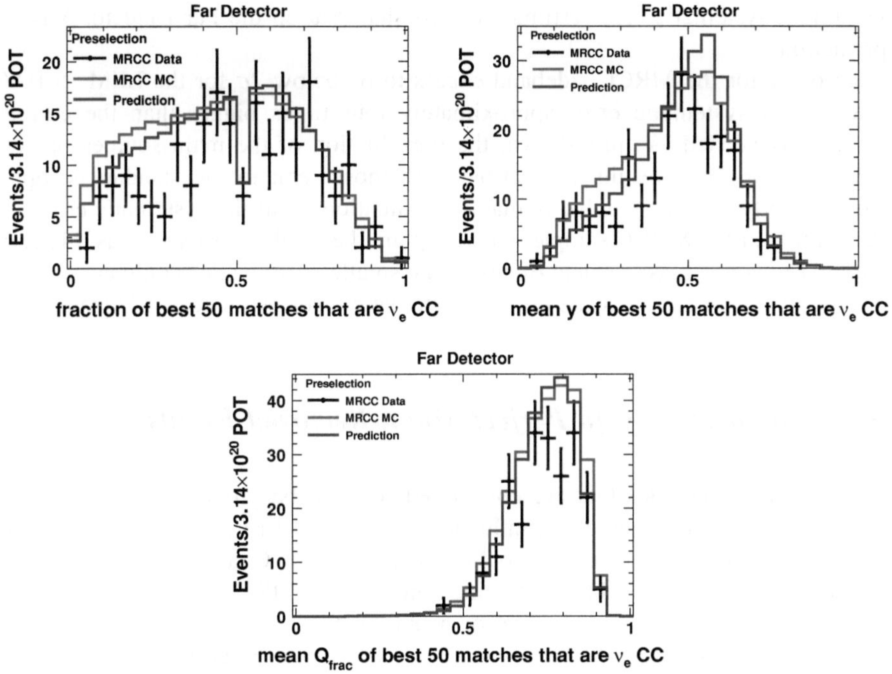

Fig. 9.7 The three variables that make up the LEM pid for events in the MRCC sideband. The data seems to agree better with the prediction than with the simulation, except perhaps in regions such as very low y_{50}

in the MRCC sample has an effect that is very similar to the one observed in the standard sample, of approximately 3.5%. This strongly suggests that crosstalk mismodeling is not the source of the discrepancy.

The largest source of systematic error in the standard sample is the uncertainty in the gains. Again, it could be that be that the MRCC procedure is more sensitive to this effect than the standard sample is. In order to evaluate the impact of this systematic, a special sample of MRCC events where the gains are shifted up and down by their uncertainties was produced, in an analogous way to what was done with the standard sample. When comparing with the normal MRCC events, a shift of approximately $\pm 8\%$ is observed in the Near Detector. This is slightly larger but still consistent (given the statistical error of the MRCC samples with the modified gains) with what is found in the standard sample.

Another large source of systematic error for the standard sample is the uncertainty in the relative energy scale. There is however no reason why this would affect MRCC events any more than it does NC events, as it is just a conversion factor that gets applied to all hits equally. In addition, the effect of the uncertainty in the hadronic model is also reassessed in the context of the MRCC sample [4]. The results are again consistent with what is found for the standard sample, yielding a systematic error estimate of less than 5% on the Far Detector MRCC prediction.

In order for the MRCC sideband excess to be below 2σ for the LEM > 0.80 selection, a systematic error approximately four times bigger than the current estimate is needed. As just shown, the reevaluation of the main sources of systematic error show results compatible with those seen in the standard sample. Consequently, it is very unlikely that a reevaluation of all the systematic errors in the context of the MRCC sample would explain the sideband's excess, as it is very hard to see how a reassessment of any of the smaller systematics could cause a big change. It thus seems that the current systematic error estimate is adequate.

9.2.2 Comparison with Perfect Muon-Removed Events

Even if the systematics that are considered seem to be properly evaluated, it is possible that an unknown effect influencing the high-LEM region has been completely missed. Such an effect could affect the MRCC and standard samples simultaneously, or it could be something unique to the MRCC process. It could be, for instance, that there is a feature or a bias in the muon-removal process that affects the two detectors differently (or, as is commonly referred to, a "Far-Near difference"). If this was indeed the case, the cause(s) of the difference would have to be located and an additional systematic error added based on how well those causes are incorporated in the simulation. If on the other hand the algorithm acts equally in the two detectors then it does not matter how good the muon removal process is, as MRCC events would be the same in the two detectors and there should be no difference in the data-MC ratios.

In order to test this hypothesis a sample with which to compare the MRCC events is needed. As suggested [5], a natural choice is to consider perfect muon-removed events, where the muon is faultlessly removed at the simulation level. In fact, a comparison between regular MRCC and perfect MRCC events in one detector has already been performed [6]. Our study reproduces the results of that work, namely that there are significant differences between perfect MRCC and regular MRCC events, as illustrated in Fig. 9.8. In average, perfect MRCC events are ~ 1.5 planes shorter and consequently ~ 0.3 GeV less energetic than regular MRCC events. The reason is simply that a part of the muon track is kept in MRCC events, which is not properly removed by the MRCC algorithm.

Given the existence of muon remnants in regular MRCC events, it is not unexpected that the LEM pid distribution differs significantly between perfect and regular MRCC events. Figure 9.9 shows the LEM pid distribution normalized to unity for these two types of events, in both detectors. The differences appear most strongly at low and high LEM values. In particular, there seem to be more than twice as many perfect MRCC events than regular MRCC events at very high-LEM values. This indicates that, as expected, the muon remnants make MRCC events

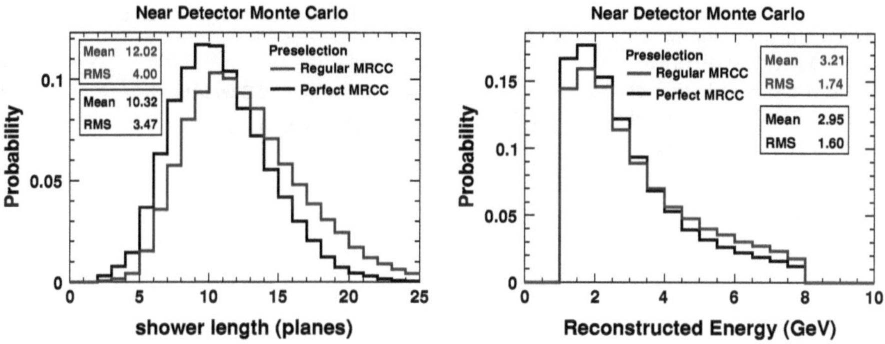

Fig. 9.8 Shower length (*left*) and reconstructed energy (*right*) distributions for perfect and regular MRCC events, normalized to unity. Regular MRCC events are longer and more energetic than perfect MRCC events

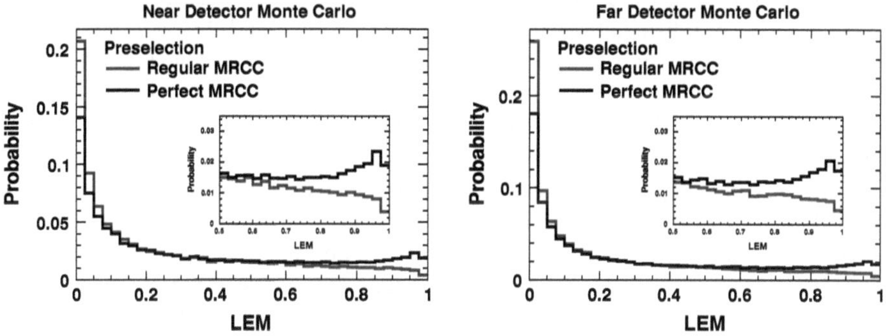

Fig. 9.9 LEM pid distribution of regular and perfect MRCC events in the Near (*left*) and Far (*right*) detectors. Significant differences exist between the two samples, particularly at high-LEM values

more similar to a high multiplicity hadronic event, thus lowering the value of the LEM pid assigned to them.

A quick aside must be made here. It is interesting to note that similar differences are observed between the NC and ν_μ CC events in the standard sample and the MRCC sample, as shown in Fig. 9.10. Once again, there is a deficit of MRCC events with respect to the hadronic showers at high-LEM values. Studies [7] have shown that those differences are not due to differences in the particle content (such as the amount of energy in the shower that is electromagnetic, or the amount of π^0's produced). Consequently, in light of this study, the most likely explanation is again that the muon remnants of MRCC events strongly affect the shower shape, making events look more background-like.

It is thus clear that the MRCC process is not perfect and that there are muon remnants that affect the LEM pid distribution. The question however is whether or not the biases in the MRCC algorithm act equally on the two detectors. In order to

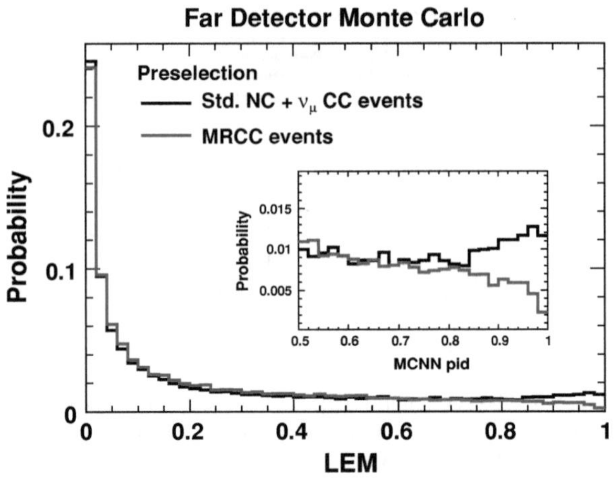

Fig. 9.10 The ratio of the LEM pid distribution of perfect MRCC events over the LEM pid distribution of regular MRCC events, for the two detectors. The differences between perfect and regular MRCC events are practically the same between the two detectors

answer this question we divide the LEM pid distribution of perfect MRCC events over the LEM pid distribution of regular MRCC events, and then we compare the results between the two detectors, as shown in Fig. 9.11. The two ratios are consistent with each other within the available statistics, except perhaps for the highest bin. The MRCC process thus seems to be acting very similarly in the two detectors.

These results argue against any bias in the MRCC process being the cause of the sideband's discrepancy. They also strengthen the case that the systematic errors assigned to the MRCC prediction must be comparable to the ones assigned to NC events in the standard sample, given that the biases in the MRCC process cancel to first order between the two detectors.[2] This suggests that the only way the systematic error on the MRCC prediction would be significantly different from the one already estimated is if there was a bias in the MRCC process that is missing from the simulation in its entirety, or an unknown Far-Near difference affecting both the MRCC and the standard samples.

9.2.3 The Energy of the MRCC Parents

It could be that the sideband discrepancy is due to a physics-related issue. For the prediction that obtained through the extrapolation to be correct the Near and Far samples of selected MRCC events must be equivalent in terms of their kinematic properties. However, this is strictly not the case for MRCC events, due to the fact that their ν_μ CC parents undergo oscillations at the Far Detector, which change their true energy distribution.

[2] If, for instance, crosstalk had a stronger effect on MRCC events than it does on the standard NC events, the comparison with the perfect MRCC events would have revealed this given that the two detectors have significantly different crosstalk patterns.

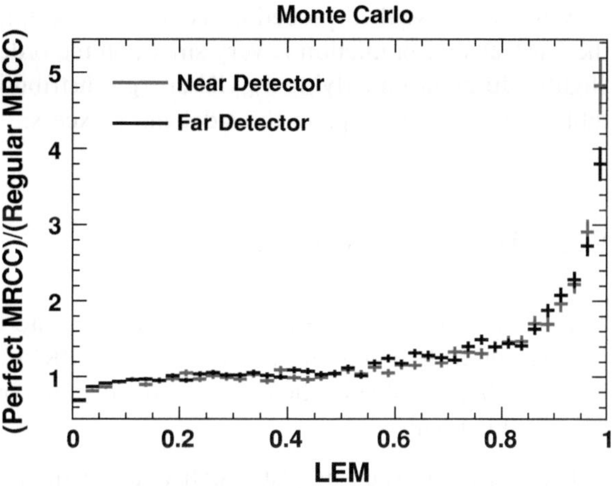

Fig. 9.11 LEM pid distributions of NC + ν_μ CC events in the standard sample and of regular MRCC events in the Far Detector, normalized to unity. Similar differences as those observed in Fig. 9.9 are observed between the standard and the MRCC samples, particularly at high-LEM values

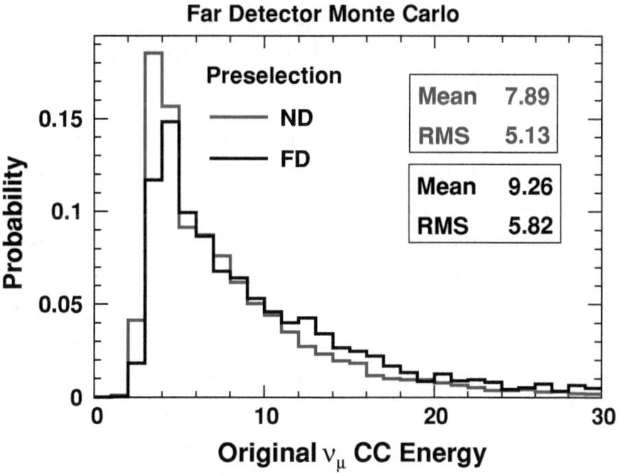

Fig. 9.12 True energy of MRCC parents in the Near and Far Detectors, normalized to unity. Due to oscillations, the average true energy of the ν_μ CC parents is $\sim$1.4 GeV higher in the Far Detector

Figure 9.12 shows the true energy of MRCC parents in the two detectors. Most of the oscillations occur below $\sim$5 GeV. Consequently, there are fewer MRCC parents with such energies in the Far Detector than in the Near Detector, while the opposite happens at higher energies. The difference is however not that large, and the average MRCC parent in the Far Detector is only $\sim$1.4 GeV higher than in the Near Detector.

In order to quantify the impact of this effect we redo the extrapolation but applying a fake oscillation weight to the Near Detector samples based on the energy of the original ν_μ CC event. This way the data-MC ratio in the Near Detector is obtained for a sample that has the same parent energy distribution as in the Far Detector. The reconstructed energy of the MRCC parent is used to determine the oscillation weight, given that the true energy is not known for the data. This is a very good approximation, as the energy resolution for ν_μ CC events is very good.

When the described procedure is applied, we find that the resulting change in the Far Detector prediction is very small, on the order of 1%. The effect of having slightly different underlying ν_μ CC energy distributions is thus practically negligible and unable to explain the sideband's excess.

9.2.4 Making a Decision

None of the studies performed to the current date of writing have revealed any problem with the main analysis or with the MRCC sideband in particular. They have also been unable to shed light as to where the problem lies. Four possibilities exist at this point:

- The observed excess is just a statistical fluctuation.
- There is a bug somewhere in the MRCC processing chain.
- There is an unknown Far-Near difference affecting only the MRCC sample.
- There is an unknown Far-Near difference affecting both the standard and the MRCC samples.

The first possibility is improbable, but not impossible. With regards to the second item, many studies have been done to check the integrity of the MRCC samples [8, 9], although the possibility still exists that there is an error somewhere in the processing chain. Furthermore, the possibility of an unknown Far-Near difference affecting only the MRCC sample is not ruled out, although diminished by some of the studies shown in this section. Although there is no other evidence for it, the last possibility is unfortunately also viable, that there is something unknown affecting the standard sample as well.

In order to make a decision as to how to proceed with the main analysis, it is useful to compare with what is seen by the other ν_e selection, which is the ANN pid [10]. Table 9.8 shows the MRCC sideband results for ANN, as obtained from [11]. The ANN selection also sees an excess, which is actually larger than the one seen by the LEM > 0.80 selection in absolute numbers. It is however less significant because of the larger background.

It is also useful to study the sideband's discrepancy as a function of cut value. Figure 9.13 shows the number of standard deviations from the prediction calculated assuming a systematic error of 10% and 7.5% for LEM and ANN respectively, as a function of cut value and of background rejection. LEM and ANN have very different shapes, and therefore considering the same cut value in both does

Table 9.8 Results of the MRCC sideband for the ANN selection. The σ diff. is the number of standard deviations from the prediction, where the systematic and statistical errors on a measurement of exactly the prediction is considered. The data is 1.8σ away from the prediction

	Sideband	Prediction	Observation	σ diff.
ANN > 0.70	MRCC	$28.6 \pm 5.3(\text{stat}) \pm 2.4(\text{syst})$	39	1.8σ

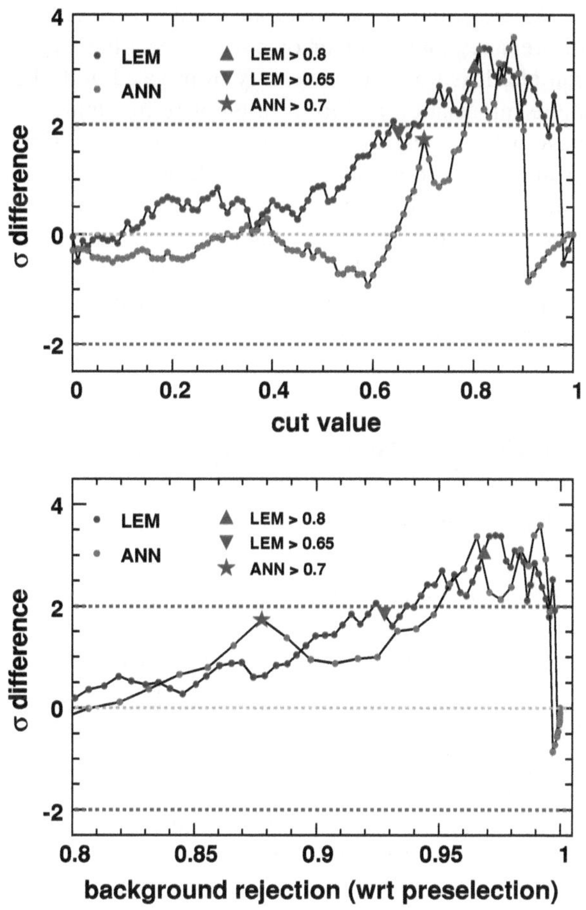

Fig. 9.13 Number of standard deviations from the prediction in the MRCC sideband, as a function of cut value (*top*) and of rejected background (*bottom*). A flat systematic error of 10% and 7.5% on the MRCC prediction is assumed for all cut values of LEM and ANN respectively when calculating the σ diff., which has a small impact given the dominance of the statistical error, particularly at medium and high cut values. Both selections see excesses of similar significance at high rejected background

not imply the same level of signal and background acceptance. Still, it can be seen that at high cut values there is also an excess seen by ANN. Furthermore, when the two pids are compared in terms of the fraction of the preselected background that they reject, there is a striking similarity between the two. In other words, if the cut at ANN had been set high enough to achieve the same amount of background rejection as LEM does, a discrepancy with an equivalent significance as the one seen by LEM would have been observed. It is however not advantageous for ANN to have such a high cut value, as the amount of accepted signal in that case is too small.

Beyond confirming that the discrepancy is not LEM specific, the above comparisons indicate that the discrepancy occurs at high background rejection. A similar conclusion was achieved when analyzing the bottom plot of Fig. 9.6, where it can be seen that in general the data agrees very well with the prediction, except for LEM > 0.80.

As previously mentioned, no study has been able to pinpoint the origin of the discrepancy and whether or not it affects the standard sample or just the MRCC sample. It is however necessary to proceed with the main analysis in order to get more information and to pave the way for future ν_e appearance analyses that will have better statistics, or else we risk stagnation. Given that the discrepancy occurs at high background rejection, we decide to move forward with the LEM > 0.65 selection instead of the optimal[3] LEM > 0.80 selection. In this way we ensure that, if the discrepancy at high background rejection is present in the standard sample, it will be absorbed to high degree by the larger and better understood background. We also satisfy the previously established requirement that all sidebands have an agreement within 2σ.

9.3 Electron Neutrino Appearance Results

9.3.1 Number of Selected Events

9.3.1.1 General Results

Having satisfied all of the pre-established criteria, we proceed to search for ν_e appearance by applying the LEM > 0.65 selection to the MINOS Far Detector data, and using the HOO decomposition as the primary one due to its better assessment of the systematics. A total of 28 ν_e-like events are observed, compared to a background-only expectation of 21.4 events in the case of the HOO decomposition and of 22.0 in the case of the MRCC decomposition. These results are summarized in Table 9.9. The observed excess is consistent with the background-only prediction at slightly more than 1 standard deviation. Figure 9.14 displays one of the ν_e candidates in the data.

The energy spectrum of LEM selected events is shown in Fig. 9.15. Most of the excess seems to be coming from the 2–3 GeV bin.

9.3.1.2 Consistency Checks

No anomalies are found in the Far Detector data or in the analysis in general. The events selected by LEM are distributed uniformly in the longitudinal and transverse directions of the Far Detector, as expected. This can be seen in Fig. 9.16. Similarly, events are consistent with being selected randomly in time (which in our case is measured in terms of POT exposure). Figure 9.17 shows the distribution of POT elapsed between selected events in the Far Detector, and alongside the

[3] in terms of sensitivity to θ_{13}

Table 9.9 Results of the LEM > 0.65 selection. The σ diff. is the number of standard deviations from the background-only prediction. The data is consistent with no ν_e appearance at 1.2σ and 1.1σ in the case of the HOO and MRCC decompositions, respectively

	Method	Bg. Prediction	Observation	σ diff.
LEM > 0.65	HOO	$\mathbf{21.4 \pm 4.6}$**(stat)** $\pm$ $\mathbf{2.6}$**(syst)**	**28**	$\mathbf{1.2\sigma}$
	MRCC	22.0 ± 4.7(stat) $\pm$ 2.5(syst)	28	1.1σ

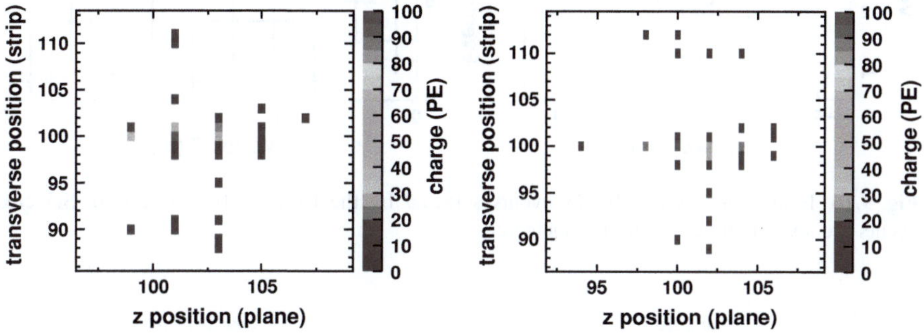

Fig. 9.14 A ν_e CC candidate in the Far Detector data (run 37980, snarl 225613) as seen in the *U* (*left*) and *V* (*right*) views. This particular event has a LEM pid value higher than 0.99. It consists of what looks like a tight electromagnetic core surrounded by crosstalk activity

Fig. 9.15 Energy spectrum of the Far Detector data alongside the background prediction as determined using the HOO method. There is an overall excess in the data with respect to the background prediction

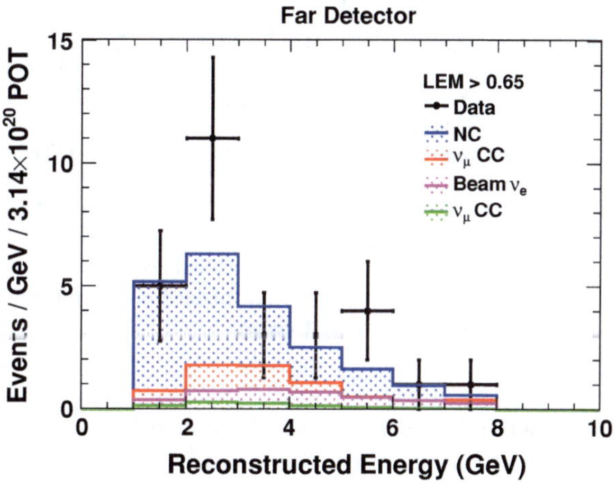

equivalent plot in the Near Detector which is shown for comparison. In both cases the distributions are consistent with the expectation.

Figure 9.18 shows the three variables that make the LEM pid for preselected events in the Far Detector. A rough prediction is obtained for each variable by reweighing the Far MC expectation by the data to MC ratio observed in the Near Detector (from Fig. 5.16). In particular, the distortion expected to occur for the y_{50} variable is indeed observed in the Far Detector, given the available statistics. Also,

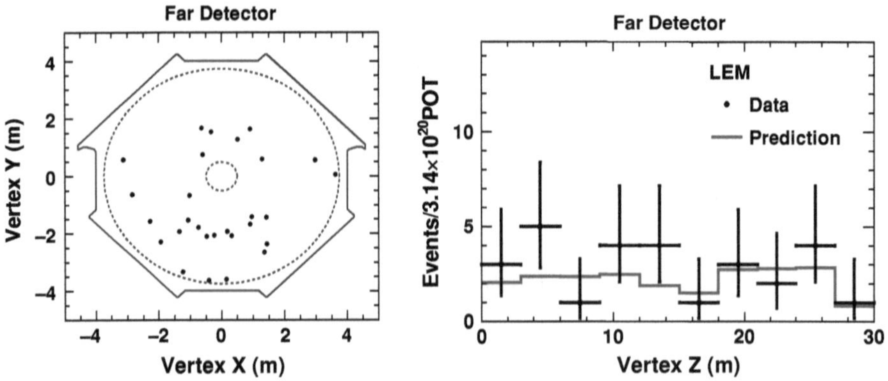

Fig. 9.16 Event vertices for the 28 events selected by the LEM > 0.65 selection. Selected events are spread all across the Far Detector

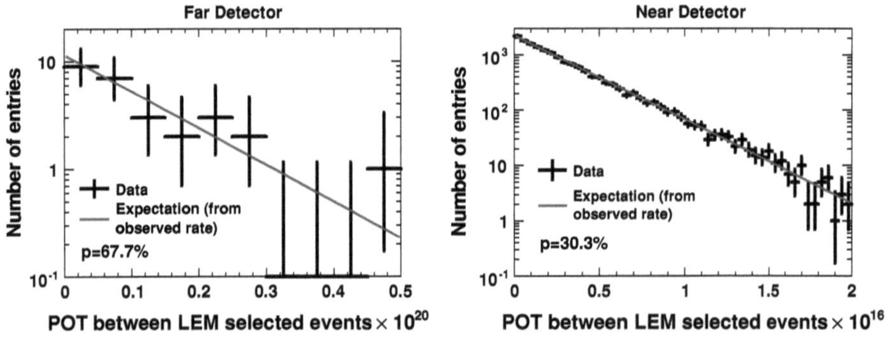

Fig. 9.17 Distributions of POT elapsed between LEM selected events in the Far (*left*) and Near (*right*) detectors. The expectations are calculated from the observed rates as described in [12] and are shown in red. In each case, a χ^2 is calculated with respect to the expectation. The probability of getting a χ^2 that is worse than the obtained is determined based on 2,500 pseudo-experiments and indicated on the lower-left corner

as shown in Fig. 9.19, the LEM pid distribution is in very good agreement with the prediction (drawn in red this time).[4]

It is also interesting to compare with what is seen by the ANN selection, as studies [13] have shown a large overlap between the two pids. In particular, $\sim 75\%$ of signal events selected by LEM are also selected by ANN, and more than $\sim 60\%$ of background events selected by LEM are also in ANN. A total of $26.3 \pm 5.1(\text{stat}) \pm 1.9(\text{syst})$ background events are expected with ANN, and 35

[4] Note that this does not have to be the case in the higher-LEM region where the signal would appear. A small indication of a signal is actually what is being observed.

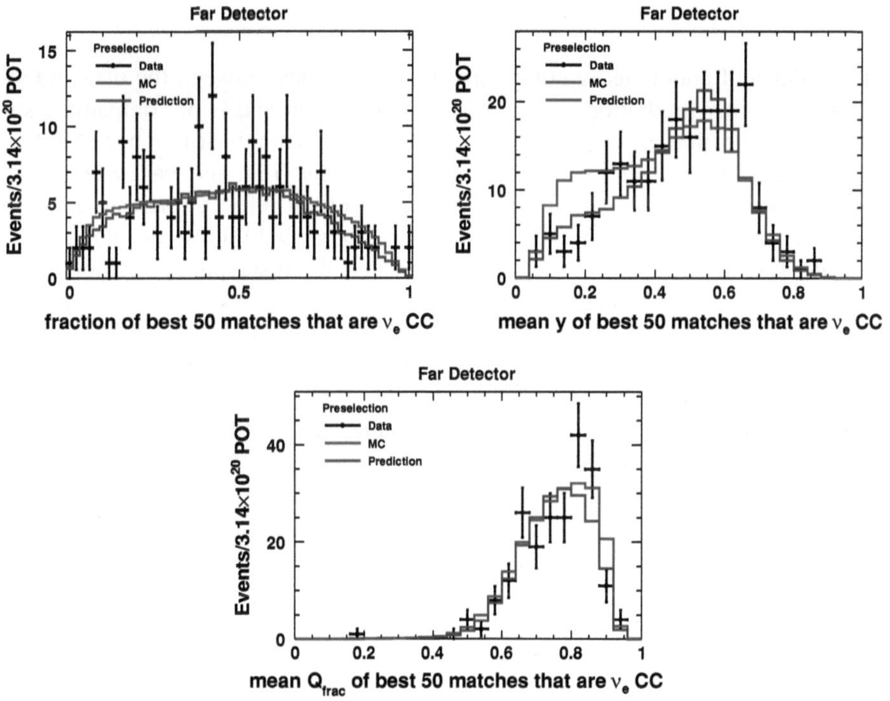

Fig. 9.18 The three variables that make up the LEM pid for preselected events in the Far Detector. The predictions are obtained by reweighing the Far MC expectation by the data to MC ratio observed in the Near Detector. The data distributions look as expected from the Near Detector

Fig. 9.19 LEM pid distribution in the Far Detector. Only the prediction and the data are shown. The events at LEM pid <0 are the preselected events that do not get any matches in the LEM library

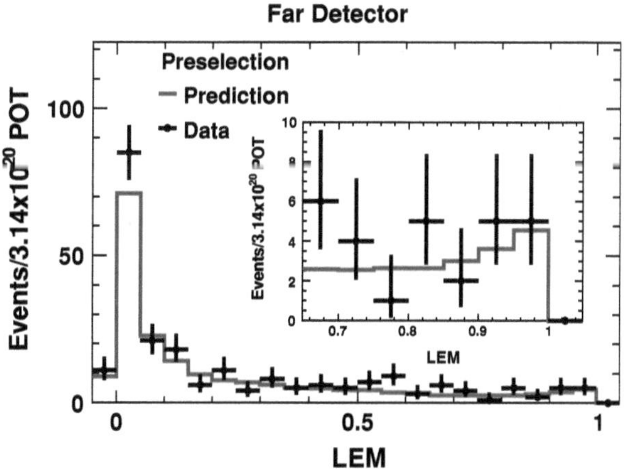

events are observed [11], giving an excess that is consistent with the background-only prediction to 1.6σ. These results thus go in the same direction and are of comparable magnitude as the ones from LEM, as expected.

9.3.1.3 Comparison with the MRCC Sideband

The reader will undoubtedly notice that the discrepancy between the data and the background-only prediction goes in the same direction and is of comparable significance as the one seen in the MRCC sideband. It is again instructive to plot the σ diff. (defined as the number of standard deviations from the background-only prediction) as a function of cut value and background rejection, and to compare with what was found with the MRCC sideband. Figure 9.20 shows this comparison for the LEM selection. There it can be seen that the discrepancy (or, more correctly, the indication of a signal) in the standard sample does exactly the opposite than what the discrepancy in the MRCC sideband does as a function of background rejection. In particular, it seems that the largest excess in the standard sample is for LEM ~ 0.55, which is a region where nothing wrong was seen with the MRCC sideband. Similarly, in the region suspected in the MRCC sideband (LEM > 0.80), the excess seen in the standard sample is not very significant.

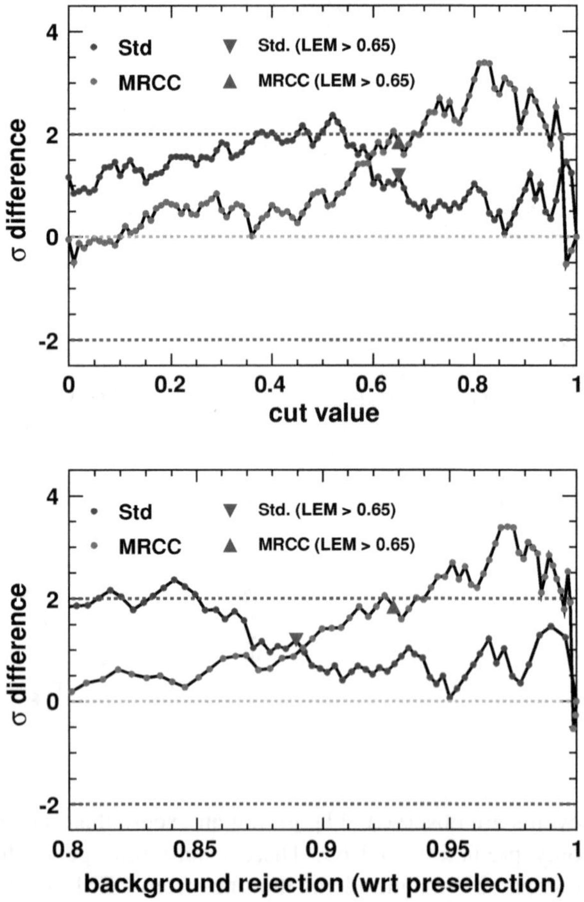

Fig. 9.20 Number of standard deviations from the background-only prediction as a function of cut value (*top*) and background rejection (*bottom*), for LEM selected events in the standard and MRCC samples. Systematic errors of 12% and 10% are used for all cut values in the standard and MRCC predictions respectively, which have a small effect given the dominance of the statistical error in both samples

These results strongly suggest that the excess we find in the standard sample is of a different nature as the one seen in the MRCC sideband. A corollary is that the MRCC sideband excess is most likely due to a problem restricted to that sample. Moreover, the excess seen in the standard sample is thus consistent with being a small indication of a signal (or a background fluctuation). It should be noted that the presence of an excess that predominantes at medium-high LEM values is not inconsistent with it being due to signal, as the corresponding signal at high-LEM values could have just fluctuated.

9.3.2 Resulting Limits on θ_{13}

In order to interpret the excess we turn to the Feldman-Cousins framework developed as described in Appendix D. Based on the observation of 28 events, a $\Delta\chi^2_{\mathrm{DATA}}$ is obtained for each point in a grid of $2\sin^2(2\theta_{13})\sin^2(\theta_{23})$ vs. δ_{CP}, allowing us to locate the best fit signal. As shown in the exclusion contours later on, this best fit signal corresponds to a value of θ_{13} slightly below the CHOOZ limit, and its energy spectra is shown in Fig. 9.21.

For each point in the $2\sin^2(2\theta_{13})\sin^2(\theta_{23})$ vs. δ_{CP} grid, the cutoff $\Delta\chi^2_{x\%}$ which encloses $x\%$ of the fake experiments is recorded. The points at which $\Delta\chi^2_{x\%} = \Delta\chi^2_{\mathrm{DATA}}$ delineate the $x\%$ confidence limit contours. The uncertainties in the atmospheric and solar oscillation parameters, as well as in the signal and the background expectations, are included in the fake experiment generation, as described in Sect. D.2.

The 98% and 68% Confidence-Level (CL) Feldman-Cousins exclusion contours for our result are shown together with the best fit signal in Figs. 9.22 and 9.23, corresponding to the HOO and MRCC decompositions respectively. The

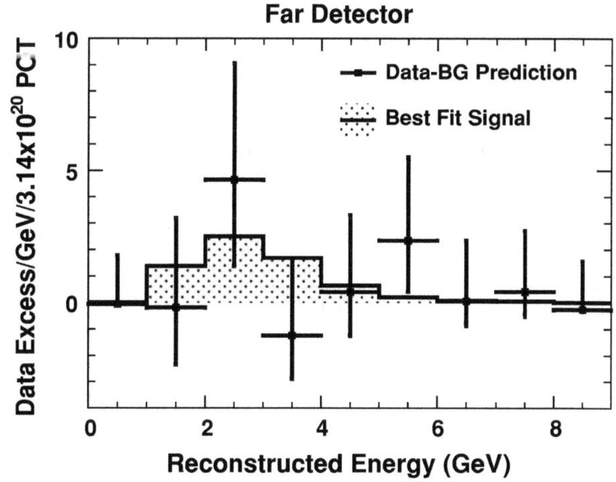

Fig. 9.21 Energy spectrum of the best fit signal, shown alongside the difference between the data and the background prediction

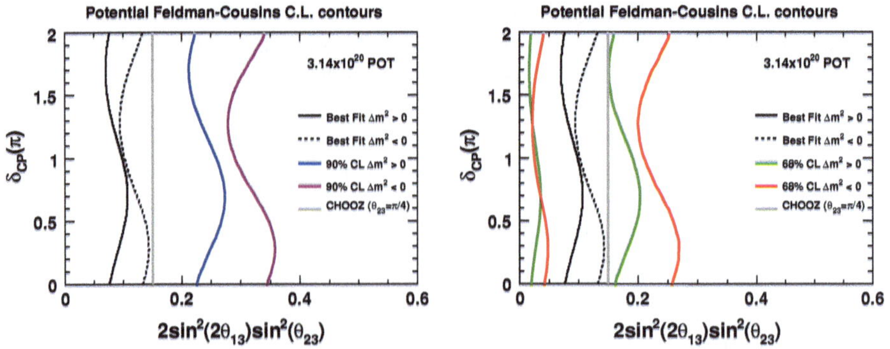

Fig. 9.22 Feldman-Cousins 90% (*left*) and 68% (*right*) C.L. contours for LEM, with the HOO decomposition, for the observation of 28 events. Since these plots constitute the main result of this thesis, they are redone under a different style and shown again in Fig. 9.24

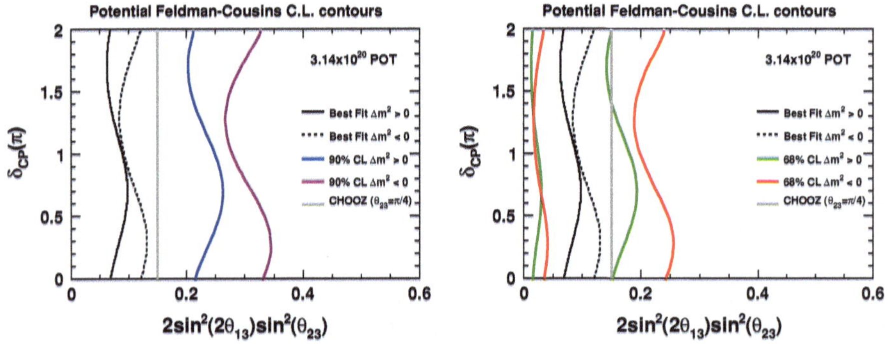

Fig. 9.23 Feldman-Cousins 90% (*left*) and 68% (*right*) C.L. contours for LEM, with the MRCC decomposition, for the observation of 28 events

Table 9.10 Best-fit values and confidence intervals for the LEM > 0.65 selection with the HOO decomposition method. The results are shown for $\delta_{CP} = 0$ as well as for the other two values of δ_{CP} giving the maximum and minimum best fit values in $2\sin^2(\theta_{13})\sin^2(\theta_{23})$

Mass Hierarchy	δ_{CP}	Best-fit $2\sin^2(2\theta_{13})\sin^2(\theta_{23})$	68% C.L. $2\sin^2(2\theta_{13})\sin^2(\theta_{23})$	90% C.L. $2\sin^2(2\theta_{13})\sin^2(\theta_{23})$
Normal	0	0.077	0.021–0.162	<0.225
	0.7π	0.107	0.036–0.206	<0.276
	1.7π	0.071	0.017–0.152	<0.213
Inverted	0	0.132	0.042–0.258	<0.345
	0.3π	0.144	0.048–0.270	<0.359
	1.3π	0.095	0.023–0.201	<0.281

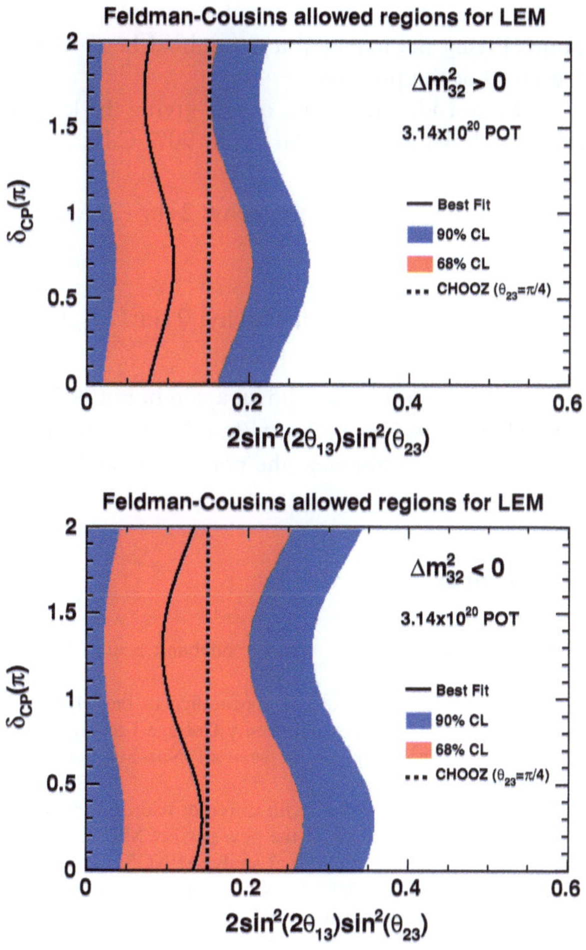

Fig. 9.24 Allowed regions and best fit results for the LEM > 0.65 selection with the HOO decomposition, for the normal hierarchy (*top*) and the inverted hierarchy (*bottom*). The best fit curves lie below the CHOOZ limit for both mass hierarchies. $\theta_{13} = 0$ is included at 90% C.L. but not at 68% C.L

MRCC contours are shown only for comparison purposes, as the primary result is the one obtained with the HOO decomposition. $\theta_{13} = 0$ is excluded with 68% confidence, but is still allowed with 90% confidence. The 90% CL limit of $\sin^2(2\theta_{13}) < 0.15$ set by the CHOOZ experiment for a value of the atmospheric mass splitting of 2.43×10^{-3}eV^2 is also shown. The CHOOZ experiment puts a limit on θ_{13} only, while the x-axis in our contours depends on θ_{23} as well. Consequently the CHOOZ limit as drawn is only valid in the case of maximal mixing ($\theta_{23} = \pi/4$). Please note also that the contours shown here and which are used to derive the final limits have been smoothed. The full procedure for obtaining the contours is described in Appendix D, where the non-smoothed contours can also be found.

Table 9.10 shows the best fit values and the 68% and 90% C.L. confidence intervals in $2\sin^2(2\theta_{13})\sin^2(\theta_{23})$ allowed by our result. In addition to $\delta_{CP} = 0$, the values are given for two other values of δ_{CP}, which correspond to the approximate[5] points giving the maximal and minimal best fit values. The results are shown for the HOO decomposition.

If we consider the values of δ_{CP} giving the less stringent limits, then our results set the following bounds on θ_{13} at 90% C.L.:

$$\text{Normal hierarchy: } 2\sin^2(2\theta_{13})\sin^2(\theta_{23}) < 0.276. \tag{9.3}$$

$$\text{Inverted hierarchy: } 2\sin^2(2\theta_{13})\sin^2(\theta_{23}) < 0.359. \tag{9.4}$$

Finally, the results are shown again in Fig. 9.24, in such a way that the 68% and 90% allowed regions in $2\sin^2(\theta_{13})\sin^2(\theta_{23})$ and δ_{CP} are more evident. This makes it necessary to distinguish the normal hierarchy from the inverted hierarchy.

References

1. G. Pawloski et al. FD AntiPID sideband results for the NUE analysis, MINOS-doc-5455 (2008)
2. A. Holin, Electron neutrino appearance in the MINOS experiment, FERMILAB-THESIS-2010-41, Ph.D. Thesis, University College London
3. T. Mann, Summary report: scan of Nue-Selected FD MRCC events, MINOS-doc-5598 (2009)
4. G. Pawloski, FD MRCC with different hadronization models, MINOS-doc-5575 (2009)
5. R.B. Patterson, A peek at the new perfect MRCC files, MINOS-doc-5925 (2009)
6. T. Yang, True muon removal study, MINOS-doc-3924 (2007)
7. J.P. Ochoa-Ricoux, MRCC vs. Std, MINOS-doc-5640 (2009)
8. J. Boehm, Status update on MR, MINOS-doc-5545 (2009)
9. J. Boehm, MRCC numbers, MINOS-doc-5552 (2009)
10. T. Yang, A study of muon neutrino to electron neutrino oscillations in the MINOS experiment, FERMILAB-THESIS-2009-04
11. J. Boehm, Measurement of electron neutrino appearance with the MINOS experiment, FERMILAB-THESIS-2009-17
12. J.P. Ochoa-Ricoux, R.B. Patterson, More on POT elapsed between selected events, MINOS-doc-5886 (2009)
13. J.P. Ochoa-Ricoux, The overlap between the LEM and ANN PIDs, MINOS-doc-5406 (2008)

[5] The chosen values of δ_{CP} are approximate, as the grid does not have infinite resolution and several points in it yield the same maximum or minimum.

Chapter 10
Pushing the Neutrino Frontier

The question of what comes after the results just shown immediately comes to mind. Accordingly, this chapter begins by reviewing the prospects for future ν_e appearance analyses. While those prospects are found to be very encouraging, we find that it is beneficial for the experiment to focus in the area of antineutrino physics after a second ν_e appearance analysis. The benefits of a small amount of dedicated antineutrino running are thus carefully explored at the end of this chapter.

10.1 The Search for a Nonzero θ_{13}

10.1.1 The Next ν_e Appearance Analysis

10.1.1.1 Reduction of Systematic Errors

The results from the first ν_e appearance results in MINOS are very exciting, as they hint at the existence of a signal slightly below the CHOOZ limit. Similar hints have been recently found by other experiments. For instance, the authors of [1] point out that the combination of the solar data with the KamLAND data suggests a slight ($\sim 1.2\sigma$) preference for $\theta_{13} > 0$. They also find a $\sim 0.9\sigma$ preference for $\theta_{13} > 0$ through a global fit to the atmospheric, long-baseline and CHOOZ data. Furthermore, a very recent low threshold global analysis by the SNO Collaboration finds a best fit value of $\sin^2\theta_{13} = 0.02^{+0.0209}_{-0.0163}$ [2], which is remarkably close to our result. Together with our measurement, all of these results constitute tantalizing first indications of a nonzero θ_{13} for now.

There is thus great motivation to carry out a second ν_e appearance analysis in MINOS as soon as possible. The first ν_e analysis allowed us to learn about the

J. P. Ochoa Ricoux, *A Search for Muon Neutrino to Electron Neutrino Oscillations in the MINOS Experiment*, Springer Theses, DOI: 10.1007/978-1-4419-7949-0_10,

intricacies of the measurement and the ways to improve it. The goal is now to either confirm the excess to a higher confidence level (preferably better than 90%), or to exclude a region of parameter space that supersedes the CHOOZ limit. In order to achieve this several improvements concerning the general infrastructure of the analysis are already in place:

- Use of a better cross-talk model and implementation of a cross-talk filter.
- Use of a corrected intranuclear rescattering model.
- More horn-off statistics.

The first two items will very significantly reduce the crosstalk and intranuclear rescattering systematics. Also, the amount of horn-off data has increased by slightly more than a factor of two, which will reduce the statistical error in the decomposition.

In addition, there are also a number of LEM-specific improvements that are in line for the next analysis:

- Use of better calibrated quantity to do the matching.
- Implementation of a more accurate matching metric.
- Use of a better compacting algorithm.

The first two items have to do with the way the event matching is done in LEM. As explained in Sect. 5.3.4, events are currently compared in terms of photoelectrons. The motivation is that it allows us to handle photostatistics directly with Poisson probabilities. We now know however that the cost of using photoelectrons is too big, given the large uncertainties in the gains. One solution is to use ADC's that have been corrected for attenuation, strip-to-strip effects, and linearity. These are much better known than photoelectrons, as they are the digitized output of PMTs and are thus directly measured in the data. The width distribution of corrected ADCs obtained when injecting a given amount of light in a strip can then be extracted empirically from the simulation, which is more precise than using simple Poisson probabilities. Preliminary results [3] show that with these changes the gains systematic can be reduced by more than a factor of two.

Also, a better compacting algorithm will be put in place, such that outliers are removed from the events [4]. This has the potential to strongly reduce the intensity systematic.

Given that the gains, intensity and crosstalk systematics are the first, third and fourth in order of importance, these improvements will significantly reduce the overall systematic error on the background prediction. In order to estimate this we assume a factor of two reduction in these three systematics, we remove the intranuclear rescattering systematic, and we take into account the increase in horn-off statistics. Recalculating the total systematic error under these rough assumptions yields estimates of 9.2 and 11.5% for the LEM >0.65 and LEM > 0.80 selections, respectively. These values are a significant improvement with respect to the 12.0 and 14.2% errors that apply in the current analysis. Their impact is assessed in the next section.

10.1.1.2 Expected Limits

The next ν_e appearance search in MINOS will be carried out with an exposure in the vicinity of 7.0×10^{20} POT. This is more than twice the data available for the current analysis and it has already been taken, as shown in Fig. 3.12.

In order to get a complete picture of what can be expected for the next analysis, we consider two scenarios:

- The excess observed in the current analysis persists.
- The observation matches the background expectation.

Figure 10.1 shows the 90 and 68% C.L. exclusion contours that would be obtained under the first scenario, where a signal corresponding to a similar[1] best fit value as the one in the first analysis is observed again. For 7.0×10^{20} POT this corresponds to an observation of 62 events. The LEM > 0.65 selection is the one considered, in the case where the systematic error in the background remains at its current value of 12% and when it is reduced to 9.2% as a result of all the expected improvements. The systematic error in the signal is left as in the current analysis, given that the improvements considered will most likely not have a strong impact on the estimation of the signal. We find that the observation of such an excess would allow us to exclude $\theta_{13} = 0$ at 90% C.L. As expected however, the limits are $\sim$5% more restrictive when the systematic error is reduced from 12 to 9.2%. If the systematic error remains at 12%, 62 events is the smallest observation that would exclude a nonzero θ_{13} to 90% confidence.

Figure 10.2 shows the 90 and 68% C.L. exclusion contours that would be obtained under the second scenario, where the observation matches the background expectation. We again round down the number of background predicted events so that the best fit corresponds to $\theta_{13} = 0$. For 7.0×10^{20} POT this gives an observation of 48 events. Under this scenario reducing the systematic error in the background has a strong impact, yielding contours that are $\sim$8% more restrictive. The expected average sensitivity for the next analysis is thus better than CHOOZ, for both mass hierarchies.

It is likely that by the next analysis the MRCC sideband excess will be fully understood, thus allowing the use of a higher cut value. With the improved systematic errors as estimated here however the difference is not very large. The contours obtained with the LEM > 0.80 selection assuming a total systematic error in the background of 11.5% are strikingly similar to the ones shown in Figs. 10.1 and 10.2 for LEM > 0.65 with a systematic error of 9.2%. This suggests that the two selections have very similar sensitivities, within $\sim$2%. The situation may change when the final systematic error estimation is completed though.

[1] Given that the number of observed events must be an integer, it is impossible to simulate the situation where exactly the same best fit value is obtained with the new exposure. In order to be conservative we round down the number of observed events expected at the new exposure. This is why, for instance, the best fit value at the top of Fig. 10.1 corresponds to a slightly lower θ_{13} than the one at Fig. 9.22.

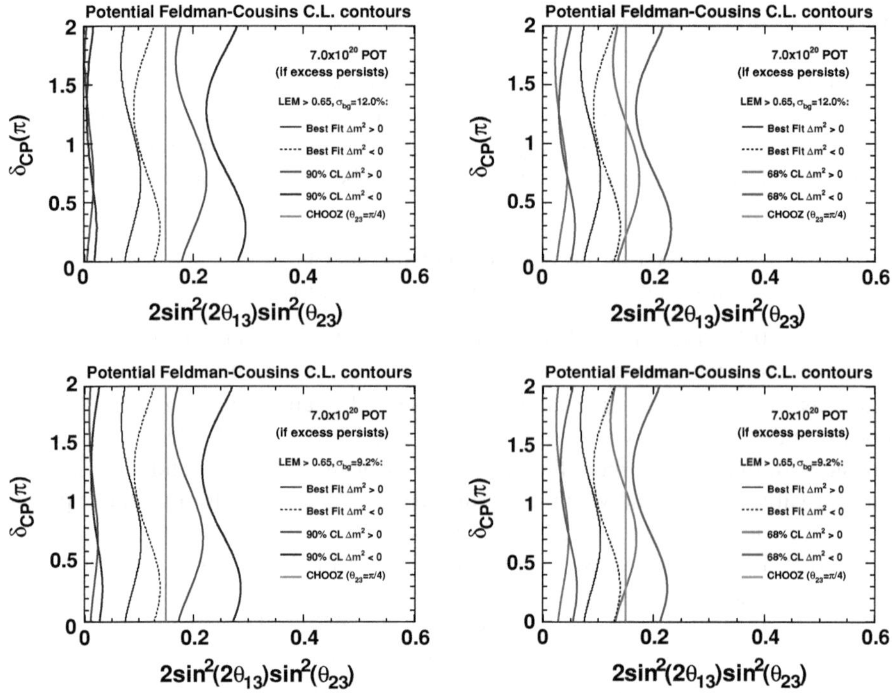

Fig. 10.1 Potential Feldman–Cousins 90% (*left*) and 68% (*right*) C.L. contours for an exposure of 7.0×10^{20} POT and an observation of 62 events. A total systematic error on the background prediction $\sigma_{\rm bg}$ of 12.0% (*top*) and of 9.2% (*bottom*) is assumed. The contours exclude $\theta_{13} = 0$ at 90% C.L

10.1.2 Beyond 7.0×10^{20} *POT*

The picture that emerges for the next analysis is thus very encouraging. If a similar-sized excess persists, then MINOS will have the first indication of a nonzero θ_{13} to 90% confidence. If, on the other hand, the observation matches (or is lower than) the background expectation, exclusion limits that are better than CHOOZ will be obtained.

A question immediately follows: what can be gained with a third analysis at, for instance, 10.0×10^{20} POT? In order to answer this question we consider the average sensitivity at the three exposures that correspond to what would be the first, second, and third ν_e appearance analyses in MINOS. As shown in Fig. 10.3, there is a very significant improvement in sensitivity when going from the first analysis to the second one. This improvement comes by reducing the systematic errors and, mostly, by increasing the data by more than a factor of two. The improvement that is obtained by performing a third analysis is not nearly as large however, as at that point we enter the regime where the systematic errors dominate.

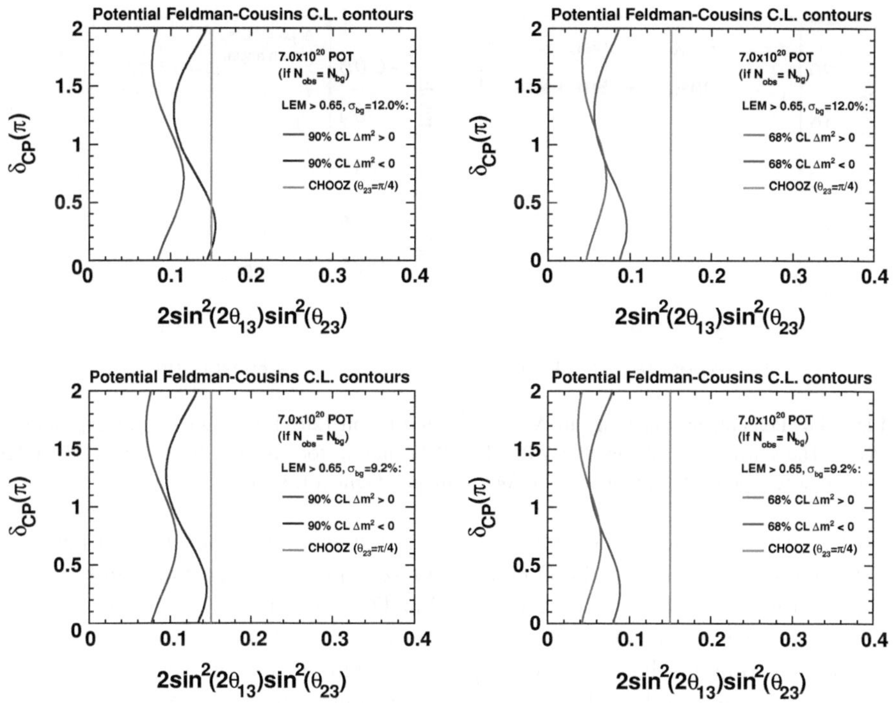

Fig. 10.2 Potential Feldman–Cousins 90% (*left*) and 68% (*right*) C.L. contours for an exposure of 7.0×10^{20} POT and an observation of 48 events, which corresponds to the expected background. A total systematic error on the background prediction σ_{bg} of 12.0% (*top*) and of 9.2% (*bottom*) is assumed. The achieved sensitivity is better than CHOOZ (in the case of maximal mixing)

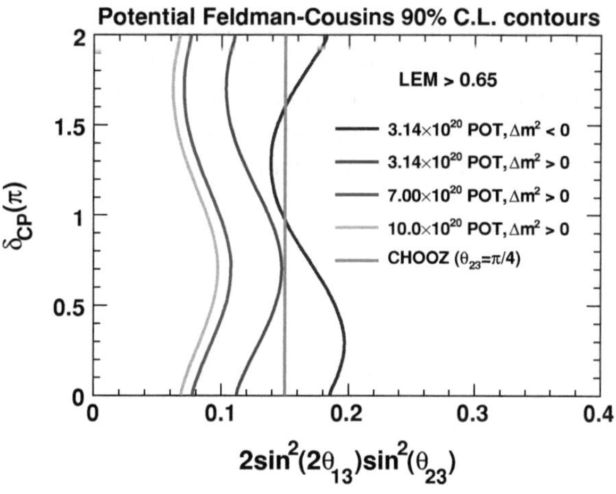

Fig. 10.3 Potential Feldman–Cousins 90% C.L. contours for the LEM > 0.65 selection at different exposures, for the case when the observation matches the background expectation. A systematic error on the background of 12.0% is used for the 3.14×10^{20} POT exposure, and of 9.2% for the other two exposures. There is a very large gain in sensitivity in going from the first analysis to the second, but not so much when going from the second to the third

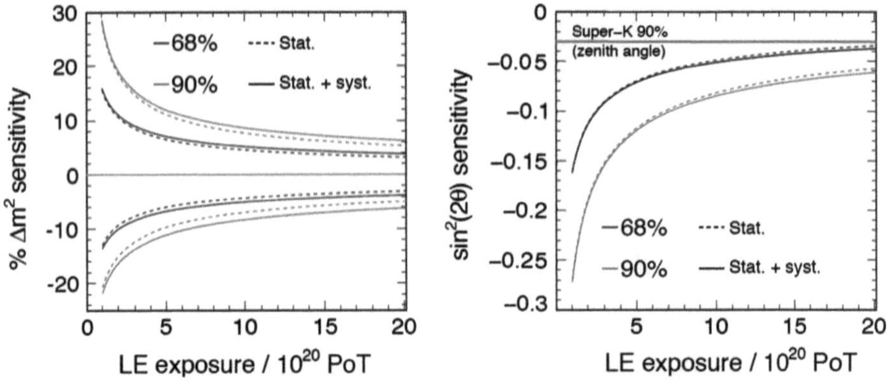

Fig. 10.4 Percentage sensitivity to Δm^2_{32} (*left*) and to $\sin^2(2\theta_{23})$ as a function of exposure in POTs. The sensitivities obtained at 7.0×10^{20} POT and at, for instance, 9.0×10^{20} POT, differ only by a few percent, particularly for Δm^2_{32}. Image obtained from [7]

It is also important to consider MINOS in the context of all the other neutrino experiments. In particular, the Double CHOOZ [5] reactor experiment in France is expected to start collecting data with its Far Detector in the first quarter of 2010 [6]. Studies show that Double CHOOZ can achieve a better sensitivity to θ_{13} in <1 year of running than what MINOS can do with all its data set.

The conclusion is thus clear that the ν_e appearance analysis in MINOS that will have the strongest impact on the neutrino community is the second one, corresponding to 7.0×10^{20} POT. The prospects are very good in that case, as MINOS could make the first observation of a nonzero θ_{13}. Beyond that however it seems that the sensitivity would not change by much and that the Double CHOOZ experiment would have surpassed MINOS in reach.

The ν_μ CC disappearance analysis is in a similar situation, in that additional data beyond 7×10^{20} POT does not improve the measurement very significantly. This can be seen in Fig. 10.4, which shows the sensitivity to Δm^2_{32} and $\sin^2(2\theta_{23})$ as a function of exposure in POTs. As we show in the next section however, the next period of data taking in MINOS is not destined to just provide the means for a refinement of the ν_e appearance and ν_μ CC disappearance measurements. It presents an outstanding opportunity to focus in a different area that has remained largely unexplored in the community: antineutrino physics.

10.2 Antineutrino Physics

10.2.1 Antineutrinos in MINOS

As explained in Sect. 3.2, the NuMI neutrino beam is produced by colliding 120 GeV protons into a graphite target. The positive particles produced in the

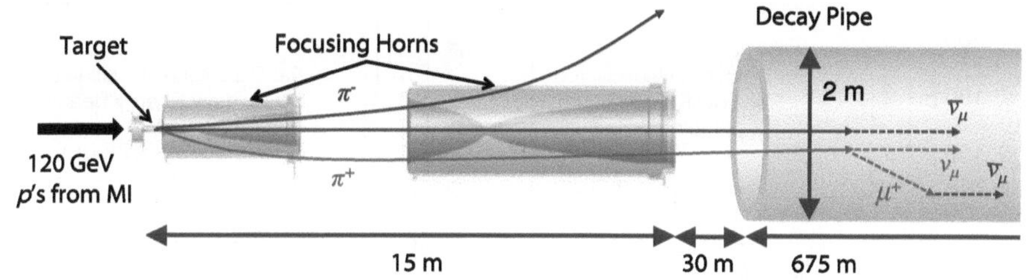

Fig. 10.5 Schematic of the NuMI beamline. Most antineutrinos are produced by those negatively charged mesons that go undeflected through the center of the horns. A small amount of antineutrinos originate from secondary μ^+ decay

collision are then focused by a parabolic horn system onto a 675 m long decay pipe. The decays of π^+ and K^+ mesons produce the ν_μ's that make up most of the beam.

Most negatively charged particles produced at the target are defocused by the horns. A few of them however avoid this fate by going straight through the horns' center, where there is no magnetic field. This is illustrated in Fig. 10.5. The decays of those π^- and K^- mesons produce most of the antineutrinos[2] in our beam:

$$\pi^- \to \mu^- + \overline{\nu}_\mu,$$
$$K^- \to \mu^- + \overline{\nu}_\mu.$$

Also, a small fraction of antineutrinos are produced in secondary muon decay alongside the beam ν_e's that constitute a background to the ν_e appearance analysis:

$$\begin{aligned} \pi^+ \to \mu^+ &+ \nu_\mu \\ &\hookrightarrow e^+ + \nu_e + \overline{\nu}_\mu. \end{aligned} \tag{10.1}$$

This is why, as explained in Appendix C, the antineutrino rate can be used to measure the beam ν_e rate.

Figure 10.6 shows the energy spectra of neutrinos and antineutrinos. Given that most of the potential antineutrino parents are deflected by the horns, the antineutrino rate is much smaller than the neutrino rate, accounting for only 6.4% of the beam at the Far Detector. Also, since most antineutrino parents are low-p_T mesons that travel through the center of the horns, antineutrinos have a small decay angle. It follows from Eq. (7.1), which relates the neutrino energy to the decay angle, that antineutrinos peak at higher energies ($\sim$8 GeV) than neutrinos ($\sim$3 GeV). This is an important point, as oscillations occur mostly at low ($\lesssim$5 GeV) energies. This means that the antineutrino spectra is not optimal for oscillation-based studies. Despite this fact, very interesting physics can still be done with antineutrinos in MINOS.

[2] Please note that by antineutrinos we refer specifically to muon antineutrinos, that is $\overline{\nu}_\mu$ events.

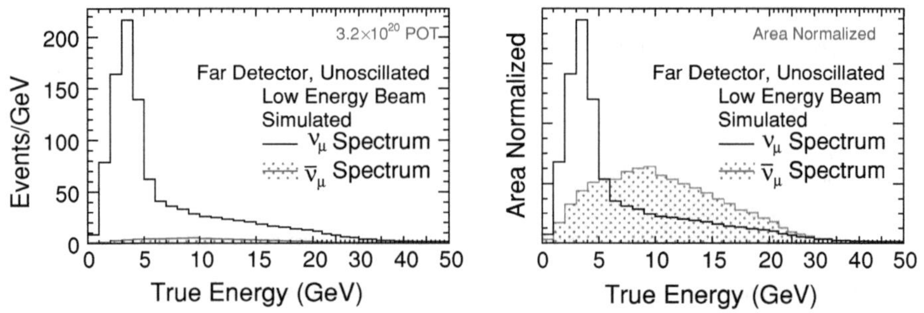

Fig. 10.6 Far Detector neutrino and antineutrino energy spectra, normalized to 3.2×10^{20} POT (*left*) and to unity (*right*). Antineutrinos make up for approximately 6.4% of all NuMI interactions at the Far Detector, and have a higher average energy than neutrinos

10.2.2 Grasping the Opportunity

10.2.2.1 MINOS' Unique Advantage

In a $\overline{\nu}_{\mu}$ CC interaction, the incoming antineutrino interacts with the nucleus N, producing an anti-muon in addition to some hadronic activity:

$$\overline{\nu}_{\mu} + \mathrm{N} \rightarrow \mu^{+} + \text{hadrons}. \tag{10.2}$$

An antineutrino CC interaction thus leaves the same unmistakeable signature that ν_{μ} CC interactions leave, consisting of a long track with some hadronic activity at the vertex. The only difference is that in the case of an antineutrino interaction the muon is positively charged.

MINOS has a privileged position with regard to other neutrino experiments in that both of its detectors are magnetized. This means that antineutrinos can be identified on an event by event basis, based on the observed curvature of the muon track. In its forward (standard) polarity, the MINOS magnetic field is such that negatively charged tracks curve inward and positively charged tracks bend outward. Figure 10.7 shows the example of a $\overline{\nu}_{\mu}$ CC candidate in the Far Detector data, where the track clearly bends toward the outside of the detector.

In sharp contrast to the situation encountered when trying to select ν_e CC events, it is possible to construct a $\overline{\nu}_{\mu}$ CC selector that achieves both high purity and efficiency. Figure 10.8 shows the contamination and the efficiency of the antineutrino selection that is used for the most recent analysis [8] as a function of energy. The selection achieves an overall efficiency of 82% with a contamination of only 3%. The selection consists of a slightly harder cut on the CC/NC discriminator used for the first ν_{μ} CC analysis [9], combined with a cut on the track fit charge sign significance and on the relative angle, which is a measure of whether the track curves toward or away from the magnetic coil hole relative to its initial

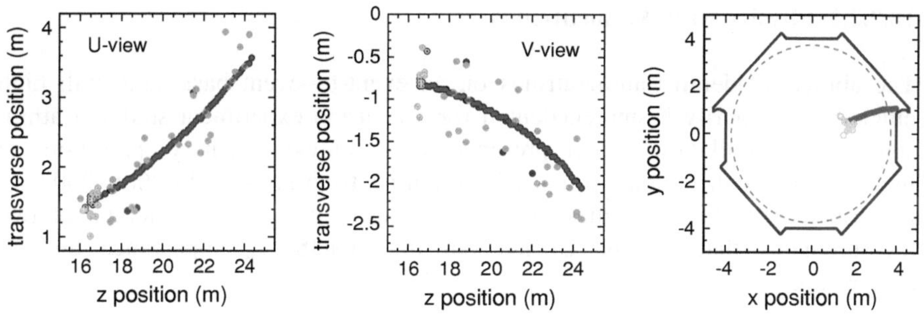

Fig. 10.7 Far Detector data $\overline{\nu}_\mu$ CC Candidate (Run 36770, Snarl 204002), as viewed from the U-direction (*left*), from the V-direction (middle) and from the front of the detector (*right*). The track clearly bends towards the outside of the detector. The low pulse height hits (in *green*) surrounding the track are due to crosstalk or bremsstrahlung

Fig. 10.8 Far Detector efficiency and contamination of the $\overline{\nu}_\mu$ CC selector used in the most recent analysis [8]

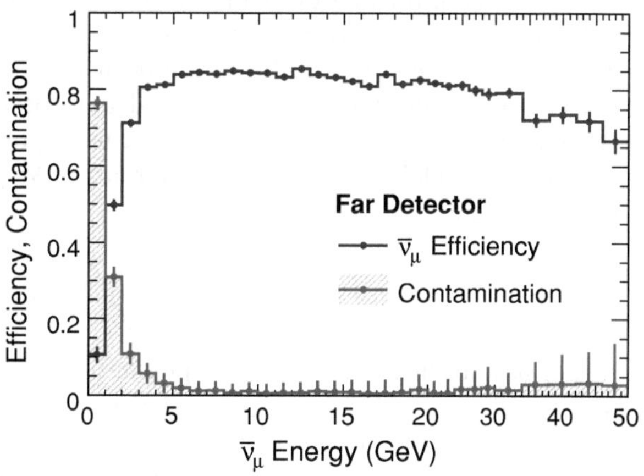

position. The contamination consists primarily of misidentified ν_μ CC events. A negatively charged track can mimic a positively charged one if, for example, the particle undergoes a hard scatter in the detector. As expected, the efficiency and the purity are smaller at low energies, where the track is not long enough to allow for the distinction between ν_μ and $\overline{\nu}_\mu$ CC events. Also, the purity and the efficiency start decreasing at high energies, due to the fact that very energetic tracks do not curve enough to unambiguously tell their sign apart.

Similarly, at the cost of some efficiency loss, it is possible to put together a $\overline{\nu}_\mu$ CC selection that has practically no contamination. Such is the case of the selection we constructed for the beam ν_e rate from antineutrinos measurement, which is described in Sect. C.4. In that case a very high purity (>99%) and a reasonably high efficiency ($\sim$45%) are achieved, by combining kinematic variables with the information of the track fit charge sign significance.

10.2.2.2 Physics with Antineutrinos

The ability to identify antineutrinos on an event-by-event basis and with high efficiency and purity is unprecedented for a neutrino experiment studying atmospheric-type oscillations, which presents us with an extraordinary opportunity to make unique contributions to the field of antineutrino physics. In 2006–2007 we performed some initial studies in order to discern the physics potential of two measurements that can be carried out in MINOS: $\overline{\nu}_\mu$ appearance and $\overline{\nu}_\mu$ disappearance.

Antineutrino Appearance

Antineutrino appearance would occur if exotic neutrino to antineutrino $\nu_\mu \rightarrow \overline{\nu}_\mu$ transitions occurred. As ν_μ CC events disappear on their way to the Far Detector, it is possible that a fraction of them transition to antineutrinos.

While this is certainly unconventional behavior, it is not completely radical either. As a matter of fact, the first time neutrino oscillations were posited by Pontecorvo in 1957 [10] neutrinos could transform into antineutrinos via mass eigenstate mixing. These transitions are actually predicted by some extensions to the Standard Model that incorporate neutrinos as Majorana particles, although at a very low level:

$$P(\nu_\mu \rightarrow \overline{\nu}_\mu) \sim \left(\frac{m_\nu}{E_\nu}\right)^2 \lesssim 10^{-18}. \tag{10.3}$$

They have been also recently [11] predicted to occur at low probability by some models with a very large neutrino magnetic moment, neutrino decay and other beyond the standard model processes.

Moreover, while ν_μ CC events have been observed to disappear with both beam [12, 9] and atmospheric [13, 14] neutrinos, they have never been observed to reappear. The Super-Kamiokande data [15] combined with the standard three-flavor oscillation model suggests that the great majority of them transition to ν_τ CC. Nevertheless, given that antineutrinos have a smaller cross-section that neutrinos, it has been speculated [16] that neutrino to antineutrino transitions could fully explain the deficit observed by all the atmospheric neutrino experiments. No search for $\nu_\mu \rightarrow \overline{\nu}_\mu$ transitions has ever been performed outside of MINOS in the atmospheric sector however.

There is no particular model that describes how these transitions would occur. There is however compelling evidence that ν_μ CC events disappear in accordance with the two-flavor oscillation expression:

$$\mathrm{P}(\nu_\mu \rightarrow \nu_\mu) = 1 - \sin^2(2\theta)\sin^2\left(\frac{1.27\Delta m^2_{32}L}{E}\right). \tag{10.4}$$

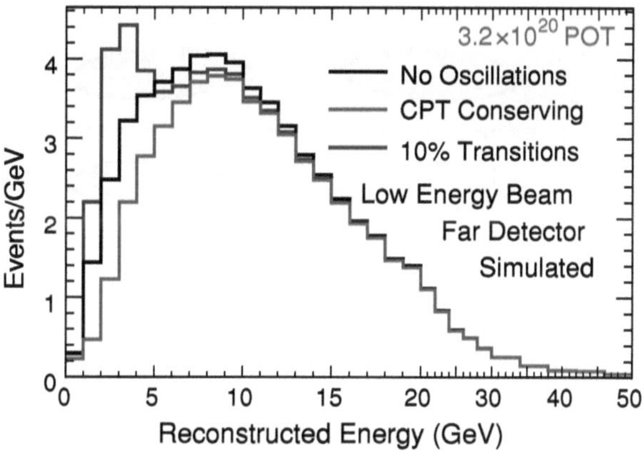

Fig. 10.9 Antineutrino simulated energy spectra for no oscillations, CPT conserving oscillations and CPT conserving oscillations with $\alpha = 0.1$ transitions. If neutrinos transition to antineutrinos an excess of low energy antineutrinos is expected to appear at low energies

This allows us to parameterize antineutrino appearance as a fraction α of the original ν_μ CC that transition to $\overline{\nu}_\mu$ CC:

$$\mathrm{P}(\nu_\mu \to \overline{\nu}_\mu) = \alpha \sin^2(2\theta) \sin^2\left(\frac{1.27\Delta m^2_{32} L}{E}\right). \tag{10.5}$$

Figure 10.9 shows what the antineutrino energy spectrum at the Far Detector would look like if 10% of neutrinos transition to antineutrinos (i.e., if $\alpha = 0.1$). As expected, transitions leave a very clear experimental signature consisting of an excess at low energies, which is where most of the ν_μ CC events disappear from.

Our studies show that if $\sim$7% or more of the disappearing neutrinos transition to antineutrinos, MINOS would be able to exclude the no-transitions scenario with 90% confidence based on an exposure of 3.2×10^{20} POT. The MINOS experiment is thus very well suited for this kind of measurement.

Antineutrino Disappearance

By isolating a pure sample of antineutrinos, a ν_μ CC-only disappearance measurement can be performed in MINOS. Such a measurement would constitute the first direct determination of the antineutrino oscillation parameters $\Delta\overline{m}^2_{32}$ and $\sin^2(2\overline{\theta}_{23})$. As such, it would also constitute a direct test of CPT conservation in the neutrino sector, and could have a very strong impact in some of the CPT violating models that are currently being invoked to, for instance, accommodate the LSND signal observation. As explained in [17], no CPT-conserving sterile neutrino scenario is able to avoid at least a 3σ discrepancy among the different experimental data sets when the LSND [18] and the latest MiniBoone results [19] are included, requiring theorists to resort to a CPT-violating spectrum.

One of the reasons these CPT violating models can still be put together is that the current uncertainty in $\Delta\overline{m}^2_{32}$ is very large. Figure 10.10 shows the 2D region of parameter space in $\Delta\overline{m}^2_{31}$ vs. Δm^2_{31} and in $\sin^2(\overline{\theta}_{23})$ vs. $\sin^2(\theta_{23})$ that is still allowed

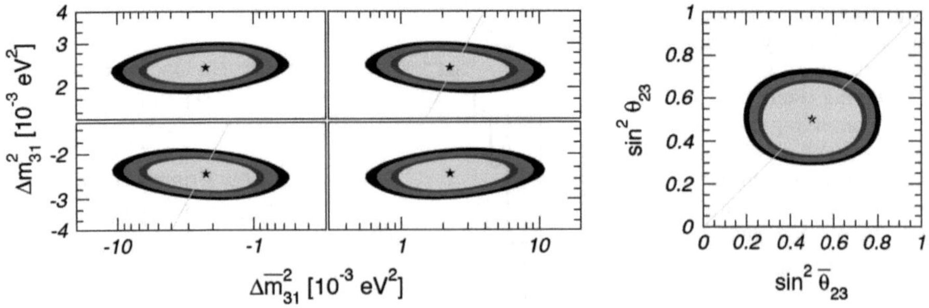

Fig. 10.10 90%, 95%, 99%, and 3σ C.L. allowed regions in the $\Delta\overline{m}^2_{31}$ vs. Δm^2_{31} parameter space (*left*) and in the $\sin^2(\overline{\theta}_{23})$ vs. $\sin^2(\theta_{23})$ parameter space (*right*), as obtained from a global fit done on all the existing neutrino data until 2008. The sign of the neutrino and antineutrino mass splittings are still unknown, which is why four different quadrants are considered in the *left* plot. The best fit point is indicated with a star, and the CPT conserving scenario with a line. The uncertainty in $\Delta\overline{m}^2_{32}$ spans more than one order of magnitude at 99% C.L. The images are obtained from [20]

by a global fit [20] of all neutrino data available up to the year 2008. There it can be seen[3] that the current uncertainty on $\Delta\overline{m}^2_{32}$ is close to one order of magnitude, depending on the confidence level considered. The reason is that the constraint on $\Delta\overline{m}^2_{32}$ comes primarily from the Super-Kamiokande data, where neutrinos and antineutrinos are indistinguishable on an individual basis and where only a very big difference between $\Delta\overline{m}^2_{32}$ and Δm^2_{32} would be apparent [21].

MINOS thus has a golden opportunity to significantly improve the current state of affairs by studying $\overline{\nu}_\mu$ CC disappearance. Our studies demonstrate however that the impact is limited if only the $\sim 6\%$ antineutrino component in the beam is used to perform the analysis. This is shown in the following section, in the context of the most recent antineutrino results in MINOS. If MINOS is to realize its full potential in terms of antineutrino oscillations, it is necessary to make the modifications which we describe in Sect. 10.2.3.

10.2.2.3 Initial Results

The results of the studies described above were first shown to the MINOS collaboration in the years 2006–2007. Since then,[4] the MINOS antineutrino group carried the $\overline{\nu}_\mu$ appearance and $\overline{\nu}_\mu$ disappearance analyses forward and, in May of 2009, the first results were released. These results are based on an exposure of 3.2×10^{20} POT.

Figure 10.11 shows the reconstructed energy spectra of selected antineutrinos in the Far Detector. As seen in Table 10.1, a total of $42\,\overline{\nu}_\mu$ CC candidates are observed in the FD data, which represents a 1.9σ deficit with respect to the

[3] We note again that $\Delta m^2_{31} = \Delta m^2_{32} + \Delta m^2_{21} \approx \Delta m^2_{32}$

[4] At that point in time the author's interests shifted to the ν_e appearance analysis.

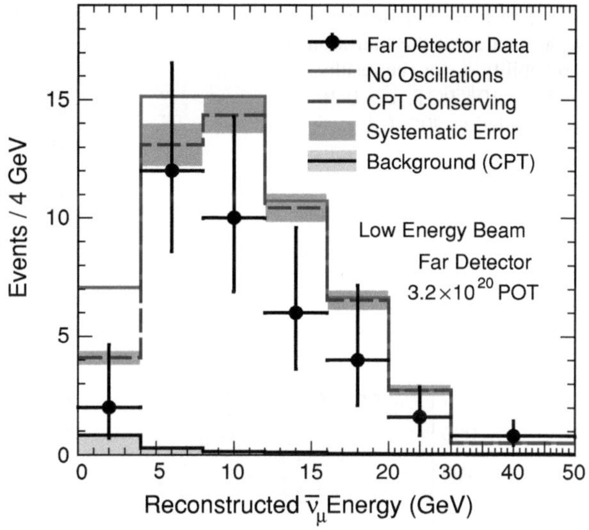

Fig. 10.11 Reconstructed energy spectrum of $\overline{\nu}_\mu$ CC candidate events in the Far Detector. A total number of 42 $\overline{\nu}_\mu$ CC candidates are observed

Table 10.1 Predicted and observed number of selected antineutrino candidates in the Far Detector, corresponding to an exposure of 3.2×10^{20} POT. There is a 1.9σ deficit in the data with respect to the CPT-conserving prediction

Prediction (no oscillations)	Prediction (CPT-conserving)	Observation
64.6 ± 8.0(stat) $\pm$ 3.9(syst)	58.3 ± 7.6(stat) $\pm$ 3.6(syst)	42

CPT-conserving expectation. No abnormalities are found in any of the FD data distributions or in the analysis in general.

With regard to the $\overline{\nu}_\mu$ appearance analysis, the observed deficit means that an even tighter limit than the one previously anticipated can be put on $\nu_\mu \rightarrow \overline{\nu}_\mu$ transitions. As explained in Sect. 10.2.2, we perform a one-parameter fit on α, the fraction of the disappearing neutrinos that transition to antineutrinos. The oscillation parameters are kept fixed at their CPT-conserving values. The exclusion level obtained for each value of α is shown in Fig. 10.12. As shown there, $\alpha > 0.026$ is excluded at 90% confidence, and $\alpha > 0.065$ at 99% confidence.

With regards to the $\overline{\nu}_\mu$ disappearance analysis, the deficit observed in the data at all reconstructed energy bins forces the best fit under the oscillations hypothesis to occur at a value of $\Delta\overline{m}^2_{32}$ that is higher than in the CPT-conserving case. This is because as the mass splitting increases, the first oscillation peak moves to higher energies as well. In order to compensate, a best fit value of $\sin^2(2\overline{\theta}_{23})$ that is not maximal is obtained. The results are consistent with it at 90% C.L., as shown on the left hand side of Fig. 10.13. The region of parameter space that is allowed at the 99.7% C.L. is very large, and spans practically all values of $\Delta\overline{m}^2_{32}$. Nevertheless, there is a small region that is excluded to 99.7% C.L.

The right hand side of Fig. 10.13 shows the 90% and 3σ exclusion contours obtained from our results and from a global fit [20] of all previous neutrino data. There is some overlap between the regions excluded by MINOS and those allowed

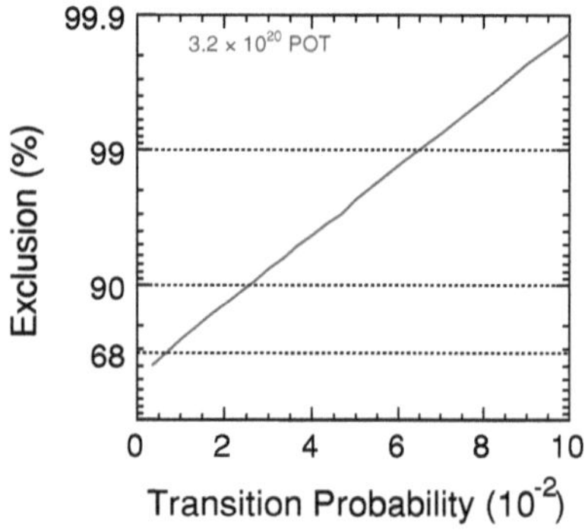

Fig. 10.12 Exclusion level as a function of the transition probability α, for the results at 3.2×10^{20} POT. $\alpha > 0.26$ is excluded at 90% C.L

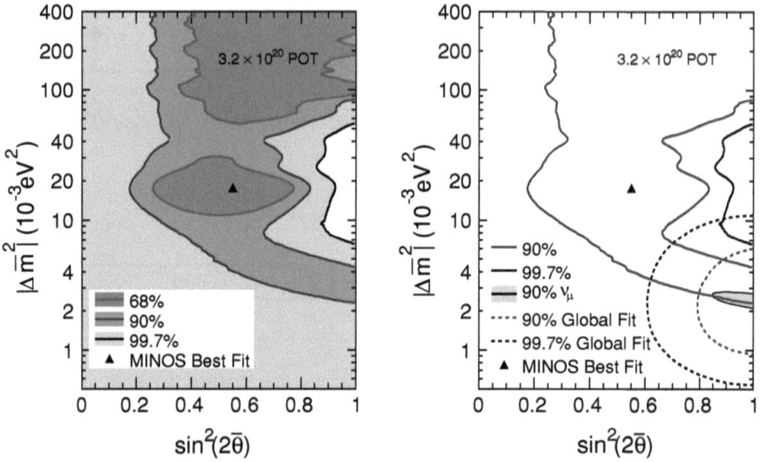

Fig. 10.13 Exclusion contours for the antineutrino disappearance results with 3.2×10^{20} POT. The *left* plot shows the regions of parameter space allowed at the 68, 90% and 3σ C.L. The unshaded region around maximal mixing is excluded at 99.7%. The *right* plot shows the 90% and 3σ contours from our results and from a global fit [21] of all previous data. The most recent ν_μ CC disappearance results from MINOS are also shown for comparison

by the global fit. The results thus exclude some previously allowed CPT-violating regions of parameter space, particularly near maximum mixing.

10.2.3 A Proposal for Antineutrino Running

10.2.3.1 Motivation

The recent antineutrino results are thus very encouraging. A very tight limit is set on $\nu_\mu \rightarrow \overline{\nu}_\mu$ transitions, such that α, the fraction of disappearing neutrinos that transition to antineutrinos, is determined to be < 0.026 at 90% C.L. Also, the $\overline{\nu}_\mu$ CC disappearance results already exclude some CPT-violating regions of parameter space that were previously allowed. However, the reduction that our results impose on the range of $\Delta\overline{m}^2_{32}$ and of $\sin^2(2\overline{\theta}_{23})$ currently allowed by global fits is probably not very large. This is particularly so for high confidence levels such as 3σ, where the right plot of Fig. 10.13 shows only a small overlap between the region excluded by MINOS and the one allowed by global fits.

As previously mentioned, the ability that MINOS has, to separate neutrinos from antineutrinos on an individual basis, is truly exceptional among other neutrino detectors. As a matter of fact, it is possible that this situation will not be repeated until the Neutrino Factory era [22, 23]. Not only is the current uncertainty in the antineutrino oscillation parameters still very large, but the recent results give a slight indication of a CPT-violating signal that must be addressed. These reasons compel us to make a better measurement of $\Delta\overline{m}^2_{32}$ and to thus take full advantage of the capabilities offered by MINOS. Taking more data in the standard beam configuration is not good enough, as will be shown shortly. The main reason is that, as mentioned previously, the energy spectrum of the antineutrino component in the beam is not optimal for studying oscillations with CPT-conserving parameters, due to its higher energy.

Having anticipated this situation, in January of 2008 we submitted a proposal [24] to the MINOS collaboration where we suggested to dedicate a significant amount of time to Reversed Horn Current (RHC) running. The concept of RHC running, as well as its benefits, are explored in the subsequent sections.

10.2.3.2 The Basics of Reversed Horn Current Running

As explained in Sect. 10.2.1, the NuMI beam is produced by focusing the positively charged particles from the target into a 675 m long decay pipe. If the current in the horns is reversed, the opposite effect is obtained: negatively charged particles are focused into the decay pipe and positively charged particles are defocused. This is illustrated in Fig. 10.14. In that situation, only those positively charged particles that go through the center of the horns and which are responsible for the high energy tail make it through.

The result of Reversed Horn Current (RHC) running is a beam composed primarily of low-energy antineutrinos. The few positively charged particles that make it through the horns create a neutrino contamination that peaks at higher energies (~ 8 GeV), in an analogous situation to what occurs with antineutrinos in the Forward Horn Current (FHC) configuration. The simulated energy spectra of neutrinos and antineutrinos in the RHC and FHC configurations are shown in Fig. 10.15. The difference between FHC and RHC running is not limited to an alternation of roles between neutrinos and antineutrinos as to which forms the main focusing peak and which the contamination. In the RHC configuration, the main focusing peak is approximately three times lower, and the contamination three times larger. This is due to the lower cross-section of antineutrinos (particularly at low energies), and to the smaller production of negatively charged mesons in the target compared to the positively charged ones.

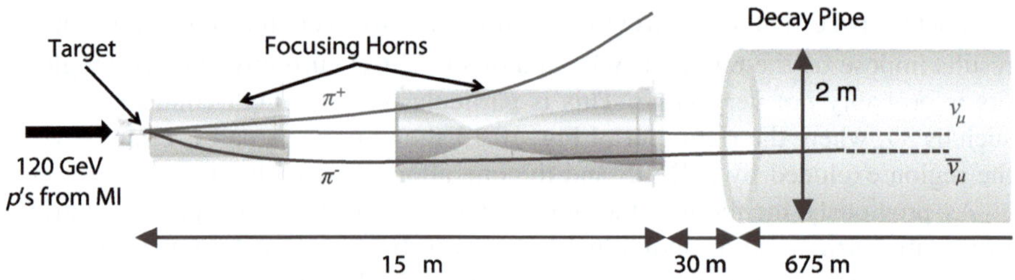

Fig. 10.14 Schematic of the NuMI beamline in Reversed Horn Current (RHC) mode. Positively charged particles from the target are defocused, and negatively charged particles are focused into the decay pipe. Only the positively charged particles that go through the center of the horns make it through. The result is a beam consisting primarily of antineutrinos

Fig. 10.15 True energy spectra of neutrinos and antineutrinos in the reverse (RHC) and forward (FHC) configurations. In the RHC configuration the beam consists predominantly of antineutrinos, although there is a significant neutrino contamination that peaks at higher energies

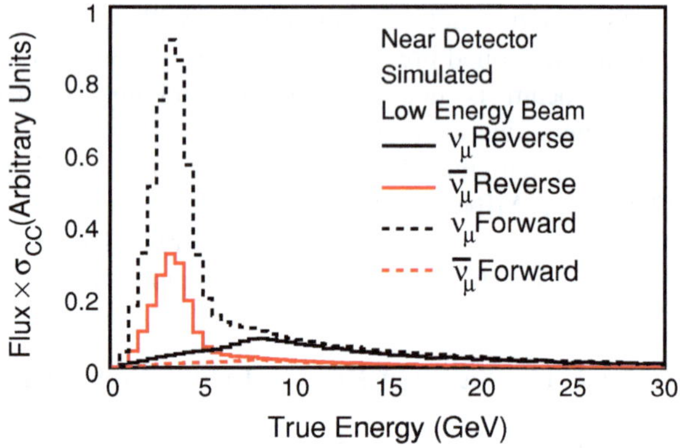

In order to get the most out of RHC running the magnetic field must be reversed as well, in such a way for μ^+ tracks to bend toward the inside of the detector and consequently maximize the selection efficiency. Reversing the magnetic field is a common procedure in both detectors which can be done very quickly. Reversing the current in the horns is a slightly more complicated issue, although well within the bounds of what can be accomplished quickly and with a relatively small cost. As shown in [25], the correct way to reverse the horn current is to invert the voltage of the power supply in order to avoid creating a very large voltage difference between the target and the horns. The equipment needed to accomplish this $10,000 to $20,000. Only a couple of months are required for preparation and one or two weeks for installation.

10.2.3.3 Expected Sensitivities

Despite the reduction in the overall event rate between the RHC and FHC configurations, the situation for antineutrinos is vastly superior in the RHC configuration than in the FHC configuration. This can be seen by comparing the two red

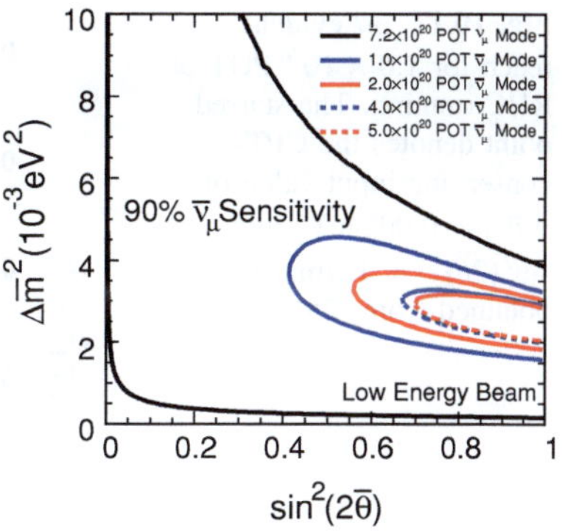

Fig. 10.16 Average 90% C.L. exclusion contours that would be obtained with different amounts of RHC data and 7.2×10^{20} POT of FHC data, in the case of CPT conservation. In all cases the area of parameter space excluded with RHC running is much larger than with FHC running

curves in Fig. 10.15. Not only is the antineutrino event rate greatly enhanced in the RHC configuration, but this enhancement occurs where it matters the most (at low energies). It should come as no surprise, then, that what MINOS can do in terms of $\overline{\nu}_\mu$ disappearance with a modest amount of RHC data is far superior to what it can do with even a large amount of FHC data.

Figure 10.16 shows the average exclusion contours that would be obtained with different amounts of RHC running, under the assumption of CPT conservation. The contours do not include systematic errors, although studies [24] show that the effect of systematic errors on the contours are very limited, as statistical errors dominate. The predicted exclusion contours obtained with RHC data are significantly better than what can be obtained with the full FHC data set existing to this date, of 7.2×10^{20}. Not surprisingly, combining the FHC and RHC antineutrino datasets has a very small impact on the resulting contours [24]. A $\sim 15\%$ measurement of $\Delta\overline{m}^2_{32}$ can be obtained with only $\sim 2 \times 10^{20}$ POT of RHC data. The 90% bounds on $\Delta\overline{m}^2_{32}$ that would be obtained in that case are approximately 4 times narrower than what is currently allowed by the global fit shown in Fig. 10.10.

The prospects for discovery are also very good with only a small amount of RHC running. The region of parameter space outside the contour shown in Fig. 10.17 gives the set of parameters that can be distinguished from the CPT-conserving scenario at 3σ significance with more than a 50% probability. In other words, for a true $\overline{\nu}_\mu$ parameter set outside the enclosed region, more than half of the experiments result in a 3σ contour that excludes the CPT-conserving input value (of $\Delta\overline{m}^2_{32} = 0.0024\text{eV}^2$ and $\sin^2(2\overline{\theta}_{23}) = 1$ in this case). This means that some of the values of $\Delta\overline{m}^2_{32}$ currently allowed by the global fit of Fig. 10.10 could be distinguished from the CPT-conserving scenario with more than 3σ significance.

It is also important to point out that such a measurement would constitute one of the most precise tests of CPT ever performed. As pointed out in [17], the best test of CPT to date outside of neutrinos comes from the neutral kaon system [26]. The difference in mass between the K^0 and $\overline{K}^0$ particles has been determined to be < 1

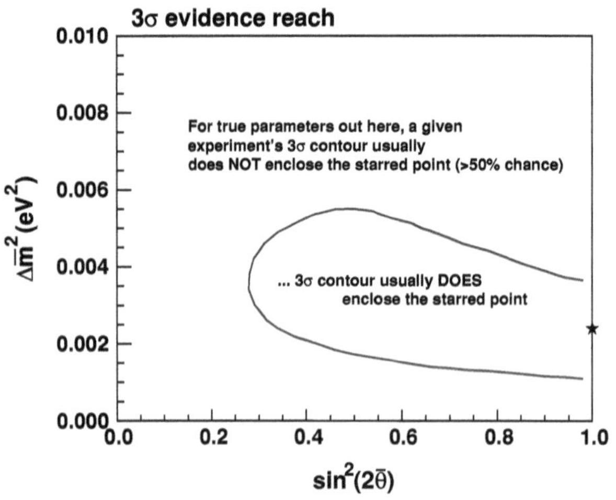

Fig. 10.17 3σ evidence reach for 1.0×10^{20} POT of RHC running. The starred point denotes the CPT-conserving input value of $\Delta\overline{m}^2_{32} = 0.0024\text{eV}^2$ and $\sin^2(2\overline{\theta}_{23}) = 1$. Image obtained from [25]

part in 10^{18}. When viewed in terms of the mass-squared difference however, the limit is equivalent to $|m^2(K^0) - m^2(\overline{K}^0)| \lesssim 0.25\text{eV}^2$. From Fig. 10.17, we see that in MINOS we could distinguish with 3σ confidence a difference in the squared masses between neutrinos and antineutrinos larger than $\sim 0.002\,\text{eV}^2$. This means that with neutrinos we can probe approximately two orders of magnitude deeper than what has been done with quarks. As pointed out in [27], the CPT test performed when comparing solar and reactor neutrino results is even more restrictive ($|m^2(\nu) - m^2(\overline{\nu})| < 1.3 \times 10^{-3}\text{eV}^2$ at 90% C.L.). Nevertheless, the test we propose here is completely independent as it probes a different region of parameter space ("atmospheric" instead of "solar").

10.2.3.4 The Time for Reversed Horn Current Running

The case for a modest ($1 - 2 \times 10^{20}$ POT) amount of RHC running in MINOS is thus very strong. Not only would it improve our knowledge of antineutrino physics by several factors, but it also allows for MINOS to have an extra set of measurements that will dominate the field for many years to come. This is not the case for the ν_e appearance and ν_μ disappearance analyses, which will be superseded once experiments such as Double CHOOZ and NOνA come online. In particular, if the NOνA collaboration decides to not run with antineutrinos in order to expand the limit on θ_{13} as far as possible, the MINOS antineutrino measurements would dominate for at least a decade.

As discussed in the first portion of this chapter, the ν_e analysis reaches its peak during the next analysis, corresponding to an exposure of 7.0×10^{20} POT. Similarly, the ν_μ CC disappearance analysis does not benefit strongly from FHC running beyond 7.0×10^{20} POT. Because of these reasons, the MINOS collaboration and the Fermilab directorate have accepted the proposal to run in RHC mode immediately after the summer 2009 shutdown. As of writing this thesis RHC running has just recently begun. The current plan is to run in RHC for approximately 1×10^{20} POT and then, based on the results from the second ν_e appearance analysis, reevaluate the run plan.

10.3 Conclusions

These are exciting times for the MINOS experiment. The first ν_e appearance results give a slight indication of a signal. With the LEM selection, the sensitivity to θ_{13} can be pushed 15–20% higher than with other methods. This will allow, in the next analysis with 7.0×10^{20} POT, to exclude $\theta_{13} = 0$ at 90% confidence if a similar excess is observed or, if no excess is observed, to produce a limit on θ_{13} that is better than CHOOZ.

The exploration of the neutrino frontier in MINOS does not stop with electron neutrinos however. MINOS plans to make full use of its capabilities by running in Reversed Horn Current mode. This will allow for the attainment of a vastly superior knowledge of the antineutrino parameters, and could even lead to very exciting discoveries, through a direct test of CPT in the neutrino sector.

We thus anticipate many cutting-edge results in these two areas and we hope that this thesis helped, in a small way, to pave the way for them.

References

1. G.L. Fogli et al., Neutrino Oscillations, Global Analysis, and Theta (13). hep-ph/0905.3549 (2009)
2. The SNO Collaboration, Low Energy Threshold Analysis of the Phase I and Phase II Data Sets of the Sudbury Neutrino Observatory. nucl-ex/0910.2984 (2009)
3. M. Orchanian, R.B. Patterson, LEM Update. MINOS-doc-6179 (2009)
4. R. Toner, LEM Intensity Effect Study, MINOS-doc-6051 (2009)
5. F. Ardellier et al., Double Chooz: A Search for the Neutrino Mixing Angle θ_{13}, hep-ex/0606025 (2006)
6. C. Palomares (for the Double CHOOZ Collaboration), Double Chooz Neutrino Experiment, in *European Physical Society Meeting* (2009)
7. J. Musser, N. Saoulidou, Report of MINOS Future Running Scenarios, MINOS-doc-5285 (2008)
8. P. Adamson et al., A Beam Muon Antineutrino Oscillation Study in MINOS (in preparation)
9. D.G. Michael et al., Observation of muon neutrino disappearance with the MINOS detectors and the NuMI neutrino beam. Phys. Rev. Lett. **97**, 191801 (2006)
10. B. Pontecorvo, Inverse beta processes and nonconservation of lepton charge. JETP **34**, 172–173 (1958)
11. P. Langacker, J. Wang, Neutrino-antineutrino transitions. Phys. Rev. D **58**(9), 093004 (1998)
12. E. Aliu et al., Evidence for muon neutrino oscillation in an accelerator-based experiment. Phys. Rev. Lett. **94**(8), 081802 (2005)
13. M. Ambrosio et al., Measurements of atmospheric muon neutrino oscillations, global analysis of the data collected with MACRO detector. Eur. Phys. J. **C36**, 323–339 (2004)

14. W.W.M. Allison et al., Neutrino oscillation effects in Soudan-2 upward-stopping muons. Phys. Rev. D **72**(5), 052005 (2005)
15. Y. Ashie et al., Evidence for an oscillatory signature in atmospheric neutrino oscillations. Phys. Rev. Lett. **93**(10), 101801 (2004)
16. E.N. Alexeyev, L.V. Volkova, Neutrino Data and Neutrino-Antineutrino Transition, arXiv:hep-ph/0504282 (2005)
17. G. Barenboim, J.D. Lykken, Status of CPT-violating neutrinos, hep-ph/0908.2993 (2009)
18. A. Aguilar et al., Evidence for neutrino oscillations from the observation of $\overline{\nu}_e$ appearance in a $\overline{\nu}_\mu$ beam. Phys. Rev. D **64** 112007 (2001)
19. A.A. Aguilar-Arevalo et al., A Search for electron neutrino appearance at the $\Delta m^2 \sim 1\,\text{eV}^2$ scale. Phys. Rev. Lett. **98**, 231801 (2007)
20. M.C. Gonzalez-Garcia, M. Maltoni, Phenomenology with massive neutrinos. Phys. Rept.**460**, 1–129 (2008)
21. S. Skadhauge, Probing CPT violation with atmospheric neutrinos. Nucl. Phys. B **639**, 281–289 (2002)
22. R. Edgecock, Potential of MINOS at a neutrino factory. Nucl. Instrum. Meth. **A451**, 192–197 (2000)
23. J.J. Gomez-Cadena, The physics case of the neutrino factory. J. Phys. Conf. Ser. **136**(2), 0220238 (2008)
24. R.B. Patterson, J.P. Ochoa-Ricoux, A. Himmel, Measuring $\Delta\overline{m}^2_{32}$ at MINOS with Reversed Horn Current Running, MINOS-doc-3978 (2008)
25. J. Hylen, Reversing Horn Current, MINOS-doc-5505 (2008)
26. C. Amsler et al.,(Particle data group), Review of particle physics. Phys. Lett. B **667**, 1(2008)
27. H. Murayama, CPT tests: Kaon vs neutrinos. Phys. Lett. B **597**, 73–77 (2004)

Appendix A
Tuning the MINOS Veto Shield

This appendix describes the service work performed on the MINOS veto shield. We begin with a brief review of the veto shield, followed by the work leading to its time calibration, which involved the development of new software tools and the resolution of a number of hardware problems. A rough estimate of the impact of the work presented here on atmospheric neutrino analyses is presented at the end.

A.1 Brief Overview of the Veto Shield

A.1.1 Motivation

In addition to recording neutrino interactions from the NuMI beam, the MINOS Far Detector also records atmospheric neutrino interactions.[1] The 5.4 kT Far Detector is not as massive as other dedicated atmospheric neutrino experiments such as Super-Kamiokande [1], which has a fiducial mass of $\sim 25\,\mathrm{kT}$. Nevertheless, there are two points in which MINOS has an advantage:

- The MINOS Far Detector is the first large deep underground detector to have a magnetic field. This permits it to make a separate measurement of ν_μ and $\overline{\nu}_\mu$ atmospheric neutrino oscillations.
- The neutrino energy and direction can be accurately determined for atmospheric neutrinos that interact inside the fiducial volume of the detector. The energy of the hadronic shower is determined from calorimetry and the energy of the muon is determined from curvature and/or range.

A measurement of atmospheric neutrino oscillations in MINOS is thus of interest, and results have already been published [2, 3]. However, the success of

[1] This refers to neutrinos produced by the interactions of cosmic rays incident on the Earth's atmosphere.

atmospheric neutrino analyses in MINOS depends heavily on the veto shield. As explained in Chap. 3, the MINOS Far Detector is a steel-scintillator sampling calorimeter with air gaps of approximately 2.41 cm between each plane. Consequently, it is very common for nearly vertical cosmic rays to travel between two planes and then interact somewhere inside the fiducial volume, mimicking an atmospheric neutrino interaction. By tagging these cosmic rays, the veto shield allows us to drastically reduce the background. For example, in the analysis published in [2], a signal to background ratio of 1:2 is obtained after the application of all the selection cuts, which increases to 20:1 after using the veto shield.

A.1.2 Veto Shield Configuration

The veto shield is a configuration of scintillator modules placed above and on the sides of the Far Detector, as shown in Fig. A.1. The "ceiling" part of the veto shield is made of two layers of scintillator, and the "walls" are made of a single layer of scintillator.

The same 8 m long modules containing 20 strips each that are used in the main body of the Far Detector are used in the veto shield, but with the orientation of strips aligned from north to south. Given that the Far Detector is approximately 30 m long, the same configuration of ceiling and wall veto shield modules is repeated four times in the longitudinal direction of the detector in order to ensure complete coverage. Each of these configurations is referred to as a "section," as illustrated on the left side of Fig. A.2. Contiguous veto shield sections have some overlap between them, to eliminate cracks.

As shown in Fig. A.2, groups of eight or four contiguous strips are read out by the same PMT pixel and are referred to as "planks." The veto shield contains 432 planks in total. The resolution of the veto shield is determined by the size of these planks and is thus given by $0.328 \times 8\,\mathrm{m}$ or $0.164 \times 8\mathrm{m}$, depending on the plank considered. Each plank is read out from its two ends. The veto shield is read out in the same manner as the main detector, as described in Sect. 3.3.2. The only difference is that the coincidence requirement of two PMTs to be hit in the same VARC in order to initiate read out is not applied. The dynode threshold is set to a

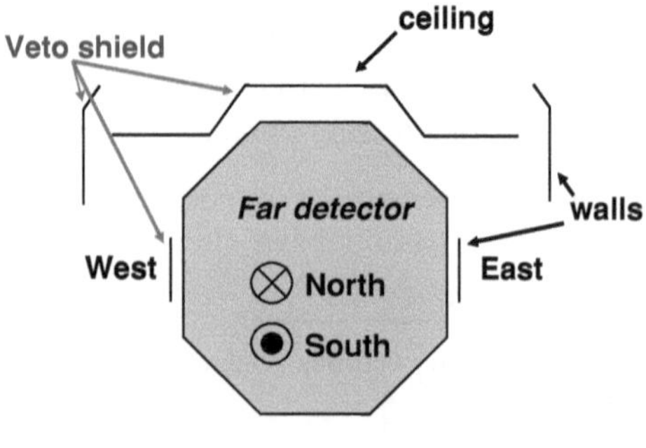

Fig. A.1 A schematic of the MINOS Far Detector and the veto shield

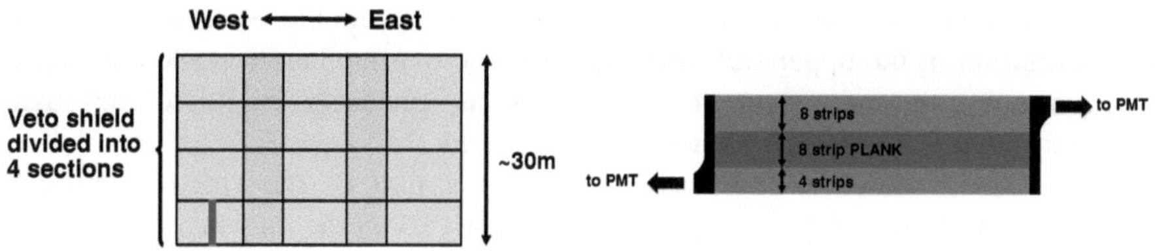

Fig. A.2 The ceiling part of the veto shield as viewed from above (*left*), and one of the 20 strip modules used in the veto shield (*right*). The veto shield planks are groups of eight or four scintillator strips that are read out by the same PMT pixel. One of the eight strip planks is shown in *red* in both plots

level corresponding to approximately one and a half photoelectrons in order to reduce false cosmic ray muon tagging that could arise from single photoelectron noise.

A.2 Implementation of New Software Tools

A.2.1 Veto Shield Geometry

The first step when tuning the veto shield was to build a package that accurately describes all of its geometric features and physical parameters. To this end a package called *ShieldGeom* was put together. With the information from the MINOS database, *ShieldGeom* models every plank in the veto shield in detail, and can thus reproduce all of its main features. Figure A.3 shows a comparison of the first section of the veto shield as modeled by *ShieldGeom* and by the previous model. Before *ShieldGeom* the shield was modeled by a series of hard-coded values obtained from the MINOS survey data. The new shield geometry allows us

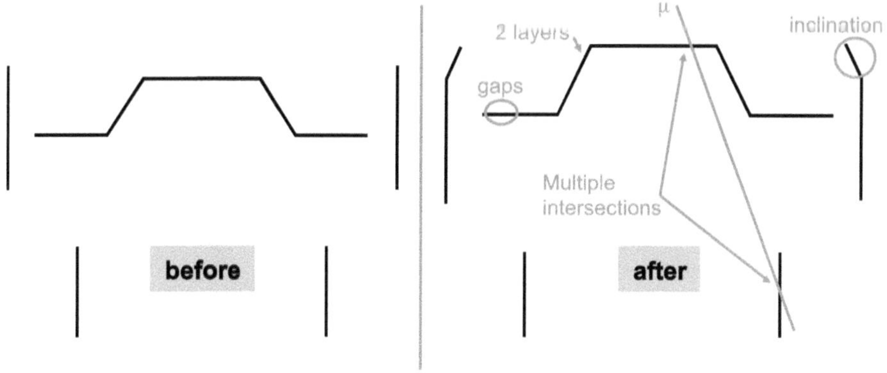

Fig. A.3 A comparison between the previous and the current geometric model of the veto shield. The new model allows us to account for many features that were previously ignored

to account for gaps between the modules, the ceiling's double layer of scintillator, the inclination of the upper wall modules, and the overlap between sections. Also, as seen next, the new geometry allows us to precisely calculate the many intersections that a track can have with the shield.

ShieldGeom, which I developed, has been a part of the standard MINOS reconstruction framework since 2006. More details about the information available in *ShieldGeom* can be found in [4].

A.2.2 *Veto Shield Traceback*

A.2.2.1 Technical Implementation

As previously explained, the spatial resolution of the veto shield is limited by the fact that groups of eight or four contiguous strips are read by the same PMT pixel. Nevertheless, from the information of reconstructed tracks in the main body of the detector, it is possible to determine the precise location in the shield where a given track went through.

Figure A.4 illustrates the way the intersection of a track with the veto shield is found. We model each plank as a perfect euclidean plane. We consider three points in this plane as obtained from *ShieldGeom*: $\vec{x_1}$ is at the plank's center, $\vec{x_2}$ is a point displaced 4 m to the north of $\vec{x_1}$, and $\vec{x_3}$ is at the center of the first strip of that plank. Also, $\vec{x_4}$ and $\vec{x_5}$ are two points along the track: $\vec{x_4}$ is at the track vertex (or end), and $\vec{x_5}$ is at the track vertex (or end) plus a unit vector given by the direction cosines as obtained from the reconstruction. The choice of whether the track's vertex or end is used is made depending on which is closer to the plank considered. For instance, given that most tracks are downward-going, their ends are typically used when considering the planks in the shield's lower walls.

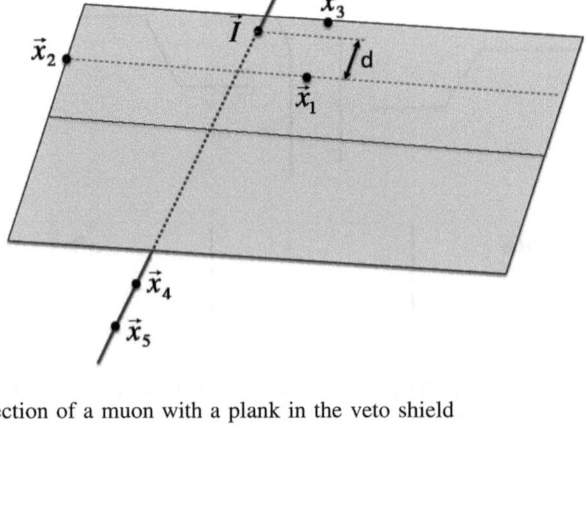

Fig. A.4 Intersection of a muon with a plank in the veto shield

The coordinates of the intersection $\overrightarrow{I}$ between the track and the plane are given by

$$\begin{aligned} I_x &= x_4 + (x_5 - x_4)t, \\ I_y &= y_4 + (y_5 - y_4)t, \\ I_z &= z_4 + (z_5 - z_4)t, \end{aligned} \tag{A.1}$$

where

$$t = -\frac{\begin{vmatrix} 1 & 1 & 1 & 1 \\ x_1 & x_2 & x_3 & x_4 \\ y_1 & y_2 & y_3 & y_4 \\ z_1 & z_2 & z_3 & z_4 \end{vmatrix}}{\begin{vmatrix} 1 & 1 & 1 & 0 \\ x_1 & x_2 & x_3 & x_5 - x_4 \\ y_1 & y_2 & y_3 & y_5 - y_4 \\ z_1 & z_2 & z_3 & z_5 - z_4 \end{vmatrix}}. \tag{A.2}$$

Once the intersection is known, the distance from it to the central line of the plank (i.e., d in Fig. A.4) is calculated in the following way:

$$d = \frac{\left|\left(\overrightarrow{x_5} - \overrightarrow{x_4}\right) \times \left(\overrightarrow{x_4} - \overrightarrow{x_1}\right)\right|}{\left|\overrightarrow{x_5} - \overrightarrow{x_4}\right|}. \tag{A.3}$$

If this distance is less than half the strip width (4.1 cm) times the number of strips in the plank, and $|I_z - z_1| < 4.0$ m, then the track is determined to have hit the plank.

A track can intersect the shield many times, as illustrated on the right side of Fig. A.3. As a matter of fact, given the overlap between shield sections, it is relatively common for a track to have eight or more intersections. Our code ensures that no intersections are missed by repeating the procedure described above with every plank in the shield. The code was optimized in order to perform these calculations as fast as possible. On a 2.8 GHz Xeon processor, the time required to calculate all of the intersections is of 15.2 ms per track [4]. This is negligible compared to all the other operations that are done in the standard MINOS reconstruction for every track.

A.2.2.2 Traceback Resolution

In order to find out how accurately the MINOS tracking allows one to trace back into the shield, we calculate a "pointing efficiency" (PE) with the MC simulation. The pointing efficiency is the probability that, when projecting a track to a given plank in the veto shield using the reconstructed track information, the same plank would be intersected if the track were projected using the true track information. In other words,

$$\text{PE} = \frac{\text{Number of times the reconstructed track is projected to plank } x}{\text{Number of times the true track is projected to plank } x}.$$

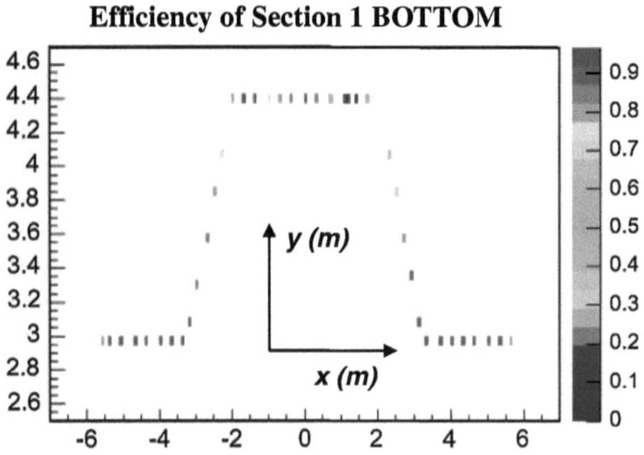

Fig. A.5 Plank efficiencies of the bottom ceiling layer of Sect. A.1. Most planks have efficiencies higher than 85%, except for a group of five contiguous planks in the top-right corner

The pointing efficiency is 96.7% which, considering the multiple scattering and energy loss of cosmic ray muons in MINOS, is very high. Also, we find that the projection using the reconstructed track differs by only 1.6 cm in one dimension from the projection using the true track information. This means that the one-dimensional traceback resolution is less than half the strip width for cosmic ray tracks.

The capability to trace back to the shield with such accuracy is very useful, and examples of this are shown in the next section. Since 2006 all tracks are automatically traced back to the shield using the method just described, as part of the standard MINOS reconstruction.

A.3 Identification of Hardware Problems

The ability to trace tracks back to the shield with high accuracy allows us to probe each component of the veto shield individually. We accomplish this by measuring each plank's tagging efficiency, defined as follows:

$$\text{Efficiency of plank}\, x = \frac{\text{Number of times plank}\, x\, \text{was hit}}{\text{Number of times a track was traced back to plank}\, x}.$$

The ability to probe each plank separately has proven to be an extremely useful tool to find and solve problems in the veto shield. There is one particular example that illustrates this very well. Once the tools described in the previous section were available, it was the first time that the efficiency of each plank could be determined. When this was done results such as the ones shown in Fig. A.5 were obtained. Most of the planks showed an efficiency higher than 85%, as expected.[2]

[2] As a matter of fact, given that most of the tracks intersect the shield at several points, the overall tagging efficiency of the veto shield is approximately $\sim$97%.

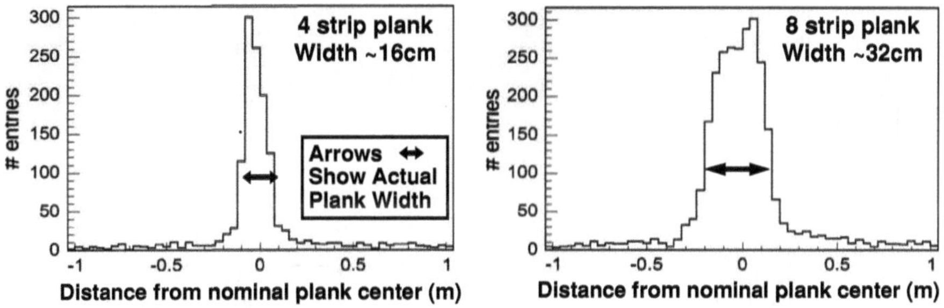

Fig. A.6 Distance from the nominal center of plank x for the interception of cosmic ray tracks with plank x. These distributions are filled every time plank x is hit. Through this technique the planks can be "seen" and the information about the plank's position in the database can be corrected

Nevertheless, the five contiguous planks at the top right corner of the ceiling portion showed medium and low efficiencies, in both layers and in all four sections. The resolution of this puzzle came by "seeing" the planks using cosmic ray tracks. Each time a given plank x is hit we project the reconstructed track into the shield and then fill a histogram of the distance from the plank's center d along the plank's plane, as described by Eq. A.3. Figure A.6 shows the resulting distributions for two normal planks.

When this technique was applied to the low efficiency planks in the top right corner of the ceiling section, it was discovered that the information about which strips formed those five planks was wrong in the database. In particular, the two modules corresponding to those five planks were effectively flipped, in such a way that the rightmost plank in each module should have been the leftmost, and so on. This is illustrated in Fig. A.7, which shows these planks as "seen" from the cosmic ray data. The problem happened for the two layers of the ceiling and in the four sections of the shield, meaning that a total of 16 modules and 40 planks were shifted in the database. This situation was immediately corrected after its discovery.

The calculation of the veto shield efficiencies on a plank-by-plank basis has been common practice since. Several other problems have been found using this technique, such as light leaks in the modules and problems with the electronics. All of them have been successfully fixed and, as of the date of writing this thesis, the shield is performing without any known hardware flaws.

A.4 Time Calibration of the Veto Shield

A.4.1 Mechanism of the Calibration

After the completion of the work described in the previous two sections, the time calibration of the veto shield was carried out. To each plank (and thus channel)

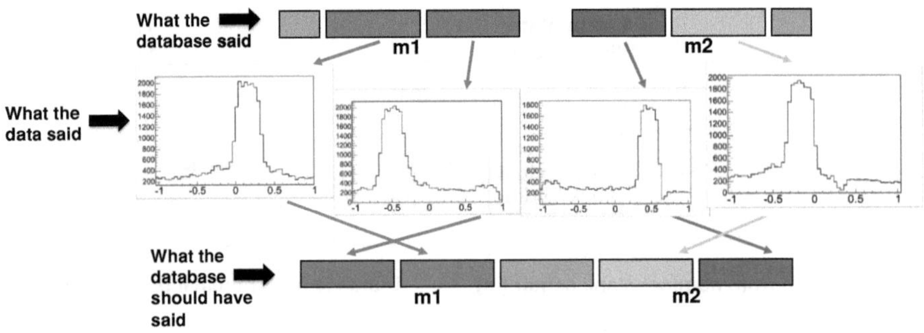

Fig. A.7 The resolution of the low efficiency puzzle for the 5 planks in the top right corner of the ceiling portion of the shield. The distributions in the middle are the same ones described in Fig. A.6, which allow us to "see" the planks. From that information it was found that the assignment of strips to planks in modules $m1$ and $m2$ was wrong. Each plank is represented by a different color. The data showed very clearly that the two modules should effectively be flipped. For instance, the plank represented in purple which, according to the database was to the left of $m2$, was actually to the right of $m2$

there is an associated time offset T_0 that is relatively constant over time and which is due mainly to

- variations in rise time of the dynode signals within a PMT,
- short time delays associated with sending the signals to the VARCs (defined in Sect. 3.3.2).

The goal of the timing calibration is to determine the T_0 constants for each channel as accurately as possible.

Figure A.8 helps illustrate how the veto shield calibration was done. For each track that intersects the veto shield at a given plank, a Δt can be calculated for each of the two channels that read out the given plank:

$$\Delta t = t_{\text{expected}} - t_{\text{measured}}. \tag{A.4}$$

The average $\langle \Delta t \rangle$ gives the T_0 offset for the corresponding channel. The expected time t_{expected} can be calculated from the information of the reconstructed track in the main body of the detector, where the timing calibration had been already performed [5]. This is achieved by determining the time at which the muon went through the vertex t_{vtx} and correcting for the time of travel between the veto shield and the vertex,

$$t_{\text{expected}} = t_{vtx} - \frac{P}{c}, \tag{A.5}$$

where P is the distance between the veto shield intersection and the track's vertex, and c is the speed of light. The time at which the track went through its vertex is the average of the position-corrected times for all strips in the track,

$$t_{vtx} = \frac{1}{N}\sum_{i=1}^{N}\left(t_i - \frac{s_i}{c}\right), \tag{A.6}$$

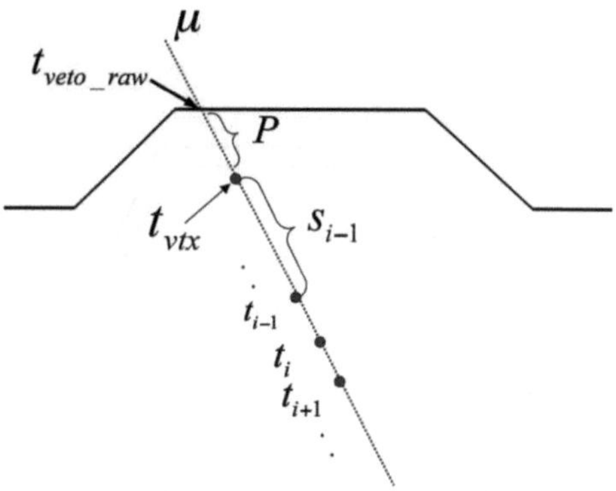

Fig. A.8 Schematic illustrating the veto shield timing calibration, as described in the text

where N is the number of strips hit by the track in the main body of the detector, t_i is the calibrated time of hit i and s_i is the distance between hit i and the vertex, as illustrated in Fig. A.8.

The time measured at the veto shield is the raw time t_{veto_raw} that corresponds to the electronic readout of the shield minus the time it took for the light to get from the track's intersection with the shield to the end of the readout t_{veto_corr},

$$t_{\text{measured}} = t_{\text{veto\_raw}} - t_{\text{veto\_corr}}, \tag{A.7}$$

where

$$t_{\text{veto\_corr}} = \frac{l}{c_f} + \frac{f}{c_f} + t_{\text{walk}}(q). \tag{A.8}$$

In the above equation l and f are the wavelength shifting fiber and clear fiber distances traveled by the light, c_f is the speed of light in the fiber ($c_f = c/n \approx 0.169\, m/ns$), and $t_{\text{walk}}(q)$ is the "time-walk" correction which is described below.

In order to maximize the precision of the calibration, only those cosmic rays whose tracks are very long and nearly undeflected by the detector's magnetic field are used. Also, only those tracks that go near the center of the planks are used, in order to ensure that the veto shield hit truly corresponds to the track.

Furthermore, there are two particular features that are specific to the veto shield and that must be taken into consideration during the calibration. The first one is that some tracks can hit two planks connected to the same PMT. This is particularly probable in the central part of the shield, where the top and bottom layers are read out by the same PMTs, as illustrated in the left side of Fig. A.9. Given that there are 5 μs of dead time after the first hit during which all hits are recorded, the timing of the second hit is not valid and must be excluded from the calibration.

The second feature is that other particles produced in the rock come in with the muon and can thus produce an early signal on some of the channels, as illustrated in the right side of Fig. A.9. Our studies show that for approximately 71% of

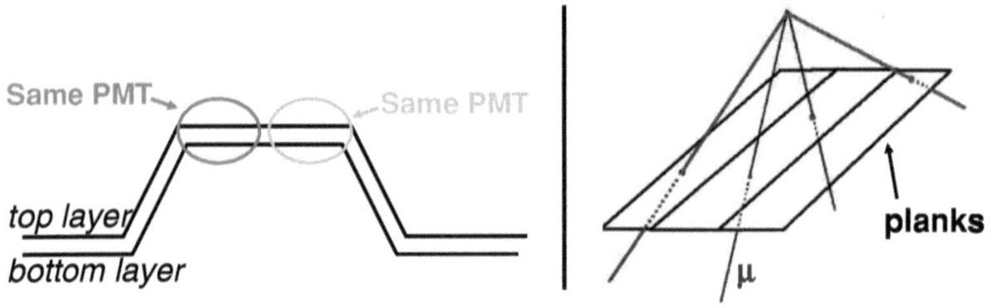

Fig. A.9 An illustration of the two additional considerations that need to be considered when performing the veto shield timing calibration. Some tracks can hit two planks connected to the same PMT, thus making the timing of only one of the hits valid (*left*). Other particles typically come in with the muon and thus affect the timing (*right*)

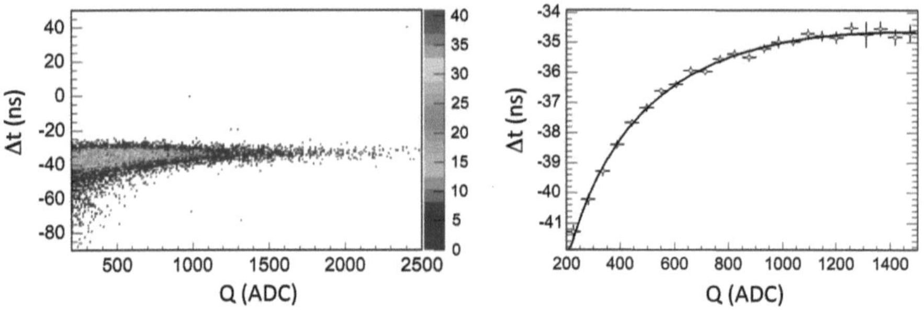

Fig. A.10 Δt vs. Q for one particular PMT. The right plot is the profile of the two-dimensional plot on the left shown on an expanded scale, and is fit to a third-degree polynomial up to 1500 ADC counts

cosmic ray events there is a nonzero number of extra planks hit simultaneously with the event and that 64% of those planks are less than 2 m away from the one intersected by the track. We reduce this effect by limiting the calibration to those events where no extra planks are hit. Nevertheless, extra particles that just hit the same plank intersected by the muon cannot be removed by this cut and cause the timing to be off. This is most likely the reason why the timing resolution in the veto shield is not as good as in the rest of the detector, as described in the next section.

Finally, a time-walk correction is needed to account for the fact that the PMTs trigger time, as determined by the arrival of the first photon that converts on the photocathode, depends on the size of the signal. For large photon yields the trigger time is likely to be small, while for low photon yields the trigger time can be large, as high as 10 ns. This is illustrated in Fig. A.10 which shows the dependence of Δt on the photon yield. The time-walk correction is extracted for each PMT by fitting the Δt vs. Q profile to a third-degree polynomial up to 1500 ADC counts.

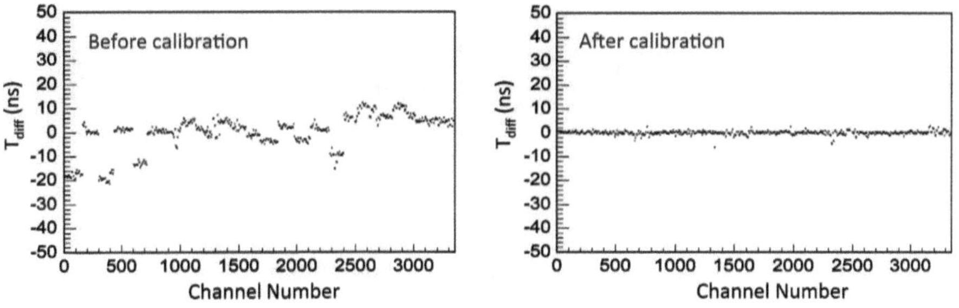

Fig. A.11 North–south time difference average as a function of channel number. The propagation delay corrections of Eq. A.8 are included in both plots. The right plot includes the T_0 offsets as determined from the calibration

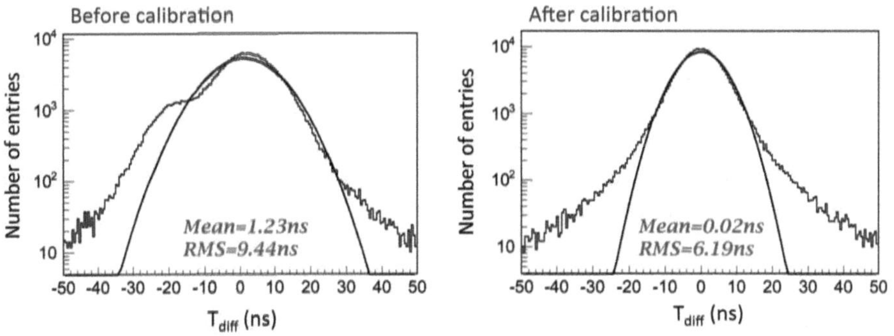

Fig. A.12 North–south time difference distribution for all channels before and after calibration. The single-hit timing resolution is $6.19/\sqrt{2} = 4.4$ ns after the calibration

A.4.2 Performance of the Calibration

As previously mentioned, each plank is read out from both its north and south ends. The precision of the calibration can thus be evaluated by computing the north–south average time difference $\langle T_{\rm diff} \rangle$. This quantity is shown before and after the calibration in Fig. A.11, as a function of channel number. A great improvement is obtained after the calibration. The width of the $\langle T_{\rm diff} \rangle$ distribution for all channels is 0.54 ns. This means that the veto shield calibration is good to $0.54/\sqrt{2}$ = 0.38 ns, which is close to the main detector's calibration precision of 0.28 ns [5].

Similarly, the non-averaged north–south time difference gives us the timing resolution of a single-hit. The distribution of $T_{\rm diff}$ for all channels is shown in Fig. A.12, before and after the calibration. The timing resolution of a single-hit improves from $9.44/\sqrt{2} = 6.7$ ns to $6.19/\sqrt{2} = 4.4$ ns with the calibration. This result is not as good as in the main body of the detector, where the resolution is 2.3 ns [5]. The most likely explanation is that other particles arrive with the muons

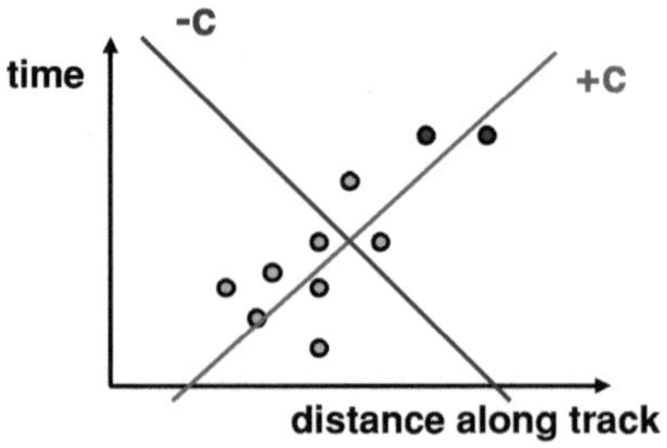

Fig. A.13 Schematic illustration of the determination of a track's direction. *Green* hits represent the hits in the main detector while *blue* hits (at the *top*) represent the veto shield hits. In order to determine if a track is upward- or downward-going, the hit times are plotted as a function of distance along the track and then fitted to the upward-going and downward-going hypotheses. The RMS deviations for each of the two hypotheses are calculated and the hypothesis with the smallest RMS is chosen

and affect the timing, as previously explained. This hypothesis is reinforced by the fact that if the single-hit resolution is recalculated ignoring hits below a certain energy deposition then it asymptotically approaches the one in the main body of the detector. The energy depositions left by the particles coming in with the muon should indeed be lower than those of the original muons.

A.5. Impact on Atmospheric Neutrino Analyses

Having presented all the work done on the veto shield, it is pertinent to evaluate the impact on the atmospheric neutrino analyses. It is impossible to quantify the impact with one single number. In order to get an idea however, we consider one of the areas where the timing of the veto shield can have a strong impact: the discrimination between upward-going and downward-going events. This can only be done with the timing information, and is an important component of atmospheric neutrino analyses as upward-going muons can only be neutrino induced. Also, the ratio of upward-going to downward-going events is an important measurement, as can be seen for instance in [2].

Figure A.13 illustrates how an event is determined to be upward- or downward-going in [2]. The hit times are plotted as a function of distance along the track and then fitted to the upward-going and downward-going hypotheses. The RMS deviations for each of the two hypotheses are calculated and the hypothesis with the smallest RMS is chosen.

As expected, the determination of a track's direction becomes easier with increasing track length, due to the fact that there are more hits and thus more information. Given that the veto shield provides us with extra hits, it is interesting to compare the efficiency to reconstruct cosmic ray muons as downward-going with and without the veto shield. This is shown in Fig. A.14, where it can be seen

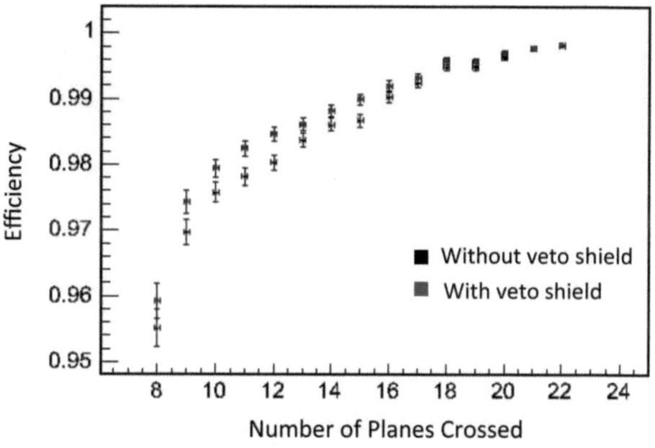

Fig. A.14 Efficiency to reconstruct cosmic ray muons as downward-going as a function of track length, with and without using the veto shield. The veto shield improves the ability to reduce the cosmic ray background when finding short upward-going muons

that the veto shield does reduce the background when finding upward-going events by as much as 25% at some track lengths. This confirms that the veto shield timing information can be very useful.

Appendix B
Some Technical Aspects of LEM

The workings and the performance of the LEM selection are described in Chap. 5. This Appendix contains the details of the more technical components of LEM. Section B.1 describes the attenuation corrections and the procedure that converts Near Detector events into Far Detector events. Section B.2 describes the technical challenges involved in the realization of the method.

B.1 Attenuation Corrections

B.1.1 Basics

As described in Sect. 5.3.3, before two events can be compared to each other they must be corrected for light attenuation in the fiber.

As the light originating from a neutrino interaction propagates through the wavelength shifting (wls) fiber in the detector it gets attenuated by a factor

$$a_w(d_w) \cong \frac{2}{3} e^{-\frac{d_w[\mathrm{m}]}{7.05}} + \frac{1}{3} e^{-\frac{d_w[\mathrm{m}]}{1.05}}, \tag{B.1}$$

where d_w is the distance traveled by the light in the wls fiber. The above expression is the result of a phenomenological fit to the measured attenuation [6], as shown in Sect. 3.3.1. Similarly, as the light travels in the clear fiber that connects the end of the module to the readout electronics it gets attenuated by a factor

$$a_c(d_c) \cong e^{-\frac{d_c[\mathrm{m}]}{10.0}}, \tag{B.2}$$

where d_c is the distance travelled in the clear fiber.

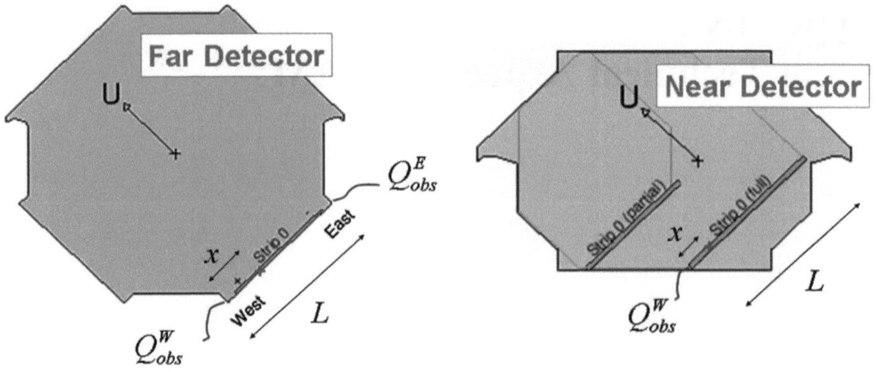

Fig. B.1 Schematics of the Far and Near Detectors. The strips are depicted in *green* and the clear fiber cables in *blue*. The location of a hit is symbolized by a *red cross*

B.1.2 Far Detector Corrections

In the Far Detector the charge[3] is read out from the east and west strip ends as $Q^E raw$ and $Q^W raw$ respectively, as shown schematically on the left side of Fig. B.1. The total charge deposited per strip Q is the sum of the charge that goes in the east direction Q^E plus the charge that goes in the west direction Q^W:

$$Q = Q^E + Q^W. \tag{B.3}$$

Thus in the Far Detector it is possible to obtain Q after correcting the observed (raw) charges for their corresponding attenuations:

$$Q = \frac{Q^E_{\text{raw}}}{a_w(L - x + w^E)a_c(c^E)} + \frac{Q^W_{\text{raw}}}{a_w(x + w^W)a_c(c^W)}, \tag{B.4}$$

where

- L is the total strip length.
- x is the distance from the interaction point to the west end.
- $w^{E(W)}$ is the length of the wls fiber (also referred to as "pigtail") that connects the east or west end of the strip end to the corresponding end of the module.
- $c^{E(W)}$ is the length of the clear fiber cable that connects the east (west) end of the module to the readout electronics.

---

[3] Clearly no charge is physically deposited on the strip by particles interacting in the scintillator, but rather energy that appears in the form of light. This light is what travels down the fibers, and not the charge. For the sake of simplicity however, since charge and light intensity are proportional at the face of the PMT, this document consistently refers to both of them as "charge."

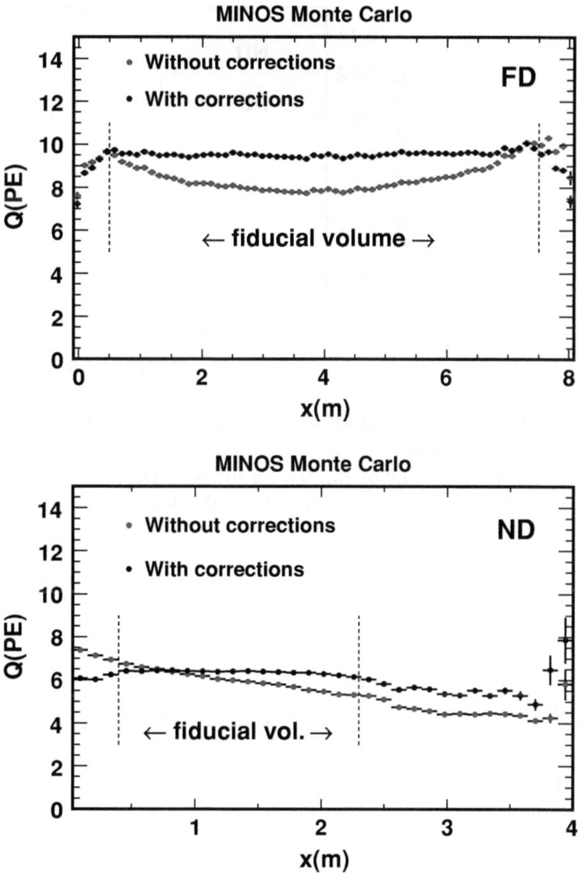

Fig. B.2 The average total charge per strip Q deposited by muons as a function of the distance from the west end x, as read from the PMTs (*red*), and after being corrected (*black*) by Eqs. B.6 and B.8 for the Far Detector and Near Detector cases respectively. Strips with zero charge are ignored. The corrected charge is to first-order flat as a function of x which indicates that the attenuation effects are well accounted for. The *dashed lines* delineate approximately the region within the fiducial volume. The drop in the corrected Near Detector charge curve occurs in regions outside the fiducial volume ($x \lesssim 0.4$ m and $x \gtrsim 2.3$ m), and is mostly due to strip edge effects (the same can be said for the Far Detector). The average corrected charge is lower in the Near Detector than in the Far Detector, where it is adjusted by the β factor included in Eq. B.9

The average attenuation for the total strip charge in the Far Detector,

$$\langle A^{FD} \rangle = \left\langle \frac{Q_{\rm raw}}{Q} \right\rangle = \left\langle \frac{Q^E_{\rm raw} + Q^W_{\rm raw}}{Q^E + Q^W} \right\rangle, \qquad \text{(B.5)}$$

has a numerical value of approximately 0.3. It is important to keep the overall charge scale close to what it is in reality, given that charge generation fluctuations at the PMTs are taken into account when comparing events, and are modelled by Poisson probabilities (see Sect. 5.3.4). Thus a standard attenuation $A^{\rm FD}_{\rm std}$ is applied to the total charge deposited per strip Q to obtain the final corrected charge:

$$Q_{final} = A^{\rm FD}_{\rm std} \left(\frac{Q^E_{\rm raw}}{a_w(L - x + w^E) a_c(c^E)} + \frac{Q^W_{\rm raw}}{a_w(x + w^W) a_c(c^W)} \right), \qquad \text{(B.6)}$$

As can be seen on the left side of Fig. B.2, the standard attenuation $A^{\rm FD}_{\rm std}$ is approximately 15% higher than the average attenuation $\langle A^{\rm FD}_{\rm std} \rangle$, which amounts to scaling the light level of all events to that of Far Detector edge events. This is

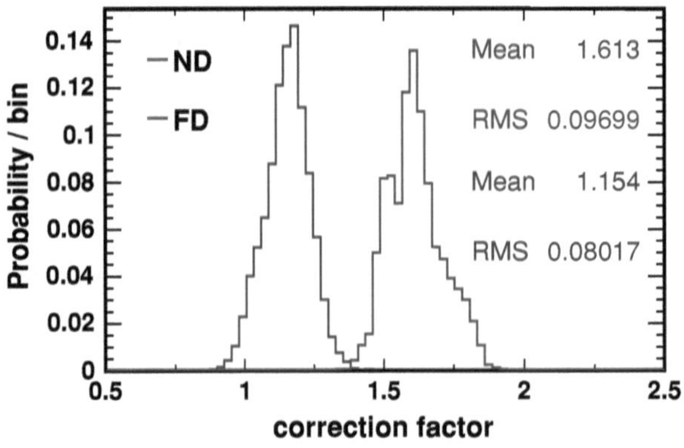

Fig. B.3 Probability distributions of the correction factor $Q^{\text{total}}_{\text{final}}/Q^{\text{total}}_{\text{raw}}$ applied to ν_e CC events in the fiducial region and with 1 GeV < E_{reco} < 8 GeV. The Q_{final} of each strip is given by Eq. B.6 for the Far Detector and by Eq. B.9 for the Near Detector, with $\beta = 1.59$ and $R = 0.8$. The geometry of the Near Detector is responsible for the double peak structure in the corresponding distribution

consistent with Fig. B.3, which shows that applying Eq. B.6 to Far Detector ν_e CC events effectively increases their charge by a factor of 1.15 ± 0.08.

B.1.3 Near Detector Corrections

B.1.3.1. Differences and Strategy

In the Near Detector the situation is different as light is reflected on the east end and only read out from the west (see the right hand side of Fig. B.1), causing the overall light level in the Near Detector to be lower than in the Far Detector. Three possible solutions to this problem can be considered:

1. Correct the Near Detector charges for attenuation and scale the light level to that of the Near Detector. Use the Far Detector libraries to do the event matching in the LEM algorithm.
2. Correct the Near Detector charges for attenuation and scale the light level to that of the Near Detector. Generate separate Near Detector libraries to use when doing Near Detector event matching in the LEM algorithm.
3. Correct the Near Detector charges for attenuation and scale the light level to that of the Far Detector. Use the Far Detector libraries to do the event matching in the LEM algorithm.

While these three solutions are all fundamentally correct, they are not all equally desirable in terms of processing time, systematic errors, and performance. The first one implies having two significantly distinct selections at each detector. This increases the dependency on the simulation which is used to account for these Near–Far differences. The second solution entails a considerable additional computational effort as Near Detector libraries need to be produced. The third option has the advantage that it requires no additional processing. In addition, while it is impossible to make the two detectors completely identical in terms of light level (see Sect. B.1.3), it is possible to make them close enough to apply the same cuts in both of them. Because of these reasons the third approach is followed, as described in the rest of this section.

B.1.3.2. Matching the Two Detectors

In the Near Detector, light emitted in the scintillator is absorbed by the wls fiber where it is re-emitted isotropically. An expression for the charge observed at the west end $Q^W raw$ can thus be obtained by having half of the total charge deposited on the strip Q going in each direction,

$$Q^W_{\text{raw}} = \frac{1}{2} Q a_w(x + w^W) a_c(c^W) + \frac{1}{2} Q a_w(2L - x + w^W) a_c(c^w) R, \tag{B.7}$$

where R is the reflectivity in the east end. Solving the above equation for Q and multiplying by the Far Detector standard attenuation for consistency we find

$$Q^{\text{prelim}}_{\text{final}} = A^{\text{FD}}_{\text{std}} \frac{2Q^W_{\text{raw}}}{[a_w(x + w^W) + a_w(2L - x + w^W)R] a_c(c^W)}. \tag{B.8}$$

The difficulty in the Near Detector case resides in the fact that the reflectivity constitutes an extra unknown in the previous equation. Determining the correct value for R is hard, as its extraction from the data is obscured by other light yield variations along the strip that make individual reflectivity measurements consistent with any value of R between 20 and 90% [7].

Based on the initial input from Near Detector scintillator experts, a tentative value of $R = 80\%$ is assumed for this parameter when solving Eq. B.8 in the LEM algorithm. The distributions of $Qfinal$ for long muon tracks can then be used to test if Eqs. B.6 and B.8 give consistent results. As can be seen in Fig. B.2, where the resulting $Qfinal$ is flat in the regions within our fiducial volume, these two equations do a reasonably good job at correcting for attenuation. This is confirmed by the fact that the LEM selection efficiency is flat as a function of position in the detector, as shown in Sect. B.1.4. There remains however a constant offset between the Near Detector and the Far Detector. This offset can be rectified by adding an extra factor β to Eq. B.8 in such a way that the mean response between the two detectors is as close as possible:

$$Q_{\text{final}} = \beta A^{\text{FD}}_{\text{std}} \frac{2Q^W_{\text{raw}}}{[a_w(x + w^W) + a_w(2L - x + w^W)R] a_c(c^W)}. \tag{B.9}$$

A value of $\beta = 1.59$ is extracted by requiring the Far Detector and Near Detector simulations of muons to match on average. Figure B.3 shows that with this value of

β Eq. B.9 increases the charge of Near Detector ν_e CC events by 1.61 ± 0.1. It is important to note that a value of $\beta = 1.59$ does not mean that the light level in the Far Detector is 1.59 times higher than in the Near Detector. As a matter of fact, Fig. B.3 shows that the Far Detector light level[4] is in average approximately $1.61/1.15 = 1.4$ that of the Near Detector. One reason why β is so high is that the light level to which Far Detector events are corrected (by Eq. B.6) is approximately 15% higher than it is on average (see the end of Sect. B.1.2). The other reason is that the original attenuation correction of Eq. B.8, while doing reasonably well in correcting for relative differences due to attenuation, does not yield the proper absolute scale for the corrected charges in the Near Detector. A good explanation for this was recently found at Caltech, where experimental tests [8] have shown that the Near Detector reflectivity R is most likely much closer to 50% than to the value of 80% that was previously assumed. Future versions of the LEM will incorporate a lower R and consequently a value of β much closer to unity.

B.1.3.3 Matching Performance

Using a more accurate value for the Near Detector east end reflectivity is not expected to render the charge matching between the two detectors significantly better than it already is. Figure B.4 shows that a very good match between the charge distributions of muons at the two detectors is obtained after all corrections have been applied. Similarly, the corrected charge distributions of νeCC events show excellent agreement for charges larger than approximately ~ 10 photoelectrons, as shown in the bottom plot of Fig. B.5.

The fact that the right plot of Fig. B.4 shows a difference between the Near Detector and Far Detector distributions for low values of the corrected strip charge is explained by threshold differences in the two detectors. Corrected strip charges of five photoelectrons correspond on average to raw charges of 4.3 photoelectrons in the Far Detector and of 3.1 photoelectrons in the Near Detector (as shown by Fig. B.3). Thus strips with these corrected charge values in the Near Detector are considerably more likely to fluctuate below the 3.0 PE threshold which is applied to both detectors, causing the observed deficit.

This threshold effect constitutes an inevitable Far–Near difference given that the two detectors are indeed different in terms of light output and that the 3.0 PE cut has to be applied before any attenuation corrections are done. The effect on the number of selected events is small however. By comparing the standard Near Detector files with a special set of Far Detector files processed with a 4.3 PE threshold the effect is found to contribute only a $\sim 5\%$ difference between the two detectors. This is because the matching in LEM is mostly driven by the strips in the core which typically possess a higher charge.

[4] As defined in this document, i.e., in terms of the total number of raw photoelectrons read out for the same original strip charge deposition in both detectors.

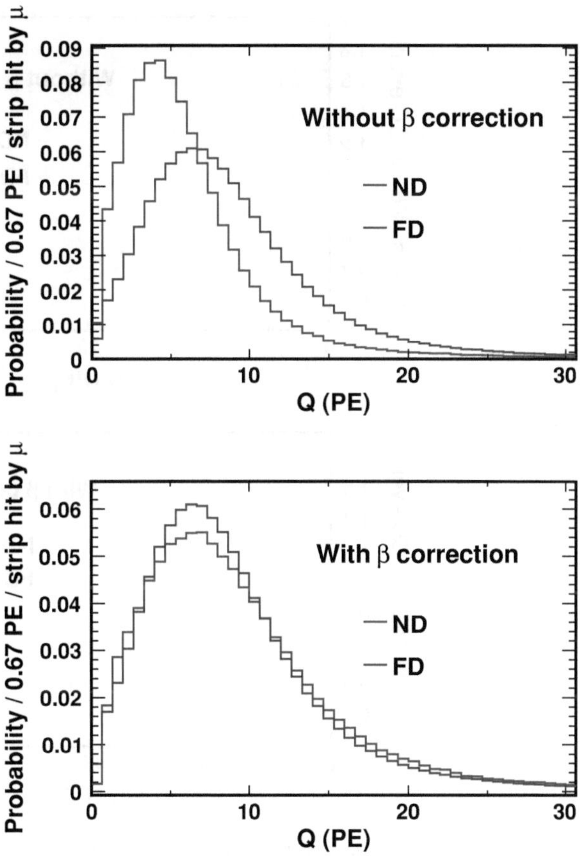

Fig. B.4 The probability distributions of the total (corrected) charge deposited per strip by muons with a track vertex in the fiducial volume. Strips with zero charge are ignored. The Far Detector charges are corrected according to Eq. B.6 in both cases. The Near Detector charges are corrected according to Eq. B.8 in the *top plot* and according to Eq. B.9 in the *bottom plot*. The shapes of the Near Detector and Far Detector charge distributions agree well after the β correction has been applied

B.1.4 Selection Efficiencies

Studying the LEM selection efficiency as a function of position in the detector can help us determine if the light level corrections are being properly handled. As shown in Fig. B.2, when no attenuation corrections are applied the light level changes by roughly 20 and 30% within the fiducial regions of the Far Detector and Near Detector respectively. Studies have shown that when the light level is artificially scaled in either of the two detectors by $\pm x\%$ the number of LEM selected events also changes by approximately $\pm x\%$, both in the data and in the MC. As can be seen in Figs. B.6 and B.7 however, the selection efficiencies[5] in

[5] These efficiencies were calculated for the finalized LEM selection (using a cut at LEM PID > 0.8).

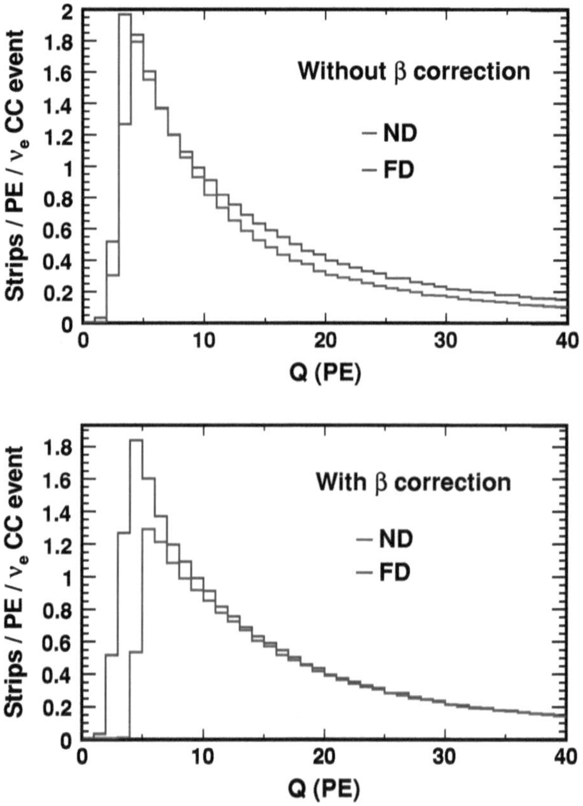

Fig. B.5 The distributions of the total (corrected) charge per strip of ν_e CC events in the fiducial volume, normalized to number of events. Events satisfy 1 GeV < E_{reco} < 8 GeV and are stripped of hits with less than 3.0 photoelectrons before any corrections. The Far Detector charges are corrected according to Eq. B.6 in both cases. The Near Detector charges are corrected according to Eq. B.8 for the *top plot* and according to Eq. B.9 for the *bottom plot*. After all corrections have been applied there remains a deficit of low charge strips in the Near Detector with respect to the Far Detector. This is explained by the lower light level in the Near Detector combined with the fact that the same 3.0 PE cut is applied in both detectors. The agreement between the Near Detector and the Far Detector charge distributions is very good for strips with corrected charge above 10 photoelectrons

both detectors are flat to first order all across the fiducial region of the detectors in both data and MC, indicating that the light level differences due to attenuation have been correctly accounted for.

Also, the overall selection efficiency is very similar in both detectors and seems to be on the order of 10% when measured with respect to the preselection. This is confirmed when the selection efficiencies are plotted as a function of reconstructed energy, as shown in Fig. B.8. There is a small difference between the two detectors, on the order of 7%. The algorithm thus acts very similarly in the two

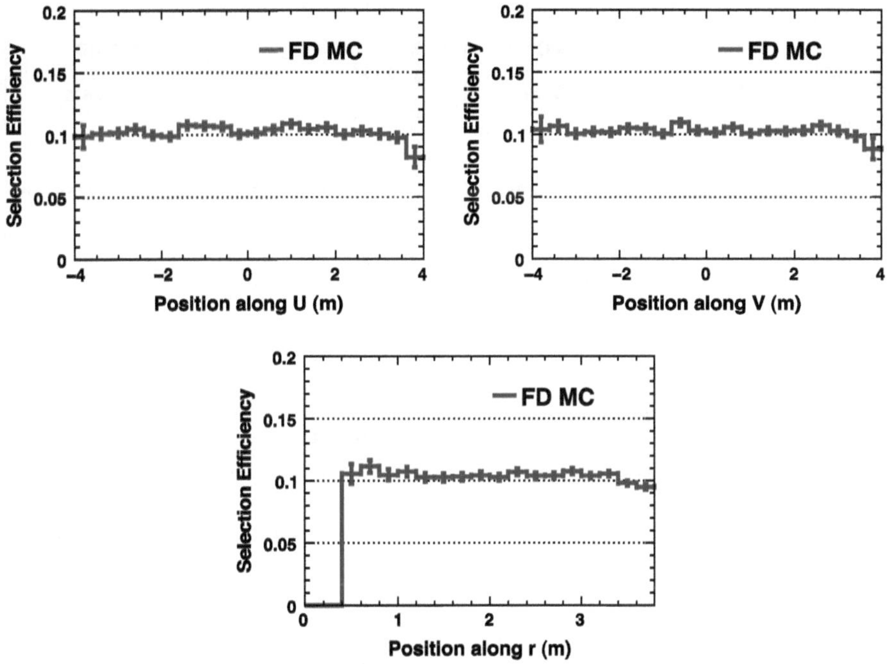

Fig. B.6 Selection efficiencies of background events in the Far Detector as a function of position in the detector. *U* and *V* are the mutually orthogonal directions along the scintillator strips, and *r* is the distance to the detector's center. The efficiencies are calculated with respect to the number of preselected events. ν_μ CC events are oscillated assuming $\Delta m^2_{32} = 2.38 \times 10^{-3} \mathrm{eV}^2$ and $\theta_{23} = \pi/4$. ν_τ CC events are not included. In the Far Detector the detector's center coincides with the coil hole, which is why the efficiency goes to zero for $r < 0.5$. No data is shown as the statistics are very limited in the Far Detector

detectors. This is important as it means that the mismodeling of hadronic showers in the simulation will have an impact on both the Near and the Far selected samples in practically the same way (the slight residual difference is quantified in Sect. 8.1.2).

These efficiencies also give us information on the impact of the slight distortion that occurs when an event's light level is scaled. According to photon statistics, if N_e photoelectrons are produced in average for a given amount of energy deposition in a strip, then the corresponding standard deviation is given by $\sqrt{N_e}$. When we scale the amount of charge in a strip by $x\%$, however, we are also effectively changing this standard deviation to a value of $x\sqrt{N_e}$, instead of to the proper value of $\sqrt{xN_e}$. This is not a problem though because the same effect occurs both in the data and in the simulation, and is just a part of how the algorithm works. But even better, the efficiency remains constant as the light level scaling x changes, which means that the selection is unaffected by this effect to first order, at least for the values of x that we deal with.

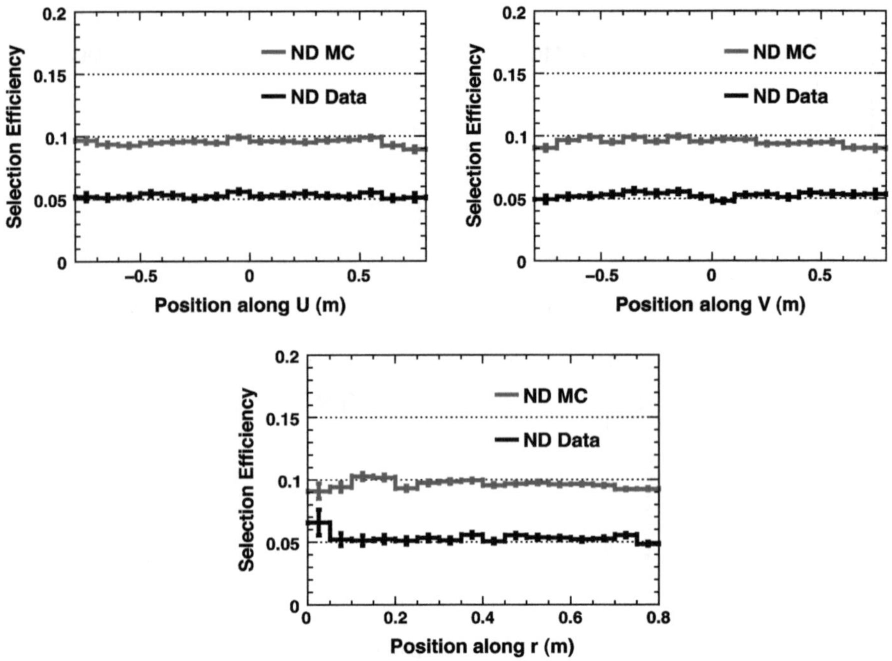

Fig. B.7 Selection efficiencies of background events in the Near Detector as a function of position in the detector. *U* and *V* are the mutually orthogonal directions along the scintillator strips, and *r* is the distance to the beam's center. The efficiencies are calculated with respect to the number of preselected events. There is a large discrepancy between the data and the MC which is addressed in Sect. 6.1

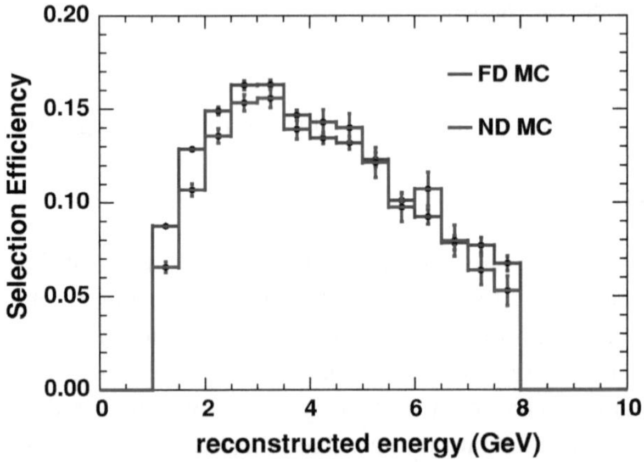

Fig. B.8 Selection efficiency of background events in the two detectors as a function of reconstructed energy. The efficiencies are calculated with respect to the number of preselected events. ν_τ CC events are not included for the Far Detector, in order to have a fair comparison between detectors. The similarity between the two curves indicates that LEM acts very similarly in both detectors. All the available MC was used to make this plot

B.2 Sorting Technical Constraints

B.2.1 Time

Due to the large number of samples that need to be processed[6] for the ν_e appearance analysis, it is important to keep the overall computation time to a minimum. A number of the elements of the LEM implementation discussed in Chap. 5 have been put in place partially or solely to reduce as much as possible the algorithm's processing time per event:

- Compacting of all events (even the library is stored and read out in compact form).
- Preselection cuts on both input events (Sect. 5.3.4) and library events (Sect. 5.3.2).
- Storage of the log likelihood values for different combinations of n_A and n_B (Eq. 5.1).

In addition, great care has been taken to ensure that no operations are repeated more than necessary, and that slow interfaces are avoided. For instance, std::vectors are used to store most of the information in the algorithm and in the compact events due to their sorting abilities, but iterators are used instead of the costly [ ] operator. Also, quantities used throughout all the algorithm such as the self-likelihood (from Eq. 5.2) are only calculated once and then stored.

Due to all these factors the speed of the LEM algorithm is remarkably good considering all the operations that are done and the volume of data to which each input event is compared. The average time the algorithm takes to process an input event that passes the precuts is slightly less than 12 s, as measured on a 2.8 GHz Xeon processor. This means that, for example, a typical beam Far Detector file with $\sim 3,000$ input events can be processed through the LEM in approximately 5 CPU-hours.[7]

B.2.2 Memory

The algorithm was constructed so that it could fit in the memory of our commonly available machines. If input events were to be compared to the entire library one after the other, the library would have to be committed to memory in order for the process to be efficient. However, the entire library in compact form occupies

---

[6] For example, the Near Detector data and MC, the Near Detector horn-off data and MC, the Near Detector muon-removed data and MC, the Far Detector data and MC, etc.

[7] Typically less than half the input events pass the precuts.

approximately 1.3 GB of space. Instead, only the compact input events are committed to memory and the main loop in the algorithm takes place over the library events, loading only one small piece of the library at a time. The lists of best matches for the input events are updated as the algorithm proceeds.

But even with this modified code structure the memory requirements are too large for files that contain tens of thousands of input events. Typically a list of the best 200 matches is kept for each input event.[8] Each best match object holds basic information such as run, snarl, interaction type, neutrino flavor and $-\Delta(\mathcal{L})$, among others, and occupies 84 bytes of memory. Some Near Detector MC files contain as many as 65,000 input events, and if we assume that approximately half of those will find matches in the LEM algorithm it means that approximately 0.5 GB of memory is needed solely to store the best-matches information. This number does not include the amount of memory needed to store the small library pieces that are loaded consecutively and the other objects needed to perform the comparisons [such as the table with the values of $\ln(S(n_A, n_B))$, as explained in Sect. 5.3.4]. To solve this problem it was necessary to modify the algorithm's infrastructure in such a way that the input file is automatically broken down into components of n_s snarls each. Each piece is processed separately and a corresponding output file produced, which is then merged with the other ones.

After all these implementations a typical LEM job divided into pieces of $n_s =$ 400 snarls each has a total memory requirement of less than 250 MB, which is well within the reach of most machines and allows for several LEM jobs to be run simultaneously.

[8] This was done originally because the optimum number of best matches was unknown; after the work of Sect. 5.4.2 it was kept as a precautionary measure, in case the extra information could be useful.

Appendix C
A Measurement of Beam ν_e's with Antineutrinos

Section 6.2 discusses the intrinsic beam ν_e component in the beam, which is an irreducible background to the ν_e appearance search. Even if this background is sufficiently well modeled in the tuned MC, it is of interest to the analysis to be able to confirm in an independent way that the actual beam ν_e rate is consistent with the simulation's prediction. This chapter presents a novel method that allows for an indirect measurement of the beam ν_e rate from μ^+ decay using antineutrinos. The measurement, although statistically limited by the amount of exposure in the pHE beam configuration (1.41×10^{19} POT), provides a cross-check of the beam ν_e flux estimation from the tuned simulation. This work was done in conjunction with David Jaffe of Brookhaven National Laboratory.

C.1 Introduction

The low energy (<8 GeV) component of the ν_e contamination of the NuMI neutrino beam is dominated by ν_e from $\mu^+ \rightarrow e^+ + \overline{\nu}_\mu + \nu_e$, as shown in Sect. 6.2. Since a $\overline{\nu}_\mu$ accompanies every ν_e from μ^+ decay and the decay matrix elements are known, a measurement of the low energy $\overline{\nu}_\mu$ flux can be used to constrain the ν_e flux. Since only the $\overline{\nu}_\mu$ flux from μ^+ decay varies significantly with the neutrino beam energy setting, we use the difference of the measured $\overline{\nu}_\mu$ spectra from low-energy (LE) and pseudo-high energy (pHE) Near Detector data to extract the component of the $\overline{\nu}_\mu$ flux from μ^+ decay, denoted by $\overline{\nu}_\mu(\mu^+)$.

C.2 Underlying Principle

Figure C.1 shows the simulated energy spectrum of $\overline{\nu}_\mu$ for the LE, pME and pHE beam configurations. As expected, changing the focusing by moving the target

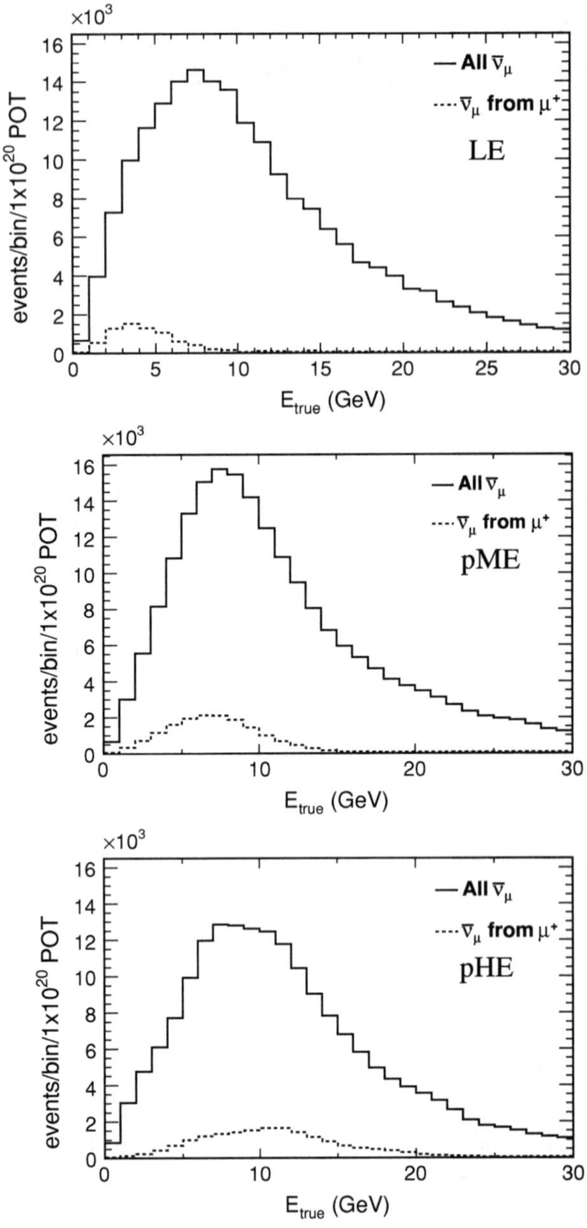

Fig. C.1 True energy spectra in the Near Detector of "all antineutrinos" and "antineutrinos from μ^+ decay" in the low-energy (LE), pseudo-medium energy (pME) and pseudo-high energy (pHE) beam configurations. All histograms correspond to 1×10^{20} POT (protons-on-target)

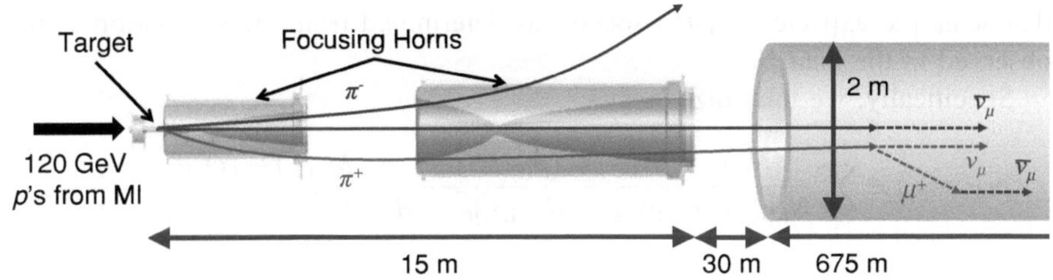

Fig. C.2 Schematic of the NuMI beamline. The antineutrinos from hadron parents $\overline{\nu}_\mu(\pi, K)$ are produced by those negatively charged mesons that go undeflected through the center of the horns. A small amount of antineutrinos originate from μ^+ decay, and the energy spectrum of those is strongly dependent on the separation between the target and the horns

upstream primarily affects the $\overline{\nu}_\mu\,(\mu^+)$ spectrum by shifting its peak to higher energy while increasing its width and magnitude. The $\overline{\nu}_\mu$ spectrum from hadrons [denoted by $\overline{\nu}_\mu(\pi, K)$] is virtually unchanged between the LE and pME settings and only slightly changed between the LE and pHE settings.

This behavior is explained by the way $\overline{\nu}_\mu$s are created in the NuMI beam. Figure C.2 illustrates how negatively charged particles are defocused by the horns, except those that go exactly through the center of them and which give rise to the $\overline{\nu}_\mu$ spectrum from hadrons $\overline{\nu}_\mu(\pi, K)$. For these antineutrinos, changing the separation between the target and the horns has very little effect, as it is always the same forward going mesons that go through the center of the horns. The situation is different for the small amount of antineutrinos originating from muon decays in the beam pipe. These muons are created in π^+ decays, which are strongly affected by the focusing. As a matter of fact, the $\overline{\nu}_\mu(\mu^+)$ spectrum varies in an analogous way to the ν_μ CC spectrum when the separation between the horns and the target is amplified (i.e., by increasing its mean and width).

C.3 Description of the Method

We endeavor to measure the $\overline{\nu}_\mu(\mu^+, \mathrm{LE})$ and $\overline{\nu}_\mu(\mu^+, \mathrm{pHE})$ spectra by taking the difference between the observed $\overline{\nu}_\mu$ spectra, scaled to POT, for the LE and pHE[9] configurations. Using the difference of normalized spectra takes advantage of the fact that the non-focusing components of the $\overline{\nu}_\mu$ spectrum change very little when the target is moved; hence, the uncertainty on the difference is relatively small. The difference in normalized spectra is fitted using two parameters p_{LE} and p_{HE}

[9] The same measurement described here can be done using pME data instead of pHE data. The amount of MINOS data taken in the pME configuration is currently very limited (less than 2×10^{18} POT). The prospects for such a measurement are nevertheless laid out at the end of this chapter, in case more pME data ever becomes available.

that scale the expected $\overline{\nu}_\mu(\mu^+)$ spectra as determined from the simulation to that observed in the data.

Specifically, we minimize

$$\chi^2(p_{\rm LE}, p_{\rm HE}) \equiv \sum_i \frac{((N_{i,\rm HE} - N_{i,\rm LE}) - (p_{\rm HE}M_{i,\rm HE} - p_{\rm LE}M_{i,\rm LE}) - (Q_{i,\rm HE} - Q_{i,\rm LE}))^2}{N_{i,\rm HE} + N_{i,\rm LE}s_{\rm LE} + M_{i,\rm HE}s_{\rm HE}^{MC} + M_{i,\rm LE}s_{\rm LE}^{MC} + Q_{i,\rm HE}s_{\rm HE}^{MC} + Q_{i,\rm LE}s_{\rm LE}^{MC}}, \tag{C.1}$$

where

- i denotes a bin in reconstructed $\overline{\nu}_\mu$ energy from 1 to 30 GeV of width 1 GeV (the bin from 0 to 1 GeV is excluded from the fit),
- $N_{i,\rm HE}$ is the observed number of $\overline{\nu}_\mu$ candidates in the ith bin in the pHE configuration for a given POT,
- $N_{i,\rm LE}$ is the observed number of $\overline{\nu}_\mu$ candidates in the ith bin in the LE configuration scaled to the same POT as the $N_{i,\rm HE}$ observation,
- $M_{i,g}$ are the expected number of $\overline{\nu}_\mu(\mu^+)$ events in the ith bin as determined from the simulation for $g = \rm LE, HE$,
- $Q_{i,g}$ are the expected number of $\overline{\nu}_\mu(\pi, K)$ events in the ith bin as determined from the simulation for $g = \rm LE, HE$,
- $s_{\rm LE} \equiv {\rm POT_{HE}/POT_{LE}}$ is the ratio of the protons-on-target for the pHE and LE data, and
- $s_{\rm LE(HE)}^{MC} \equiv {\rm POT_{HE}/POT_{LE(HE)}^{MC}}$ is the ratio of protons-on-target in pHE data with respect to LE(pHE) Monte Carlo.

The expected statistical precision and systematic biases on $p_{\rm LE}$ and $p_{\rm HE}$ are assessed with studies using "fake" data corresponding to the statistics of the currently available samples of LE and pHE data and MC (Table C.1). The $\overline{\nu}_\mu(\mu^+)$ and $\overline{\nu}_\mu(\pi, K)$ spectra were fitted with analytic functions to obtain $M_{i,g}$ and $Q_{i,g}$. We generated a fake experiment and calculated $\chi^2(p_{\rm LE}, p_{\rm ME})$ at every point in a 200×200 grid for $p_{\rm LE}$ and $p_{\rm ME}$ in the range (0, 2). Figure C.3 shows the result of one such experiment with "fake" data and with the fitted result superimposed. One thousand such fake experiments are combined to obtain the χ^2 behavior of an average experiment as shown in Figs. C.4, C.5 and C.6. The fitted values of p_{LE} and p_{HE}, although correlated as expected, are unbiased and can be determined to a precision of $\sim$35 and $\sim$19%, respectively, on average. With infinite statistics for the LE and pHE MC and LE data, the precision on $p_{\rm LE}$ would be $\sim$19%. We used similar procedures to determine the bias induced in $p_{\rm LE}$ due to the systematic effects described in Sect. C.5.

Table C.1 The data and MC samples used in this analysis. The POT values for data are after "good beam" cuts

Configuration	Data (10^{19}POT)	MC (10^{19}POT)
LE runI (runII)	2.46 (2.21)	4.44
pHE	1.41	1.19

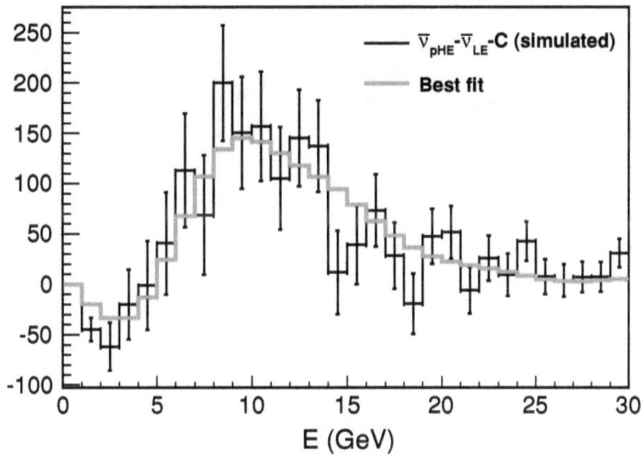

Fig. C.3 The fitted results of a single experiment with "fake" data corresponding to the statistics of Table C.1. The *black* histogram with error bars represents the 'fake' data and the *green* histogram represents the best fit result. C is the correction due to the antineutrinos from hadron parents, $C \equiv Q_{\text{HE}} - Q_{\text{LE}}$

C.4 Antineutrino Selection

Antineutrino candidates were first required to pass a preselection consisting of five cuts:

- Fiducial volume: The reconstructed event vertex must satisfy $1.0 < z < 5.0$ and $r < 1$, where r is the radial distance from the event vertex to the beam axis and z is the z-coordinate of the event's vertex.
- Track quality cuts: A track must exist in the reconstructed event which passed the fit, with a difference between the number of U and V planes being less than 6, and with $\chi^2/\text{ndf} < 20$.
- Track Fit Probability cut: Prob $(\chi^2, \text{ndf}) > 0.1$.
- Track length > 25 planes.
- abPID[10] > 0.85 (to further remove Neutral Current events).

Then a multivariate likelihood-based discriminant (PID) was constructed to finalize the selection. Four reconstructed variables were used in the construction of the PID, which are (i) the track's curvature divided by its error $(q/p)/\sigma_{q/p}$, (ii) the reconstructed y (defined to be the ratio between the reconstructed shower and neutrino energies), (iii) the track length (measured in planes) and (iv) the cosine of the track's angle with respect to the beam direction. The selection takes advantage of the correlations of the curvature distribution with event length. Figure C.2 shows the distributions of these input observables in data and simulation after preselection.

The resulting PID distribution can be seen in Fig. C.4. The simulated antineutrino spectra obtained in the different beam configurations after applying a cut at PID > 0.75 is shown in Fig. C.1. Table C.2 shows the achieved purities in the LE and pHE configurations.

[10] The abPID is the NC/CC discriminator described in [9].

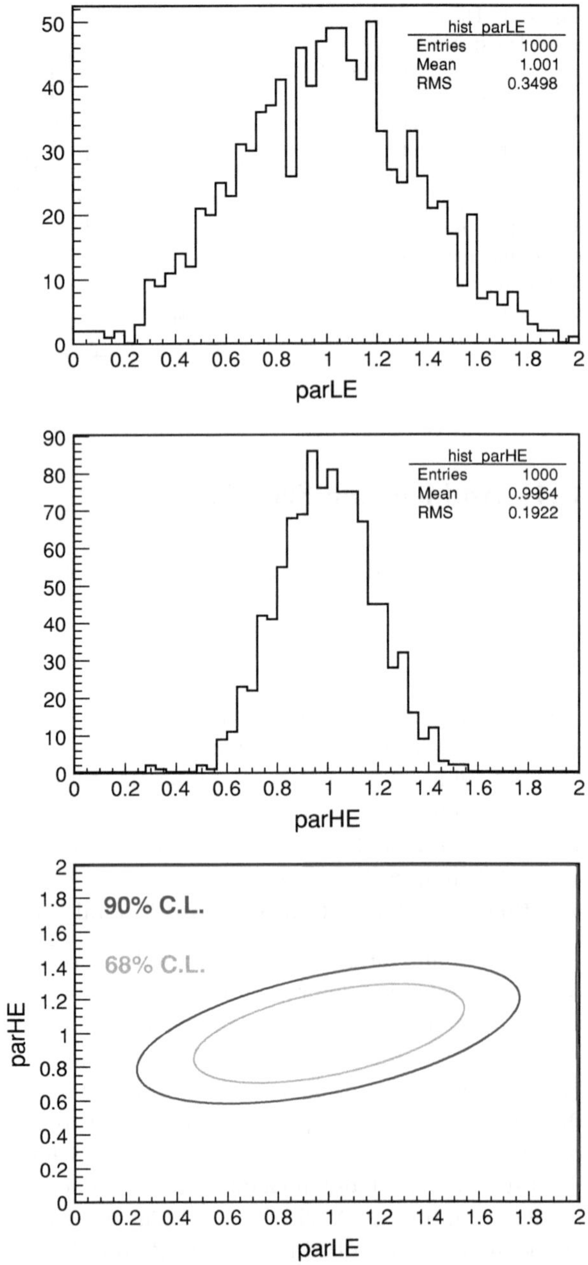

Fig. C.4 The results of 1000 simulated experiments using "fake" data for the parameters p_{LE}(upper) and p_{HE}(middle) given the data and MC statistics in Table C.1. The lower plot shows the expected correlation and accuracy of the measurement of p_{LE} and p_{HE}

Table C.2 Obtained purities for the $\overline{\nu}_\mu$ selection in the 0–30 GeV range as a function of the beam configuration

Configuration	Purity (%)
LE	99.3
pHE	96.8

C.5 Systematic Uncertainties

We estimate the contribution of systematic uncertainties to the measurement from the following sources:

1. Beam optics,
2. Hadron production,
3. Background,
4. $\overline{\nu}_\mu$ selection efficiency,
5. $\overline{\nu}_\mu$ cross-section uncertainty, and
6. Uncertainty on the ratio of the ν_e and $\overline{\nu}_\mu$ cross-sections.

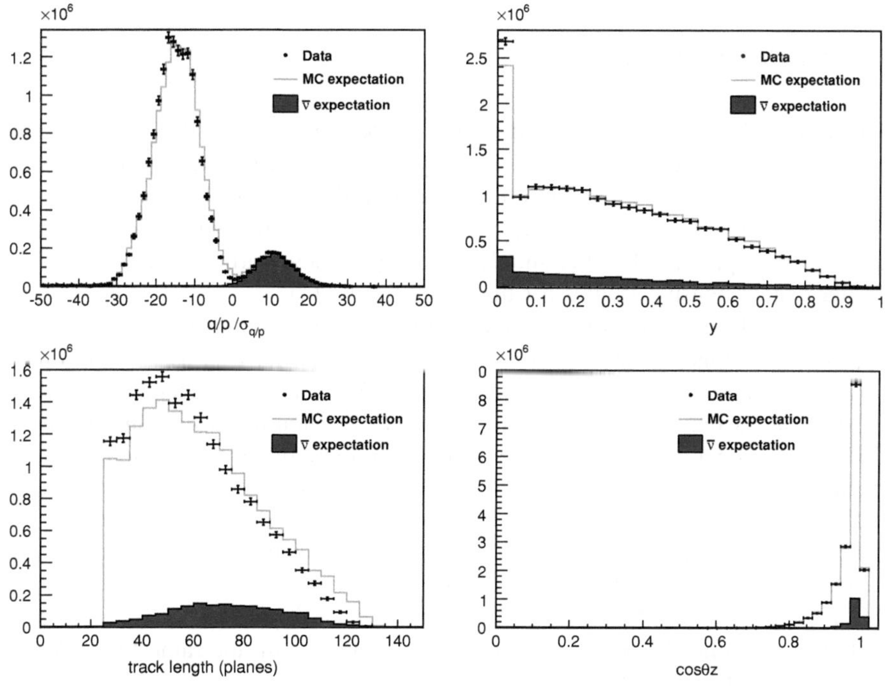

Fig. C.5 Input observables to PID are shown after preselection for ND events in data (*points*) and expected MC (*grey*) in the low energy (LE) beam configuration. The expected $\overline{\nu}_\mu$ contribution as obtained from the simulation is shown by the *shaded* histograms

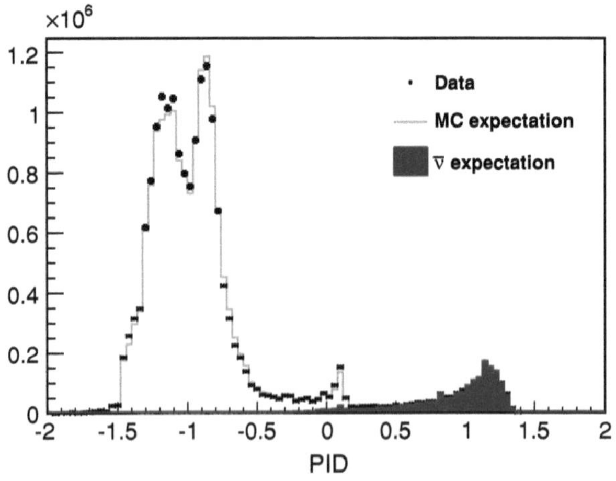

Fig. C.6 Resulting PID distribution used for $\overline{\nu}_\mu$ selection is shown for ND events in data (points) and expected MC (grey) in the low energy (LE) beam configuration. The expected $\overline{\nu}_\mu$ contribution as obtained from the simulation is shown by the shaded histogram

We note that use of the pHE–LE method (as opposed to the pME–LE method) suppresses the contributions from background and $\overline{\nu}_\mu$ cross-section uncertainties.

C.5.1 Beam Optics

Study of the beam elements that affect the $\overline{\nu}_\mu$ spectrum showed that horn1 had the most profound influence [10]. Specifically the horn1 offset and current had the greatest effect on the $\overline{\nu}_\mu$ spectrum, largely due to the solid angle subtended by the horn1 aperture. The target z position was also varied as it contributes to the solid angle of the aperture. The magnitude of the systematic variations was guided by earlier comprehensive studies for the ν_μ spectrum [11].

The NuMI beamline simulation was used to assess the change in the ND $\overline{\nu}_\mu$ spectra for changes in the offset and current of horn1 as well as changing the z position of the target in the LE configuration. Spectra were generated for horn1 current values that were $0, \pm1$ and $\pm3\%$ times the nominal value and binned in 2 GeV bins of neutrino energy. A second order polynomial was fitted to the spectra in each bin as a function of the horn1 current. The fitted function was evaluated at $\pm1\%$ to obtain the expected change in the $\overline{\nu}_\mu$ spectra. A similar procedure was followed to assess the effect of a 1 mm vertical displacement of horn1 with displacements of 0, 1, 2 and 4 mm as well as the effect of a ±1 cm uncertainty in LE z position using MC with z-displacements of 0, ±1 and ±3 cm. At least $\sim 1 \times 10^8$ simulated POT were used for each setting with the same FLUKA files used as input for all settings.

The results of the study [12] are summarized in Table C.3. To assign a systematic uncertainty for each component, we take the magnitude of the largest measured bias and symmetrize it.

Table C.3 The measured bias in the parameters p_{LE} and p_{HE} in percent for various changes in the beam geometry. The second and third columns show the bias when the variation is applied to the LE and pHE spectra simultaneously. The fourth and fifth (sixth and seventh) columns show the bias when the variation is applied to the LE (pHE) spectra only. The rightmost column contains the assigned uncertainty in p_{LE}

Variation	Parameter bias (%)						Assigned uncertainty in p_{LE}(%)	
	Vary both		Vary LE only		Vary pHE only			
	p_{LE}	p_{HE}	p_{LE}	p_{HE}	p_{LE}	p_{HE}	Min	Max
+1% horn1 current	−15	−14	−5	−9	−3	−5	−15	15
−1% horn1 current	−9	−14	−4	−5	−10	−9	−15	15
+1 mm y_{horn1}	+3	+3	+6	+6	−3	−2	−6	6
+1 cm z_{tgt}	NA	NA	−6	−6	NA	NA	−6	6
−1 cm z_{tgt}	NA	NA	+1	+1	NA	NA	−6	6
Sum in quadrature(%)							−17.2%	17.2%

C.5.2 Hadron Production

The fit done simultaneously on ν_μ CC and $\overline{\nu}_\mu$ CC data taken in different beam configurations in order to address hadron production at the target, which is described in Sect. 3.4.2, is commonly referred to as SKZP. The systematic uncertainty in p_{LE} due to hadron production uncertainties was evaluated by replacing the shape of $M_{i,g}$ and $Q_{i,g}$ (Eq. C.1) with SKZP-reweighting by analogous shapes without SKZP-reweighting. Reweighting only has a significant effect on $Q_{\mathrm{HE}} - Q_{\mathrm{LE}}$ as shown in Fig. C.7. Lack of the beam reweighting causes a relative bias of 15% in p_{LE} which we assign as the systematic uncertainty due to hadron production. This procedure may overestimate the uncertainty due solely to hadron production as only 16 of the 26 parameters of the SKZP fit are directly related to hadron production [13].

C.5.3 Background

The background to the selected $\overline{\nu}_\mu$ sample is composed of μ^- reconstructed as μ^+ ("misidentified") and hadrons reconstructed as μ^+ ("fake") at a rate of approximately 2:1. The uncertainty in the size of these background components is estimated by comparing data and simulation.

The "fake" background rate is estimated by comparing the rate of $\overline{\nu}_\mu$ candidates selected by the nubar-PID in MRCC data and Monte Carlo samples. The analysis showed that the MC estimate of the "fake" background rate should be scaled by 0.77 ± 0.25 to obtain the rate of "fake" background in the data. We assign a relative uncertainty of ±25% to the $\overline{\nu}_\mu$ background from this source.

The "misidentified" background rate is estimated from the results of the "muon-chopper" [14] applied to data and MC. The muon chopper identifies

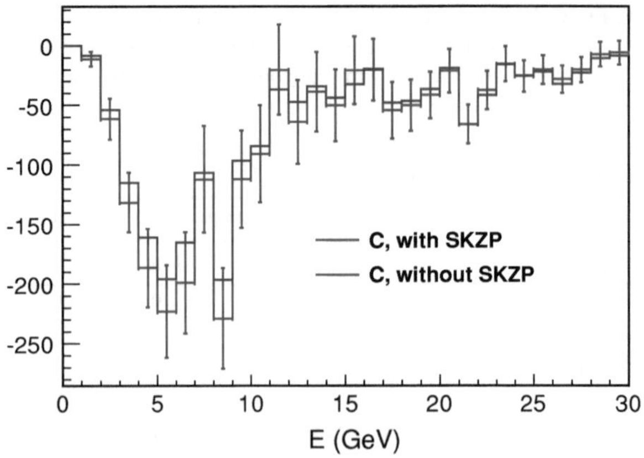

Fig. C.7 The expected difference $C \equiv Q_{HE} - Q_{LE}$ in the $\overline{\nu}_\mu$ flux from pions and kaons with (*red line*) and without (*blue line*) SKZP-reweighting. The error bars represent the statistical uncertainty due to the size of the available MC samples

stopping muons in the ND and removes all but the last X GeV of the track to create a "chopped muon" where X can be selected by the user. New "raw data" is generated using the chopped muon. The charge of the chopped muon is known by placing stringent cuts on the original muon. The rate of 900 MeV/c chopped μ^- being reconstructed as μ^+ was 3.0 ± 0.1 and 3.4 ± 0.2% in data and MC, respectively. From this result we assign a relative uncertainty of ±10% to the rate of "misidentified" $\overline{\nu}_\mu$ background.

The overall relative uncertainty in the background is the weighted combination of the two sources or $1/3 \times 25\% + 2/3 \times 10\% = 15\%$. The systematic bias in the measurement of p_{LE} is $<1\%$ when the background contribution of both LE and pHE is shifted coherently by ±30%. Thus we assign a relative uncertainty on p_{LE} of $<1\%$ due to the uncertainty in the background contribution.

C.5.4 Antineutrino Detection Efficiency

We can obtain a reasonable estimate of the uncertainty using the results of the muon-chopper [14]. The rate of 900 MeV/c chopped μ^- being reconstructed as μ^- was $70.1 \pm 0.2\%$ in data and 65.8 ± 0.5% in MC from which we infer an estimated relative uncertainty on the $\overline{\nu}_\mu$ reconstruction of ±8%.

C.5.5 Antineutrino Cross-Section

Studies show that varying the parameters[11] M_A(QE), M_A(res) and KNO$_r$ by 15, 15 and 10%, respectively, introduced a bias in p_{LE} of ±1.9% when p_{LE} was extracted

[11] More information about these parameters and how cross-section uncertainties are handled in the main analysis can be found in Sect. 8.1.2.

from the difference in pME and LE $\overline{\nu}_\mu$ spectra. We expect similar results for the pHE–LE method and assign a relative systematic uncertainty to p_{LE} of $\pm 1.9\%$ due to uncertainty in the $\overline{\nu}_\mu$ cross-section.

C.5.6 Relative ν_e and $\overline{\nu}_\mu$ Cross-Sectional Uncertainty

In order to translate the $\overline{\nu}_\mu$ rate measurement to a measurement of the ν_e rate, we must take into account the relative uncertainty in the ratio of $\nu$and $\overline{\nu}$ CC cross-sections and invoke lepton universality. As the relevant energy range is roughly that of the ν_μ disappearance analysis [15], we use the assessment of contributions to the cross-sectional ratio of 4, 8, and 8%, and from the total, quasi-elastic and resonant contributions, respectively, to estimate an overall uncertainty of $\pm 12\%$ in the ratio. This estimate is consistent with an earlier evaluation of the uncertainty $\sigma(\overline{\nu}_\mu)/\sigma(\nu_\mu)$ of $\pm 13.5\%$ for the atmospheric neutrino sample which is in the relevant energy range [16]. We note that the primary method for estimating the beam ν_e contribution is derived from the measured ν_μ CC rate and thus does not suffer from this uncertainty.

C.6 Results and Conclusions

The results of minimizing χ^2 as defined in Eq. C.1 are shown in Fig. C.8 and yield $p_{LE} = 1.53 \pm 0.37$ and $p_{HE} = 0.52 \pm 0.19$ where the uncertainties are statistical only. The minimum value of χ^2 is 25.4 indicating a good fit for 27 degrees of freedom. If we replace the fitted values of of p_{LE} and p_{HE} with the nominal values of unity, the χ^2 increases by 16.9 and the agreement is qualitatively worse as seen in Fig. C.9

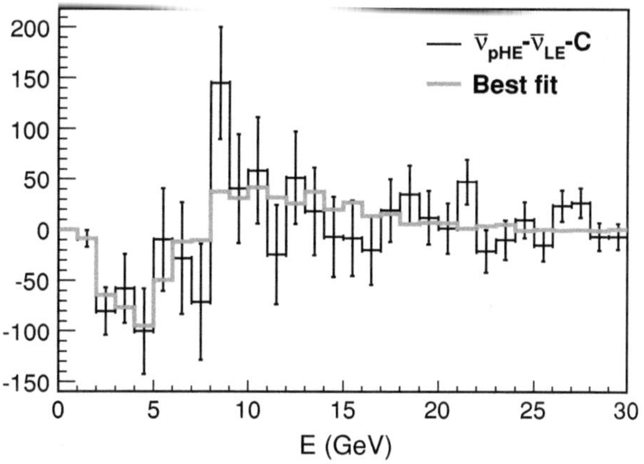

Fig. C.8 The histogram with statistical error bars is the difference in the measured normalized pHE and LE $\overline{\nu}_\mu$ spectra in data corrected by the difference $C \equiv Q_{\mathrm{HE}} - Q_{\mathrm{LE}}$ derived from simulation. The *green solid line* is the best fit result

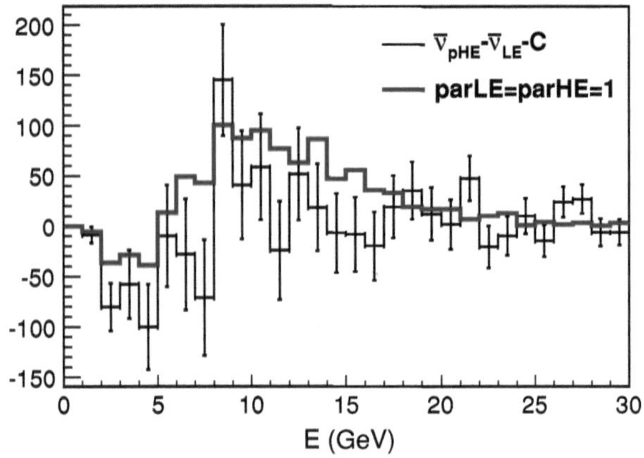

Fig. C.9 The histogram with statistical error bars is the difference in the measured normalized pHE and LE $\overline{\nu}_\mu$ spectra in data corrected by the difference $C \equiv Q_{\mathrm{HE}} - Q_{\mathrm{LE}}$ derived from simulation. The *blue solid line* represents the expected shape when p_{LE} and p_{HE} are set to the nominal value of unity

Table C.4 Summary of the components of the uncertainty in p_{LE}

p_{LE} relative uncertainty (%)	Component
17.2	Beam geometry
15	Hadron production
<1	Background
8	Efficiency
1.9	$\overline{\nu}_\mu$ cross-section
24.3	Sum in quadrature

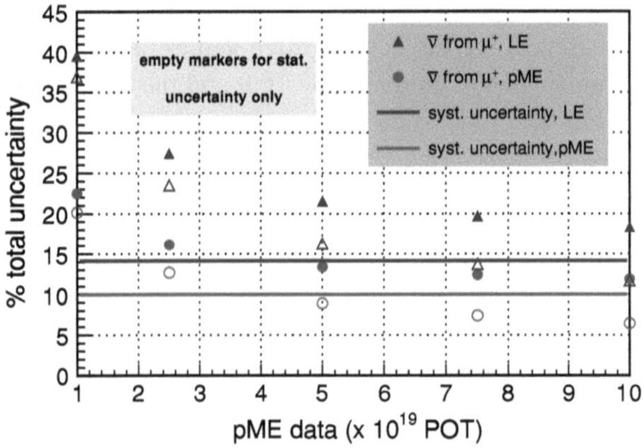

Fig. C.10 The expected uncertainty on p_{LE} (*blue triangles*) and p_{ME} (*red circles*) as a function of pME exposure. The total uncertainty is indicated by the *filled markers* and the *empty markers* represent the statistical uncertainty only. The *horizontal lines* indicate the total systematic uncertainty

The relative systematic uncertainty in p_{LE} is 24.3% (Sect. C.5.6) giving a measured rate of $\overline{\nu}_\mu$ from $\mu^+ \rightarrow e^+ + \overline{\nu}_\mu + \nu_e$ of $p_{\mathrm{LE}} = 1.53 \pm 0.37 \pm 0.37$. Taking into account the relative uncertainty in the ratio of ν_e and $\overline{\nu}_\mu$ cross-sections yields an estimate of the beam ν_e rate from $\mu^+ \rightarrow e^+ + \overline{\nu}_\mu + \nu_e$ of $1.57 \pm 0.37(\mathrm{stat}) \pm 0.41(\mathrm{syst})$ times the expectation based on the SKZP-tuned simulation. If the three sources of uncertainty are added in quadrature, this measurement confirms the tuned MC expectation at 1 SD.

We note that the statistical uncertainty in this measurement can be reduced to $\sim$20% with approximately five times more pHE MC. Also, this measurement can be carried out with pME data instead of pHE data. In that case, the impact of the main sources of systematic uncertainty, the beam geometry and hadron production, would be greatly reduced. The reason for this is that the correction $Q_{\rm ME} - Q_{\rm LE}$ is significantly smaller than the one used with the pHE data $Q_{\rm HE} - Q_{\rm LE}$. Figure C.10 shows the sensitivity to which the measurement could be carried out as a function of pME exposure. We surmise that the addition of sufficient pME data ($\sim 2.5 \times 10^{19}$) to the available pHE data would enable the effect of the two main sources of uncertainty to be virtually eliminated.

Finally, we remark that this method could be exploited to great advantage by future long baseline neutrino experiments that have a near detector capable of muon charge-sign determination and an adjustable beam focus.

Appendix D
Feldman–Cousins Framework for the $\nu_\mu \rightarrow \nu_e$ Oscillations Analysis

The ν_e appearance search probes a region of space that is close to the $\sin^2(2\theta_{13}) = 0$ physical boundary. Moreover, the expectation at the Far Detector is on the order of tens of events, which leads to non-Gaussian behavior. Because of these reasons, a Feldman–Cousins approach is used to derive the confidence intervals in this thesis. The general features of our Feldman–Cousins implementation are described first, followed by a discussion concerning the incorporation of the uncertainties in the oscillation parameters.

D.1 General Implementation

D.1.1 Motivation

A confidence interval at $x\%$ confidence level (C.L.) is a range of the measured parameter that has a $x\%$ probability of containing the true value of the parameter. The approach introduced by Feldman and Cousins [17] to set a confidence interval on a parameter near the boundaries of a physical region presents several advantages with respect to the Bayesian and frequentist alternatives described in [18].

- It defines a true confidence interval in the statistical sense, even close to physical boundaries.
- The confidence interval is always in the allowed region of the parameter.
- The method is well defined before looking at the data.

D.1.2 Procedure

Because of the previous reasons we chose to implement a Feldman–Cousins approach to derive the confidence intervals resulting from the ν_e appearance analysis. This is achieved by executing the following steps:

- Divide the parameter space into a grid. In our case the parameter space consists of a pair of two-dimensional planes of $\sin^2 2\theta_{13}$ vs. δ_{CP} or, as discussed later, of $2\sin^2 2\theta_{13}\sin^2\theta_{23}$ vs. δ_{CP}. One plane is for the normal mass hierarchy, and the other one is for the inverted mass hierarchy. Each plane is divided into 401 points in each dimension, giving a total of $2 \times 401 \times 401 = 321{,}601$ grid points.
- For each point in the grid calculate the expected total (signal plus background) number of events N_{exp}. This is done by repeating the entire process described in Chap. 7, but changing the oscillation parameters as appropiate.
- Also for each point in the grid perform 10,000 fake experiments. This is done by shifting the number of expected background N_B and signal N_S events by their corresponding total uncertainties σ_B and σ_S with a Gaussian random number generating function. The sum of the shifted signal and background events is then fluctuated with a Poisson random number generating function to simulate the statistical uncertainty of the measurement. This procedure is summarized by the following equation,

$$N_{obs} = R_P(R_G(N_B, \sigma_B) + R_G(N_S, \sigma_S)), \tag{D.1}$$

where R_P and R_G denote the Gaussian and Poisson random generating functions respectively. In order to save computational effort, the fractional systematic errors σ_B/N_B and σ_S/N_S as derived in Chap. 8 are assumed to be independent of θ_{13}, δ_{CP} and any other oscillation parameters.[12]

- Calculate the $\Delta\chi^2$ between each fake experiment and the expectation:

$$\Delta\chi^2 = \chi^2 - \chi^2_{min}. \tag{D.2}$$

The χ^2 is calculated as

$$\chi^2 = -2\ln P = 2\left(N_{exp} - N_{obs} + N_{obs}\ln\left(\frac{N_{obs}}{N_{exp}}\right)\right), \tag{D.3}$$

where P is the poisson probability of observing N_{obs} events with a mean of N_{exp}. χ^2_{min} is the value of χ^2 at the best fit point. In our situation, the calculation of χ^2_{min} is straightforward. As shown in Fig. D.1, for a fixed δ_{CP}, the number of expected events increases monotonically with θ_{13}, with a minimum N^{min}_{exp} reached at the $\theta_{13} = 0$ boundary. Furthermore, N^{min}_{exp} is the same for all values of δ_{CP}. The value of χ^2_{min} is thus exactly zero if $N_{obs} \geq N^{min}_{exp}$, given that in that case a best fit point where N_{exp} exactly matches the observation can always be found in the parameter space. If, on the other hand, $N_{obs} < N^{min}_{exp}$, then the best fit point occurs at

[12] This is a good approximation, as the background is dominated by NC events, which are insensitive to oscillations. Moreover, the fractional error on the signal would change only due to strong variations in the shape of the energy spectrum. While all the oscillation parameters affect the ν_e appearance probability, the variations in θ_{13} and θ_{23} considered here are the ones that have the strongest impact on the ν_e yield. The probability of ν_e appearance is proportional to $\sin^2 2\theta_{13}\sin^2\theta_{23}$ to first order, which affects all energy bins equally.

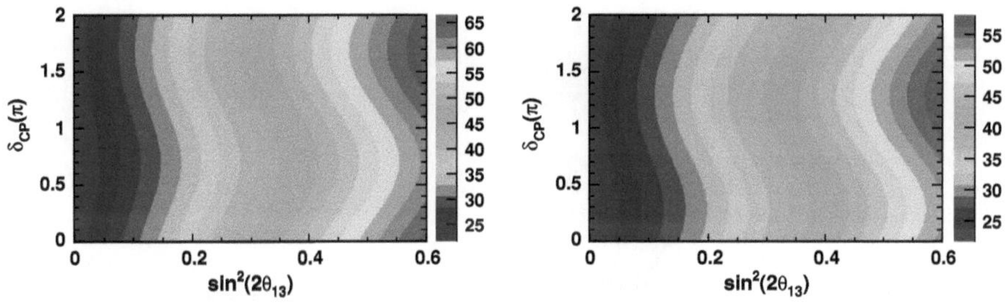

Fig. D.1 Number of expected events N_{exp} selected with LEM > 0.65 as a function of $\sin^2 2\theta_{13}$ and δ_{CP}, for the normal mass hierarchy (*left*) and the inverted mass hierarchy (*right*). N_{exp} increases with θ_{13} and depends on δ_{CP}. At the $\theta_{13} = 0$ boundary however there is no dependence on δ_{CP}

the $\theta_{13} = 0$ boundary, and χ^2_{min} is obtained with Eq. D.3 but setting $N_{exp} = N^{min}_{exp}$.

- Store the $\Delta\chi^2$ values for each experiment in a one-dimensional histogram. Then find the value $\Delta\chi^2_{x\%}$ that encloses $x\%$ of all the fake experiments.
- Calculate the $\Delta\chi^2_{DATA}$ as done for the fake experiments but with $N_{obs} = N_{DATA}$.
- If $\Delta\chi^2_{DATA} < \Delta\chi^2_{x\%}$ then the particular grid point is inside the $x\%$ C.L. contour. The entire procedure is summarized in Fig. D.2.

D.1.3 Obtaining Smooth Contours

The 90% Feldman–Cousins contours for LEM > 0.65 corresponding to an observation of 28 events are shown on the left side of Fig. D.3. Our Feldman–Cousins implementation gives results that are consistent with other implementations. The fact that the contours show a certain "thickness" has been reproduced elsewhere [19] and is an unavoidable feature of the method, stemming from the fact that we measure small, integer numbers. The right side of Fig. D.3 shows the $\Delta\chi^2$ distribution of 10,000 fake experiments at the $\sin^2 2\theta_{13} = 0.15$, $\delta_{CP} = 0$ point in the normal hierarchy grid. Due to the fact that N_{obs} can only be an integer, and that N_{exp} is small and constant (at each grid point), the resulting $\Delta\chi^2$ distribution is discrete. This in turn creates a discreteness in the contours.

The "thickness" of the contours is thus a real feature of our measurement. It is however an inconvenient one, for both aesthetic and practical reasons. If a limit is set on $\sin^2 2\theta_{13}$ from these contours, the limit might not accurately reflect the limit that is set by values of δ_{CP} near the one that was used. In other words, the limit might be too optimistic or too conservative depending on the choice of δ_{CP} down to several decimal places. This also means that the quoted limit would be dependent on the binning that was chosen for the grid. It is thus advantageous to bypass this issue by artificially "smoothing" the contours. We achieve this through the use of the *Smooth* function [20] which is incorporated in the ROOT software package [21]. For convenience, we apply this function to the two-dimensional grid

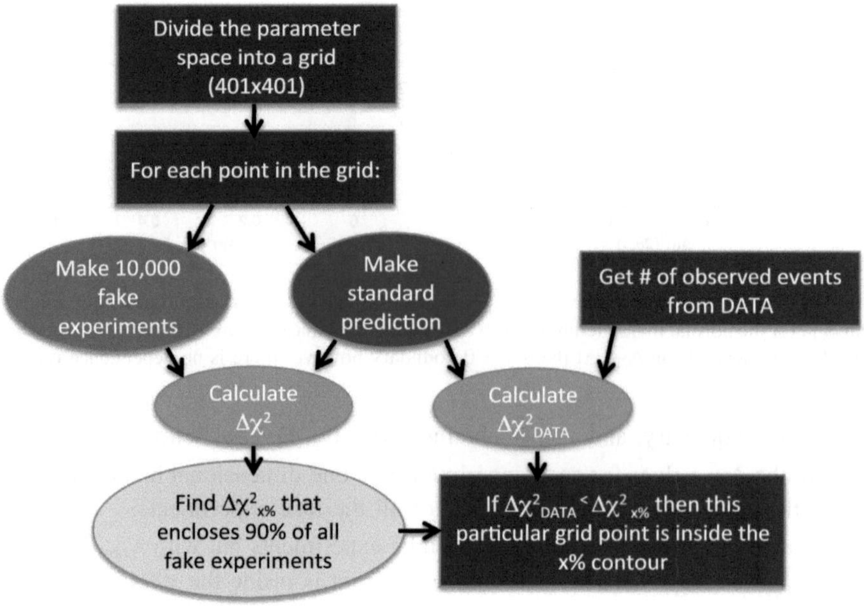

Fig. D.2 Schematic of our Feldman–Cousins implementation. The details of each step are given in the text

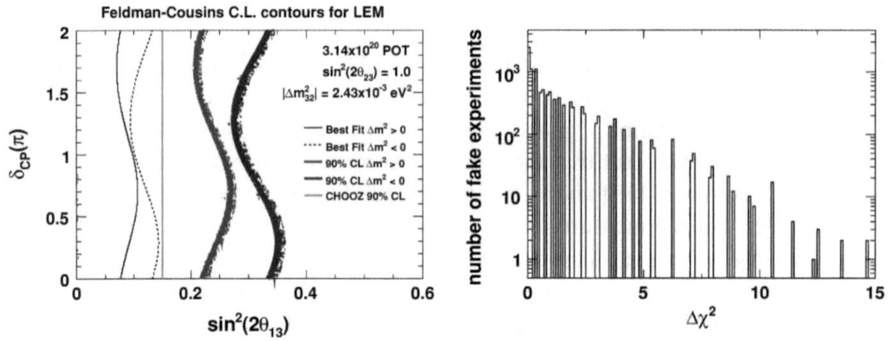

Fig. D.3 Feldman–Cousins 90% C.L. exclusion contours with the LEM > 0.65 selection for an observation of 28 events (*left*), and $\Delta\chi^2$ distribution of the $10,000$ fake experiments for the $\sin^2 2\theta_{13} = 0.15, \delta_{CP} = 0$ normal hierarchy grid point (*right*). The $\Delta\chi^2$ distribution is discrete due to the fact that only an integer number of events can be observed and that such observation is low statistics. Because of this the contours have a certain "thickness" to them. For the right plot, the values that enclose 68 and 90% of all fake experiments are 1.15 and 3.55 respectively

$\Delta\chi^2_{x\%-\mathrm{diff}}$, which is the difference between the $\Delta\chi^2_{x\%}$ grid (containing the cutoff values that enclose $x\%$ of the fake experiments) and the $\Delta\chi^2_{\mathrm{DATA}}$ grid. Figure D.4 shows the $\Delta\chi^2_{90\%-\mathrm{diff}}$ grid before and after smoothing. The edge of the region for which $\Delta\chi^2_{90\%-\mathrm{diff}} < 0$ delineates the 90% C.L. contour.

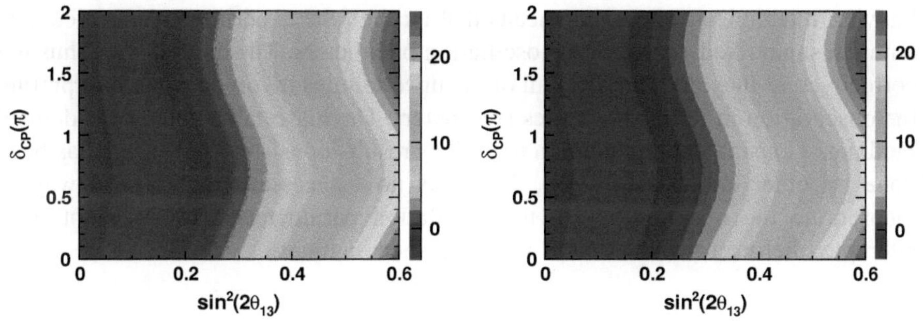

Fig. D.4 $\Delta\chi^2_{90\%-\mathrm{diff}}$ grid before and after smoothing. The edge of the region for which $\Delta\chi^2_{90\%-\mathrm{diff}} < 0$ delineates the 90% C.L. contour

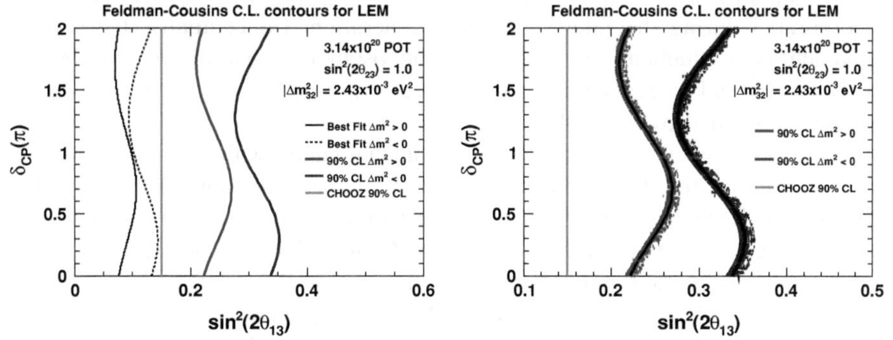

Fig. D.5 *Left* smooth 90% C.L. contours with the LEM > 0.65 selection for an observation of 28 events. *Right* a comparison between the raw (in *blue* and *purple*) and the smooth contours (superimposed in *black*). Please note that the x-axis scale is different for the two plots. The smooth contours fall approximately in the middle of the raw contours

Figure D.5 shows the smooth contours that result from the process just described. The contours are shown by themselves but also imposed on top of the raw contours. The smooth contours fall approximately in the middle of the raw contours. From this point onward and in the entire thesis only the smooth contours are considered, unless specified.

D.2 Incorporating the Uncertainties on the Oscillation Parameters

D.2.1 Motivation

As indicated in the legend of Figs. D.3 and D.5, the contours presented so far are generated under the assumption that the oscillation parameters are perfectly known. In technical terms, this means that all the fake experiments are calculated with the same fixed values of the oscillation parameters. The contours are thus not correct yet, as they do not reflect all of the uncertainties involved when interpreting the observation in terms of physics parameters. Having established a well-defined Feldman–Cousins implementation that gives results consistent with what has been done by other collaborators [19, 22], it is now necessary to devise a way to incorporate the uncertainties in the oscillation parameters. This had not been attempted before for the MINOS ν_e appearance analysis.

D.2.2 Technical Challenges

In our implementation, the uncertainties in the oscillation parameters have to be incorporated during the fake experiment generation. Conceptually, the simplest approach is to fluctuate the values of the oscillation parameters within their uncertainties and to regenerate a Far Detector prediction that is then fluctuated again as described by Eq. D.1

There are however significant technical challenges with this approach. The regeneration of the Far Detector prediction has to be done for every fake experiment. Each regeneration takes on the order of 0.15 s on a 2.8 GHz xeon processor, given that many of the steps of Chap. 7 have to be repeated.[13] This means that to generate two Feldman–Cousins grids of 401×401 size an estimated amount of 67,000 CPU-hours are needed just to regenerate all the Far Detector predictions. Even with a farm consisting of 400 slots such as the one existing at Caltech, this corresponds to approximately a week of wait time, which is prohibitively long.

We propose two solutions to this problem which are described next.

D.2.3 First Solution

One solution is to treat the effect of varying the oscillation parameters as an additional error that can be folded in with the systematic errors. In this case the magnitude and the characteristics of such an effect have to be determined as a function of $\sin^2 2\theta_{13}$ and δ_{CP}. This can be done with a lower resolution than the one used for the Feldman–Cousins grids.

We consider a 201×81 grid of $\sin^2 2\theta_{13}$ vs. δ_{CP}. For each point in this grid, the oscillation parameters are simultaneously and independently varied within their

[13] This is remarkably fast given all of the steps involved, and is the result of an optimization performed by the author of [19]. It is thus unlikely that the code could be made significantly faster than it already is.

Table D.1 Oscillation parameters and their uncertainties, as incorporated in the calculation of the Feldman–Cousins contours

Δm^2_{32}	θ_{23}	Δm^2_{21}	θ_{12}
$(2.43 \pm 0.13) \times 10^{-3}\text{eV}^2$	$\frac{\pi}{4} \pm 0.122$	$(8.0 \pm 0.6) \times 10^{-5}\text{eV}^2$	0.59365 ± 0.041

uncertainties using Gaussian random number generating functions for a total of 2,000 times, each generating a new predictions N_{fluct}. Each grid point's one-dimensional "fractional variation distribution" [i.e., the distribution of $(N_{\text{fluct}} - N_{\text{exp}})/N_{\text{exp}}$] is then saved to file, where N_{exp} is the Far Detector prediction obtained with the best fit values of the oscillation parameters and is calculated once per grid point. The number of CPU-hours required for these calculations is now reduced to $\sim 13,000$, which is achieved in a few hours with a farm such as the one available at Caltech. Also, the advantage of saving the information in terms of these fractional variation distributions is that they do not have to be recomputed when the POT exposure changes.

Table D.1 shows the oscillation parameters and their uncertainties as considered in the calculation. The uncertainty on Δm^2_{32} is obtained from the latest MINOS results [23]. The uncertainty on θ_{23} is obtained from the results published by Super-Kamiokande [24]. The uncertainties on the solar parameters are obtained from the combination of the solar and KamLAND results [25]. In some cases these do not correspond to the most recent bounds available on the oscillation parameters, but they are consistent with the values used throughout the rest of the analysis.

The fractional variation distributions are, in most regions of parameter space, well fitted by a Gaussian distribution with a mean at zero. Figure D.6 shows the RMS of the fractional variation distributions for signal and background events selected by LEM > 0.65, in the case of the normal hierarchy. Varying the oscillation parameters has a negligible impact on the background, of 1.2% at the

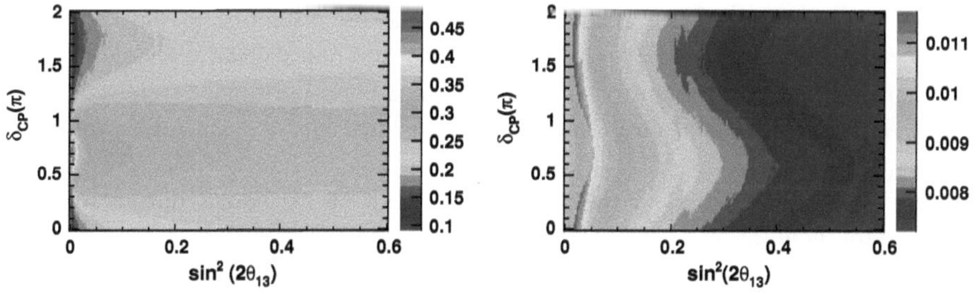

Fig. D.6 Fractional error due to the uncertainties in the oscillation parameters on the number of signal (*left*) and background (*right*) events, as a function of $\sin^2(2\theta_{13})$ and δ_{CP}. The errors correspond to the RMS of the $(N_{\text{fluct}} - N_{\text{exp}})/N_{\text{exp}}$ distributions, with the LEM > 0.65 selection and in the case of the normal hierarchy. The error is negligible for the background, but quite important for the signal. The same conclusions apply in the case of the inverted mass hierarchy

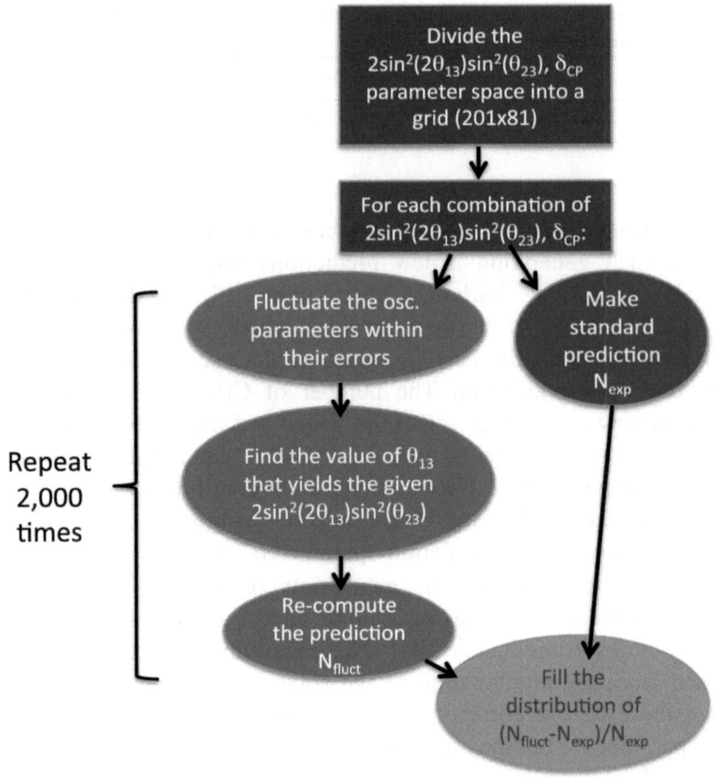

Fig. D.7 An illustration of the procedure followed in order to quantify the error due to the uncertainties in the oscillation parameters when the parameter space is divided in terms of $2\sin^2(2\theta_{13})\sin^2(\theta_{23})$ vs. δ_{CP}

most. As previously explained, this is because the background consists mostly of NC events which are insensitive to oscillations. The situation is different for the signal however, where the error induced by the uncertainties in the oscillation parameters is higher than 20% in most regions and can exceed 40%. This is not surprising, since the probability for ν_e appearance is proportional to $\sin^2\theta_{23}$ to first order. The $\sim$15% uncertainty on θ_{23} thus translates into a $\sim$28% uncertainty in $\sin^2\theta_{23}$ to first order.

The size of the error on the signal that needs to be added during the fake experiment generation is thus larger than the systematic and statistical errors combined for $\sin^2 2\theta_{13} \gtrsim 0.20$. This is undesirable, as this additional error is obtained with fewer fake experiments and with a lower resolution than the Feldman–Cousins grids. Furthermore, the impact on the resulting contours is quite large, as could be expected. The contours of Fig. D.5 degrade by roughly 15% with the implementation of this error. Because of these reasons, the first solution described here is not the preferred one. Nevertheless, the results obtained so far are

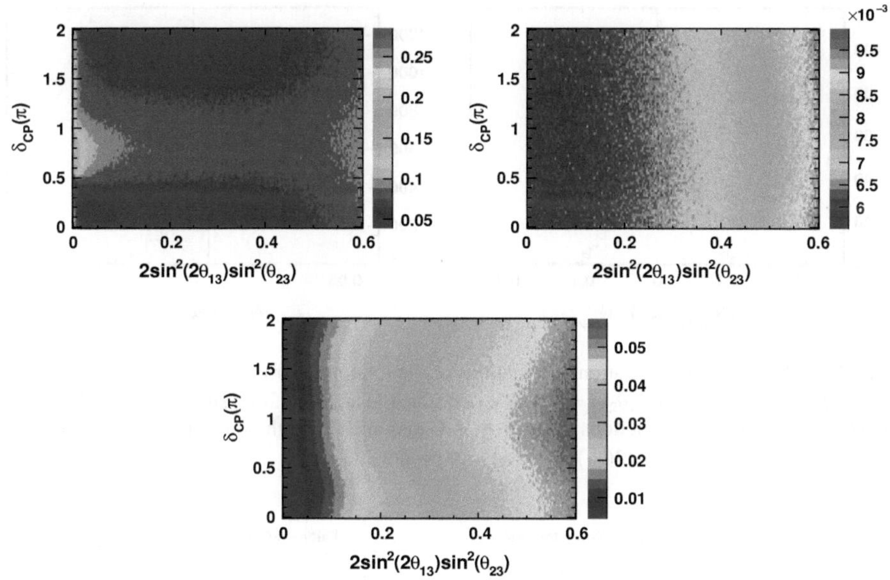

Fig. D.8 fractional error due to the uncertainties in the oscillation parameters on the number of signal (*left*), background (*right*) and total (*bottom*) events, as a function of $2\sin^2(2\theta_{13})\sin^2(\theta_{23})$ and δ_{CP}. The errors correspond to the RMS of the $(N_{fluct} - N_{exp})/N_{exp}$ distributions, with the LEM > 0.65 selection and in the case of the normal hierarchy. The error in the total number of expected events is less than 5%. The same conclusion applies in the case of the inverted mass hierarchy

very valuable, as they indicate that the contours calculated with fixed oscillation parameters are $\sim 15\%$ too optimistic.

D.2.4 Second Solution

To first order, the ν_e appearance probability in MINOS is given by

$$P(\nu_\mu \rightarrow \nu_e) \approx \sin^2(2\theta_{13})\sin^2(\theta_{23})\sin^2\left(\frac{\Delta m^2_{31}L}{4E}\right). \quad \text{(D.4)}$$

Therefore, a second solution to the problem of incorporating the uncertainties in the oscillation parameters is to parameterize the parameter space in terms of $2\sin^2(2\theta_{13})\sin^2(\theta_{23})$ vs. δ_{CP}. This way, most of the uncertainty in θ_{23} is absorbed by the $2\sin^2(2\theta_{13})\sin^2(\theta_{23})$ term.

Even with this approach there remains an additional error due to the uncertainties in the oscillation parameters that needs to be accounted for. This error is partly due to the fact that there are second-order terms in the ν_e appearance probability. Even if two combinations of θ_{13} and θ_{23} give the same value of

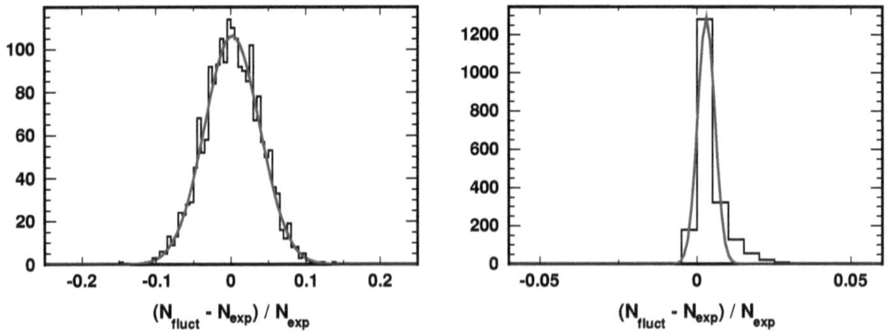

Fig. D.9 The fractional variation distributions of the total number of expected events with LEM > 0.65, at $2\sin^2(2\theta_{13})\sin^2(\theta_{23}) = 0.3$, $\delta_{\mathrm{CP}} = 0.97\pi$ and at $2\sin^2(2\theta_{13})\sin^2(\theta_{23}) = 0.012$, $\delta_{\mathrm{CP}} = 0.97\pi$ (*right*). Most distributions fit well to a Gaussian, although some distributions near the physical boundary of $2\sin^2(2\theta_{13})\sin^2(\theta_{23}) = 0$ do not

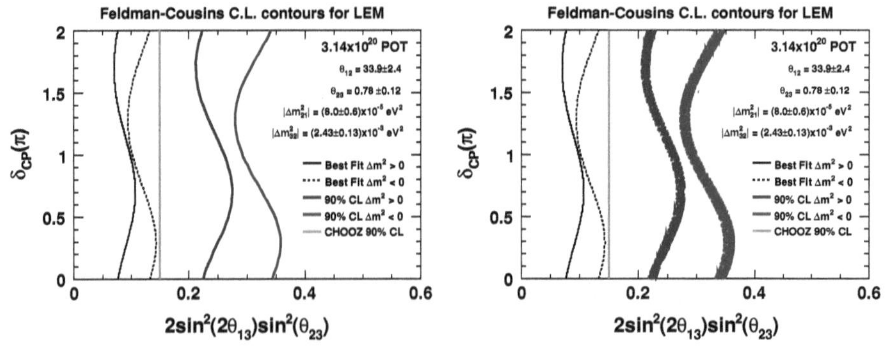

Fig. D.10 Smooth (*left*) and raw (*right*) Feldman–Cousins 90% C.L. contours for the LEM > 0.65 selection, for an observation of 28 events. The uncertainties in the oscillation parameters are included

$2\sin^2(2\theta_{13})\sin^2(\theta_{23})$, the corresponding ν_e appearance probabilities can be slightly different. Also, there are other oscillation parameters in addition to θ_{13} and θ_{23} that are involved in ν_e appearance.

This additional error is quantified as described by Fig. D.7. The procedure is similar to the one described for the first solution, except that now the value of θ_{13} is slightly different at every iteration, being affected by the fluctuated value of θ_{23}. Figure D.8 shows the RMS of the fractional variation distributions for signal, background and total events selected by LEM > 0.65, in the case of the normal hierarchy. The error in the signal is now much smaller and exceeds 20% only near the $\theta_{13} = 0$ physical boundary. Near that boundary however the signal is very small compared to the background. Because of this, the error on the total number of expected events does not exceed $\sim$5% in the region of parameter space

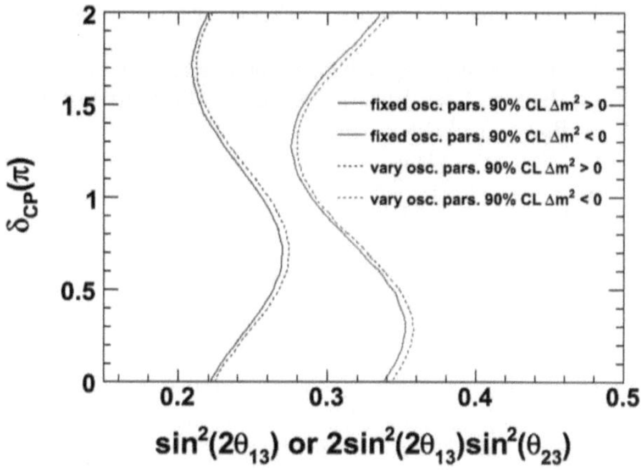

Fig. D.11 A comparison between the contours of Fig. D.10 with those of Fig. D.5. When the x-axis parameterization is changed to $2\sin^2(2\theta_{13})\sin^2(\theta_{23})$, the impact of varying the oscillation parameters is very small. If the parameterization is not changed however the contours degrade by approximately 15%, as explained in Sect. D.2.3

considered. This is small compared to the statistical error in the measurement.

Figure D.9 shows two fractional variation distributions. In some cases the distributions do not match well to a Gaussian. Consequently, they are stored for later use during the Feldman–Cousins grid generation. There, the distribution corresponding to a particular point in the Feldman–Cousins grid is randomly sampled at each fake experiment in order to shift the total number of expected events. The systematic and statistical fluctuations are applied in the same way as described in Sect. D.1.

The results are shown at Fig. D.10 for the LEM > 0.65 selection. Given the small size of the error due to the uncertainty in the oscillation parameters in comparison with the systematic and statistical errors inherent to the measurement, the contours look practically identical to the ones of Fig. D.5, except for the label of the x-axis. An explicit comparison is done in Fig. D.11, where it can be seen that the differences are smaller than a few percent.

The raw contour is also shown on the right side of Fig. D.10 for completeness. The contours included throughout the rest of this thesis however are the smooth contours with the error in the oscillation parameters incorporated as just described.

References

1. Y. Ashie et al., Measurement of atmospheric neutrino oscillation parameters by Super-Kamiokande I. Phys. Rev. D **71**(11), 112005 (2005)
2. P. Adamson et al., First observations of separated atmospheric ν_μ and $\overline{\nu}_\mu$ events in the MINOS detector. Phys. Rev. D **73**, 072002 (2006)
3. P. Adamson et al., Charge-separated atmospheric neutrino-induced muons in the MINOS far detector. Phys. Rev. D **75**, 092003 (2007)
4. J.P. Ochoa-Ricoux, Users guide to the candshield package, MINOS-doc-1156, 2005
5. A. Blake, A study of atmospheric neutrino oscillations in the MINOS far detector, FERMILAB-THESIS-2005–77

6. J. Ling, MINOS detector WLS fiber attenuation calibration documentation, MINOS-doc-4850, 2008
7. L. Mualem, N. Tagg, Personal communication, 2007
8. L. Mualem, Near detector-style fiber reflectivity test, MINOS-doc-4388, 2008
9. A. Blake, A likelihood method for CC/NC separation, MINOS-doc-2720, 2007
10. D. Jaffe, J.P. Ochoa-Ricoux, Systematics of beam ν_e measurement using pHE-LE10 beams, MINOS-doc-2841, 2007
11. D. Indurthy, S. Kopp, Ž. Pavlović, Systematic uncertainties in the NuMI beam flux, MINOS-doc-1283, 2005
12. D. Jaffe, J.P. Ochoa-Ricoux, Horn & target systematics for beam nue's from antineutrinos measurement, MINOS-doc-3230, 2007
13. Z. Pavlovic, Observation of Disappearance of Muon Neutrinos in the NuMI Beam, FERMILAB-THESIS-2008–59
14. C. Howcroft, Numubar Summary, MINOS-doc-2643, 2007
15. C. Andreopoulos et al., Updated Cross-Section Model Uncertainties for the Charged Current Analysis, MINOS-doc-2989, 2007
16. H. Gallagher, Nubar/nu cross section uncertainties at atmos nu energies, MINOS-doc-1424, 2006
17. G.J. Feldman, R.D. Cousins, Unified approach to the classical statistical analysis of small signals, Phys. Rev. D **57**(7), 3873–3889 (1998)
18. C. Amsler et al., (Particle Data Group), Review of particle Physics. Phys. Lett. **B**(667), 1 (2008)
19. J. Boehm, Measurement of Electron Neutrino Appearance with the MINOS Experiment, FERMILAB-THESIS-2009–17
20. D. McKee, http://root.cern.ch/root/html/TH2.html#TH2:Smooth
21. R. Brun, F. Rademakers, ROOT: An object oriented data analysis framework. Nucl. Instrum. Meth. A **389**, 81–86 (1997)
22. T. Yang, A study of muon neutrino to electron neutrino oscillations in the MINOS experiment, FERMILAB-THESIS-2009–04
23. P. Adamson et al., Measurement of neutrino oscillations with the MINOS detectors in the NuMI beam. Phys. Rev. Lett. **101**, 131802 (2008)
24. Y. Ashie et al., Evidence for an oscillatory signature in atmospheric neutrino oscillations. Phys. Rev. Lett. **93**(10), 101801 (2004)
25. B. Aharmim et al., Electron energy spectra, fluxes, and day-night asymmetries of ^{8}B solar neutrinos from measurements with NaCl dissolved in the heavy-water detector at the Sudbury Neutrino Observatory. Phys. Rev. C **72**(5), 055502 (2005)

Index

MIX
Papier aus verantwortungsvollen Quellen
Paper from responsible sources
FSC® C105338

If you have any concerns about our products,
you can contact us on
ProductSafety@springernature.com

In case Publisher is established outside the EU,
the EU authorized representative is:
Springer Nature Customer Service Center GmbH
Europaplatz 3, 69115 Heidelberg, Germany

Printed by Libri Plureos GmbH
in Hamburg, Germany